T0141959

Studies in Big Data

Volume 44

Series editor

Janusz Kacprzyk, Polish Academy of Sciences, Warsaw, Poland
e-mail: kacprzyk@ibspan.waw.pl

The series "Studies in Big Data" (SBD) publishes new developments and advances in the various areas of Big Data- quickly and with a high quality. The intent is to cover the theory, research, development, and applications of Big Data, as embedded in the fields of engineering, computer science, physics, economics and life sciences. The books of the series refer to the analysis and understanding of large, complex, and/or distributed data sets generated from recent digital sources coming from sensors or other physical instruments as well as simulations, crowd sourcing, social networks or other internet transactions, such as emails or video click streams and other. The series contains monographs, lecture notes and edited volumes in Big Data spanning the areas of computational intelligence incl. neural networks, evolutionary computation, soft computing, fuzzy systems, as well as artificial intelligence, data mining, modern statistics and Operations research, as well as self-organizing systems. Of particular value to both the contributors and the readership are the short publication timeframe and the world-wide distribution, which enable both wide and rapid dissemination of research output.

** Indexing: The books of this series are submitted to ISI Web of Science, DBLP, Ulrichs, MathSciNet, Current Mathematical Publications, Mathematical Reviews, Zentralblatt Math: MetaPress and Springerlink.

More information about this series at http://www.springer.com/series/11970

Sanjiban Sekhar Roy · Pijush Samui
Ravinesh Deo · Stavros Ntalampiras
Editors

Big Data in Engineering Applications

Springer

Editors
Sanjiban Sekhar Roy
School of Computing Science
    and Engineering
Vellore Institute of Technology
Vellore, Tamil Nadu
India

Pijush Samui
Department of Civil Engineering
National Institute of Technology Patna
Patna, Bihar
India

Ravinesh Deo
University of Southern Queensland
Springfield, QLD
Australia

Stavros Ntalampiras
Polytechnic University of Milan
Milan
Italy

ISSN 2197-6503   ISSN 2197-6511   (electronic)
Studies in Big Data
ISBN 978-981-13-4162-5   ISBN 978-981-10-8476-8   (eBook)
https://doi.org/10.1007/978-981-10-8476-8

© Springer Nature Singapore Pte Ltd. 2018
Softcover re-print of the Hardcover 1st edition 2018
This work is subject to copyright. All rights are reserved by the Publisher, whether the whole or part
of the material is concerned, specifically the rights of translation, reprinting, reuse of illustrations,
recitation, broadcasting, reproduction on microfilms or in any other physical way, and transmission
or information storage and retrieval, electronic adaptation, computer software, or by similar or dissimilar
methodology now known or hereafter developed.
The use of general descriptive names, registered names, trademarks, service marks, etc. in this
publication does not imply, even in the absence of a specific statement, that such names are exempt from
the relevant protective laws and regulations and therefore free for general use.
The publisher, the authors and the editors are safe to assume that the advice and information in this
book are believed to be true and accurate at the date of publication. Neither the publisher nor the
authors or the editors give a warranty, express or implied, with respect to the material contained herein or
for any errors or omissions that may have been made. The publisher remains neutral with regard to
jurisdictional claims in published maps and institutional affiliations.

Printed on acid-free paper

This Springer imprint is published by the registered company Springer Nature Singapore Pte Ltd.
part of Springer Nature
The registered company address is: 152 Beach Road, #21-01/04 Gateway East, Singapore 189721,
Singapore

Contents

Applying Big Data Concepts to Improve Flat Steel Production Processes

Jens Brandenburger, Valentina Colla, Silvia Cateni, Antonella Vignali, Floriano Ferro, Christoph Schirm and Josef Melcher

Abstract In this chapter we present some results of the first European research project dealing with the utilisation of Big Data ideas and concepts in the Steel Industry. In the first part, it motivates the definition of a multi-scale data representation over multiple production stages. This data model is capable to synchronize high-resolution (HR) measuring data gathered along the whole flat steel production chain. In the second part, a realization of this concept as a three-tier software architecture including a web-service for a standardized data access is described and some implementation details are given. Finally, two industrial demonstration applications are presented in detail to explain the full potential of this concept and to prove that it is operationally applicable. In the first application, we realized an instant interactive data visualisation enabling the in-coil aggregation of millions of quality and process measures within seconds. In the second application, we used the simple and fast HR data access to realize a refined cause-and-effect analysis.

J. Brandenburger (✉)
VDEh-Betriebsforschungsinstitut GmbH, BFI, Düsseldorf, Germany
e-mail: jens.brandenburger@bfi.de

V. Colla · S. Cateni · A. Vignali
Scuola Superiore Sant'Anna, SSSA, Pisa, Italy
e-mail: colla@sssup.it

S. Cateni
e-mail: s.cateni@sssup.it

A. Vignali
e-mail: a.vignali@sssup.it

F. Ferro
ILVA S.P.a, Novi Ligure, Italy
e-mail: floriano.ferro@gruppoilva.com

C. Schirm · J. Melcher
Thyssenkrupp Rasselstein GmbH, Andernach, Germany
e-mail: christoph.schirm@thyssenkrupp.com

J. Melcher
e-mail: josef.melcher@thyssenkrupp.com

© Springer Nature Singapore Pte Ltd. 2018
S. S. Roy et al. (eds.), *Big Data in Engineering Applications*,
Studies in Big Data 44, https://doi.org/10.1007/978-981-10-8476-8_1

1

Keywords Big data · Manufacturing · Flat steel production
Data visualization · Analytics

1 Introduction

According to the German ICT industry association BITKOM.

Big Data means the analysis of large amounts of data from a variety of sources at high speed aiming to generate economic benefit [1].

In other words, the aim of Big-Data is not to create vast data pools, but to make use of the data in a target-oriented manner.

From the viewpoint of manufacturing industries, this means that it is much more important how to combine the data with production knowledge than to process huge amounts of data in real-time. In this context, the focus must be set on the usability of Big-Data and not only on the technological limits of data processing. Therefore, in [2] the concept of the domain expert is introduced. Contrary to the technological expert, he does not care about the well-known Vs,[1] but demands 3 Fs to Big-Data applications:

For the domain expert the usage of Big Data should be F̲ast, F̲lexible and F̲ocused.

Thus, in this chapter we present a solution that tries to maximize data usability, already before storage, by means of a multi-scale data representation. Accompanied with the knowledge about the production history of each single product this can be used to implement ETL-procedures providing tailored data for any kind of through process analysis, enabling fast, flexible and focused Big-Data applications.

2 Problem Definition

Today modern measuring systems support the production of high quality steel and provide an increasing amount of high resolution (HR) quality and process data along the whole flat steel production chain. Although this amount appears to be quite small compared to the huge amount of data that occurs e.g. in the world wide web, especially the complexity of the flat steel production chain makes detailed analytic tasks difficult for classical relational database management systems (RDBMS).

---

[1]V̲olume, V̲ariety, V̲elocity, V̲eracity

2.1 Spatial Querying

First reason impeding data analytic tasks in flat steel production is the fact, that the product (steel coil) basically implies a 2-dimensional spatial object. Whereas on the one hand it is no problem for a standard RDBMS to access the whole HR data for each single coil produced quite fast, it is much more difficult to retrieve only partial data.

If the task is:

> Retrieve all available HR data within a certain coil-region over one whole production period containing thousands of coils.

Standard data-warehouse concepts are approaching their limits very fast. To formulate spatial queries using standard SQL leads to extremely complex statements. Spatial extensions of common RDBMS like MS SQL Server 2008 [3] or postGIS for PostgreSQL [4] are able to cope with spatial objects, apply spatial indices and execute spatial queries, but they are devoted to supply geographical data.

For instance in case of the MS SQL Server the spatial indices are based on regular grid structures in different resolutions approximating spatial objects [3]. The main problem of this approach is that the regular grid structure should preferably be small compared to each single spatial object described by this index to perform well. However, in case of single surface defects measured by an automatic surface inspection system (ASIS) this relation is just the opposite. There are many small objects detected on the coil that would need a very small grid structure for the spatial index and thus cannot be efficiently covered. The underlying index structure is dedicated to less and larger objects as usually the case in geographic applications but not for industrial data.

The addressing and aggregation of a multiple-coil request, as foreseen for HR data evaluation is not supported directly and thus not of high-performance. Figure 1 shows the result of a trial setup using a geospatial database (PostgreSQL + PostGIS) and GeoServer [5] as frontend, which is an open source geospatial web service engine dedicated to process geographical data. The applicability and performance of this architecture was tested by means of Apache JMeter [6] measuring the querying and aggregating performance of surface defects of 5 coils at a time (different for each request).

In this trial, an average response time of 550 ms on the basis of 1500 requests was obtained, but the response time increased exponentially with the number of queried coils, leading to unacceptable results if more coils are analysed [7].

However, for quality monitoring and improvement of the production processes a more statistical view on the data is mandatory. The absolute coil-position of a measurement looses importance and a normalized view on the data should be used to compare not only single coils but also full production cycles and/or material groups regarding suspicious data distributions. This involves not only 5 but several

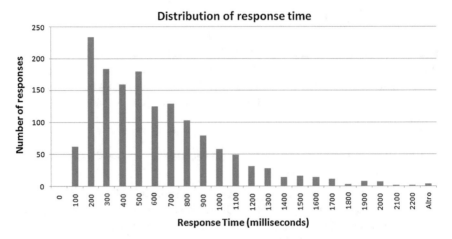

Fig. 1 Distribution of response time for spatial querying the data of 5 coils

hundred coils thus a new concept of HR data representation had to be found allowing the fast aggregation of spatial information over a massive amount of individual coils.

2.2 Product Tracking

To be able to locate a quality problem in the flat steel production chain it is mandatory to be able to track each individual coil or part of a coil through the full production process. As exemplary shown in Fig. 2 for the tinplate production, such a production chain can be quite complex.

Moreover, it is very likely that an effect of a quality problem is measured at a process step different from the step that causes that problem. Consider an ASIS installed at a tinning line located at the end of the processing chain that detects surface defects emerging at a degreasing line.

To be able to correlate process parameters with such a quality problem, it must be possible to calculate the emerging position p^* of each single surface defect from the position p of the ASIS detection at the tinning line as

$$p^* = \left(T_{d \to t}^c\right)^{-1}(p) \tag{1}$$

where $T_{d \to t}^c$ is the individual position transformation of coil c from the degreasing line to the tinning line. This transformation is a composition of the following base transformations occurring in the flat steel production process:

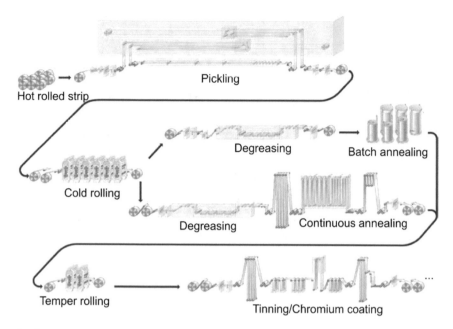

Fig. 2 Exemplary production chain for tinplate production [8]

- **winding**—always causes a switch of coil start and end. Furthermore a switch of top/bottom and left/right side is possible
- **production process**—rolling processes cause coil elongation and thus linear position shifts
- **cutting/welding**—due to continuous production, customizing and repairing operations coils may be cut and new coils assembled from multiple coil parts of preceding process steps

In Fig. 3 each blue dot on the most right picture represents the relative position of one or more surface defects as detected by an ASIS installed at the tinning line. These defect positions are transformed to the preceding lines according to the available tracking information for the individual coil. Consequently, the quality data of each coil has to be transformed individually before they can be combined with process data from preceding lines making this kind of through-process investigations very complex and time-consuming for standard data-warehouse concepts. A dedicated storage concept should be able to consider tracking information already in the ETL-procedures to provide fast query response times for data analytic tasks.

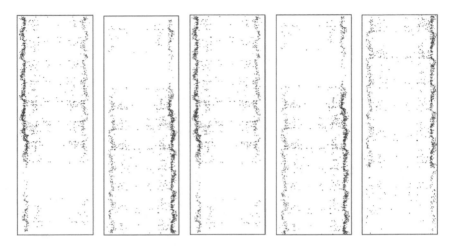

Fig. 3 Position of surface defects tracked through the production chain

3 How to Apply Big Data Concepts in Flat Steel Production?

The HR data that occurs in flat steel production are values provided by measuring systems installed at different steps in the production chain. Thus for a suitable production model, each processing step of a single product has to be considered as one product instance.

Furthermore, all gathered data has to be assignable to an individual product and each measurement to a single position on this product. This means in particular, that time-series must be synchronized with the production and consequently all data becomes spatial. To allow synchronization between different production steps additionally the full tracking information of each coil should be available. Figure 4 shows the different data types that have to be considered as HR-data in flat steel production.

Usually for further processing this kind of information is aggregated based on constant length segments (e.g. 1 m) and stored in a factory-wide quality database (QDB). Thus, already today common applications based on the available data in a QDB can monitor the current production, support quality decisions or allow dedicated investigations in case of customer claims [9]. However, to realize a solution supporting as well fast in-coil aggregations as displaying data from the viewpoint of different production steps by considering material tracking information, a tailored data model is needed.

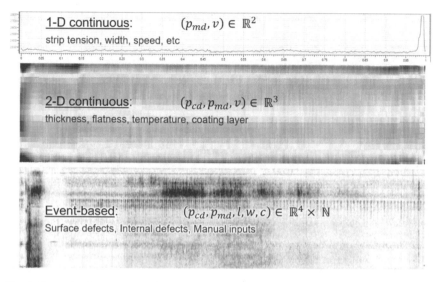

1-D continuous: $(p_{md}, v) \in \mathbb{R}^2$

strip tension, width, speed, etc

2-D continuous: $(p_{cd}, p_{md}, v) \in \mathbb{R}^3$

thickness, flatness, temperature, coating layer

Event-based: $(p_{cd}, p_{md}, l, w, c) \in \mathbb{R}^4 \times \mathbb{N}$

Surface defects, Internal defects, Manual inputs

Fig. 4 Example of measurement data entries in flat steel production, with coil length-position p_{md}, coil width-position p_{cd} and measurement value v or event attributes: length l, width w and class c

3.1 Production Data Model

A suitable data model that is able to provide efficient HR data access for flat steel production was developed in the European research project 'EvalHD' [7]. The general idea is to address the problem from the domain expert point of view. To visualize HR-data on the screen an image has to be created representing a pixel matrix $I := [0, N_x] \times [0, N_y]$. Each pixel $p_{x,y} \in I$ represents a rectangle $R_{x,y}$ on the coil that is located relative to each pixel position in I [10]. The color (resp. value) of each pixel is calculated from the HR measurements within the corresponding rectangle $R_{x,y}$.

Thus, for visualisation and a given image I it is sufficient to store only the position x, y together with the aggregated value $p_{x,y}$ in a production data model without loosing any information. By means of a bijective function

$$\mu: [0, N_x] \times [0, N_y] \to [0, N_x N_y] \tag{2}$$

furthermore a unique *TileID*: $= \mu(x, y) \in [0, N_x N_y]$ can be assigned to each pixel position simplifying the aggregation over multiple coils significantly as this aggregation can be directly performed over equal *TileID*s.

Moreover, this means that using a grid structure fitted to the size and resolution of the intended visualisation I is a minimal data representation as it stores exactly the data that is required. On the other hand there are some applications where a resolution much lower than $N_x \times N_y$ is reasonable. One example is the

Table 1 Visualisation of surface defect distribution over about 8000 coils in different resolution stages

Stage 0	Stage 1	Stage 2
n = 2	n = 8	n = 32
Stage 3	Stage 4	Stage 5
n = 128	n = 512	n = 2048
Stage 6	Stage 7	Stage 8
n = 8192	n = 32768	n = 131072

cause-and-effect analysis described in 5.2, but also for visualisation it may be beneficial to use lower resolutions. They can be used to realize a 'coarsest first' visualisation and a user-experience similar to other modern rendering engines as applied e.g. by the virtual globe "Google Earth" and in detail described in [11]. A parallel querying of the desired information over multiple resolutions and the immediate visualisation of the finest available data as soon as it is completely processed leads to a low response time and high user-acceptance of the system [12].

This idea leads to a multi-scale grid representation of measurement data as shown in Table 1. According to Eq. (2) in this representation each grid cell can be uniquely addressed by means of the pair (*Stage, TileID*), thus keeping the fast aggregation capabilities of the single stage model.

For the final production data model a grid of 1 cell in cross direction (CD) times 2 cells in production direction (MD) was chosen as coarsest stage 0 resolution. The different number of cells was chosen because of the unequal length to width ratio of a steel coil (often < 1:10000). For the next stage, the resolution is multiplied by 2 in each dimension leading to the final grid hierarchy shown in Table 2. In this setup, exactly 4 grid cells of stage $i+1$ fit in a grid cell of stage i and thus each *TileID* of stage $i+1$ can be uniquely assigned to a single stage i grid cell.

To extract the raw data from the productive databases, transform and load them into the common HR data model at first each coil has to be normalized to a length and width of 1. This means that each point $P_c := (x, y)$ on a coil c is converted to a new point

Table 2 Grid definitions and exemplary sizes of grid cells for a coil length of 7500 m and a coil width of 1500 mm

Stage	Tiles CD	Tiles MD	$\Delta x_{c,s}$ ($c_w := 1500$ mm) (mm)	$\Delta y_{c,s}$ ($c_l := 7500$ m) (m)
0	1	2	1500	3750
1	2	4	750	1875
2	4	8	375	937.5
3	8	16	187.5	468.75
4	16	32	93.75	234.38
5	32	64	46.88	117.19
6	64	128	23.44	58.59
7	128	256	11.72	29.3
8	256	512	5.86	14.65

$$\overline{P_c} := \left(\frac{x}{c_w}, \frac{y}{c_l} \right) \in [0,1]^2, \text{with } c_w := \text{coil width and } c_l := \text{coil length.} \quad (3)$$

Consequently, the reachable synchronization accuracy can be calculated dependent on the coil c, the coil dimensions c_w, c_l and the resolution stage $s \in [0, 8]$ as

$$\Delta x_{c,s} = \frac{c_w}{s} \text{ resp } \Delta y_{c,s} = \frac{c_l}{s} \quad (4)$$

Some exemplary values for $\Delta x_{c,s}$ and $\Delta y_{c,s}$ are also given in Table 2.

Once each coil position is normalized, the transformation of point-based raw data into the grid structure can be performed quite easy by simple cell-based aggregation of all measurements falling into one specific grid cell. Regarding 1D and 2D continuous measurements, the aggregations stored in the grid structure are minimum, maximum, mean and count of the measuring values. Event-based data (like surface defects) are usually stored as rectangular regions combined with a certain identifier describing the type of the event (e.g. defect class) and can either be aggregated as absolute counts per grid cell or overlapping area relative to the full cell area.

Given this multi-grid data representation, the question remains how to enable the combination of data across production stages. This again can be easily solved by not only simultaneous storage of data across different resolutions, but also across different perspectives. Assumed that the information about all coil transformations is given during data transformation, the data can be tracked upstream and/or downstream and further grid data can be created and stored for each measurement from the perspective of other production steps. The data is stored for each plant separately according to the available material tracking information. Thus, the data is available simultaneously in different plant coordinates enabling fast HR data access by means of redundant data storage.

Finally, to analyse production data and find causes of quality problems it is essential to be able to filter data according to different production parameters, like material, thickness, production period, etc. Thus, further filter conditions have to be added to each grid entry of the same HR-type (see Fig. 4) to keep filter capabilities of the data representation. The grid attributes finally stored can be classified into five different categories:

- **Coil Filter**—Unique coil identifier that allows filtering grid entries by coil attributes like material type, thickness, process parameters, etc.
- **Identifier**—Unique grid cell identifier needed for fast aggregation (Stage, TileID)
- **Sub Filter**—Further type specific filter conditions (defect class, measuring device, etc.)
- **Data**—Per grid cell aggregated measuring data (min, max, mean, count)

This production data model is able to synchronize and aggregate HR data of different kind from different perspectives very fast. Therefore, it acts as a kind of database index on the available HR raw data supporting dedicated querying of grid data.

4 Implementation

The production data model described above was implemented as classical three-tier architecture as shown in Fig. 5. This architecture has the benefit that it separates presentation, application processing, and data management functionality.

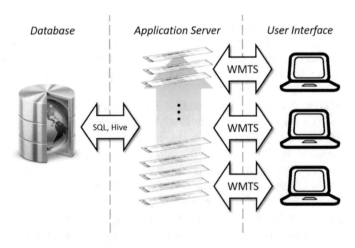

Fig. 5 Schema for HR data access

At the bottom of this architecture, a database management system (DBMS) implements the HR-data model. In this approach, it is not relevant if the database is a standard RDBMS or a Hadoop cluster. The application server has to cope with it and use the correct query syntax to provide the desired grid data by means of parallel querying the employed database. On top of this architecture, a browser application communicates with the application server following a unified web-service definition that is based on the Web-Map-Tile-Service (WMTS) standard provided by the Open-Geospatial Consortium [13]. In the implemented setup the querying of the data follows a two-step approach:

1. Query all coils meeting certain filter conditions applied by the user
2. Query grid data according to the selected coils

The resulting grid data can be provided either aggregated (for visualisation) or per-piece (for cause-and-effect analysis). If material tracking should be considered one important detail of the final implementation is, that each coil queried in the first step knows its own production history. This allows switching the viewpoint to another process step without re-querying the selected coil-set. Furthermore, it is possible to select only coils that where processed at a certain line being another important aspect when searching for quality problem causes. For further details on the web-service definition, please refer to [14].

5 Application

To proof the usability of the architecture depicted in Fig. 5 it was finally implemented at two industrial sites. The production data was transformed to grid data and continuously imported in the HR data model. Based on the available data a solution for the fast data visualisation was realized supporting instant-interactive data analysis and a solution for refined cause-and-effect analysis was implemented.

5.1 Visualisation

The system implemented at thyssenkrupp Rasselstein GmbH finally involved 1137 HR-measurements from 24 main aggregates of the complete tinplate production chain together with the full material tracking information. This includes data from the hot strip mill to the finishing lines at the end of the production. As database, an MS SQL Server 2012 has been chosen with a capacity of 20 TB being sufficient to cover approx. 1 year of full production grid data.

It was necessary to put a lot of effort in the implementation of the import services to be able to store the available HR data to the server without flooding. Extensive use of methods like bulk inserts, parallel processing and index-free temporary tables

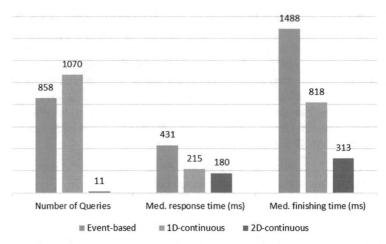

Fig. 6 Performance statistics of HR server over 2 months usage

were required to finally achieve 'coil-realtime', meaning that the time required for data storage can follow the production. It can be reasonable assumed that this will be no issue using a database system dedicated to Big-Data processing. On the other hand, it has to be investigated if the query performance of such a system can be competitive with the index structures provided by the standard RDBMS.

Figure 6 shows a performance statistic over two months of system usage. In this period the median response time of the system, providing defect data was 215 ms. This response time refers to the first visualisation of the lowest resolution stage queried. The querying process was implemented by means of parallel SQL-queries for 8 equally sized full width stripes distributed over the full coil length. In this trial the multi-scale visualisation started with stage 2 and refined over stage 6 before finally stage 8 results were presented.

On average (median) the browser application was able to provide the full resolution defect data to the user (8 stripes at stage 8 resolution) in less than 1.5 s. Furthermore, it can be seen that the usage of the system in the testing period was mainly focused on the analysis of 1D and event-based data, whereas 2D-continuous data played only a minor role.

5.1.1 Paw-Scratch Example

The following example clearly demonstrates the benefit of the developed solution as it shows how a quality problem could be successfully solved using the interactive visualisation solution presented in this chapter. The quality problem investigated was the so-called 'paw-scratch'—defect that often looks like a paw print of an animal. This defect is well detected by ASIS and can be classified very reliable by using context information in post-processing rules [9]. Thus, it is a good choice for

Relative defect positions as detected by ASIS at finishing line

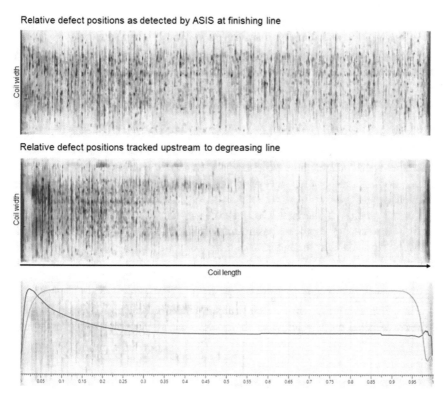

Relative defect positions tracked upstream to degreasing line

Fig. 7 Relative defect positions of paw-scratch defects at the finishing (top) and tracked to the degreasing line (middle). Bottom: in-coil aggregated mean values of related process variables (light: line speed, dark: strip tension)

a detailed ASIS data analysis as no manual data verification is required [15]. The investigation started with the analysis of paw-scratch defects as detected by an ASIS installed at the finishing line. The visualisation on top of Fig. 7 shows the distribution of this type of defect over a set of more than 2000 coils affected by this defect and combines more than 500.000 single defects in one image. Herein the most blue grid positions represent more than 500 single paw-scratch detections.

The picture in the middle of Fig. 7 shows the same result as the top picture after each single ASIS result of each individual coil has been tracked to one of the two degreasing lines located at the thyssenkrupp site in Andernach. In this case, a characteristic distribution of the paw-scratch defects becomes visible and it appears that significant more paw-scratches were located at the beginning of the coils.

This example impressively shows what happens if no tracking information is considered for data analysis. Due to the individual coil transformations as described in Sect. 2.2, the characteristic defect distribution at the causative line gets completely lost throughout the production chain. Thus, in this case no reasonable correlation analysis can be performed by means of the defect positions as measured by the ASIS.

Generally, an almost uniform distribution as shown on top of Fig. 7 is a strong indication that the cause of this defect is not located at the specific line. On the other hand once there is a kind of characteristic distribution visible as in the middle, it makes sense to correlate the relative positions of the surface defects with process parameters of the plant.

The bottom picture of Fig. 7 shows an overlay of the paw-scratch distribution at the degreasing line and the stage 8 grid mean values of the 1-dimensional process data 'line speed' (light) and 'strip tension' (dark). It can be seen that there is a strong correlation between the dark strip tension graph and the paw-scratch locations.

For the quality engineers this correlation was taken as reason to perform some trials, how to adapt the process parameters of the degreasing line in such a way that a lower strip tension at the beginning of the coil can be achieved. To evaluate the trial results again the ASIS data of the finishing lines had to be tracked back to the degreasing line to see if the new control strategy led to the desired result. This again could be done pretty easy using the developed visualisation system.

Thus, iteratively the paw-scratch problem could be solved and today 90% less coils are affected by paw-scratches than before the implementation of this system.

5.2 Cause-and-Effect Analysis

The occurrence of ripple defects in the course of the Hot Dip Galvanizing (HDG) process on thick coils (i.e. thickness ≥ 1.5 mm) with low zinc coating (i.e. in the range 50–71 g/m m^2) has been examined at ILVA s.p.a. Ripples are vertical line shaped defects that could be designed as diffuse coating ruffles so that they are identified by ASIS systems without difficulty at the end of HDG lines [16]. The process parameters, which mainly affect the occurrence and the significance of ripples, are the air blades configuration, cooling techniques, process speed and wiping medium. The real effects of each process variable deviation is still not very clear; skilled personnel control ripple defects by employing nitrogen as wiping medium in air blades but it is not always an effective method and this action could increase costs uselessly. Moreover, a greater understanding of the phenomenon under observation can improve the quality by decreasing reworked or scrapped material.

The above-described problem has been dealt by analysing data from a HDG line at ILVA, including 1D-continuous HR measurements of 20 process attributes that can be categorised into four categories:

1. Air blades;
2. Temperatures (zones before and after the zinc bath, top-roll, water bath);
3. Line speed;
4. Fan coolers.

The case study has been divided into two analysis considering the use of nitrogen in the air blades. The first analysis is devoted to study the process conditions that minimize the ripple presence despite only air is blown and the second analysis regards the knowledge of process conditions that lead to a high defectiveness while employing Nitrogen. This analysis is important as the Nitrogen is expensive and it is interesting to minimize its use maximizing its effectiveness.

When only air is blown, the target is to find a set of process conditions that allows minimizing the ripple occurrence, while, on the other side, when nitrogen is employed, the target is to avoid the occurrence of ripples at all. Nitrogen is in facts expensive, thus, its use should be minimized and its effectiveness maximized.

Due to this reason, two datasets were organized for air and nitrogen blowing, respectively, which comprised HR measurements of the process attributes highlighted above as inputs, and a binary classification of the tiles (null value for tiles without ripple defect and unitary values for highly defective tiles) has been carried out. Dataset is composed by about 360 coils that are developed through the HR data model dealt in 3.1 and pre-processed in order to remove outliers performing a multivariate Fuzzy-based method (FUCOD) that is in detail designated in [17, 18].

Another issue regards the fact that the available variables are 1D-continuous, while defects are 2D-continuous. In order to aggregate, input and target tiles are combined into so-called 'slices' along the coil width by summing ripple defects along that direction. An example of the stage 2 slices is shown in Fig. 8.

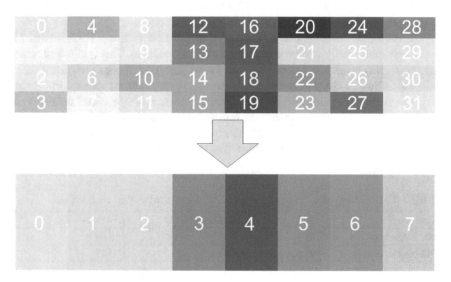

Fig. 8 Tile aggregation along coil width (top: tiles for stage 2, bottom: aggregation of tiles to create associated slices)

A binary classification based on Decision Tree (DT) has been developed; class 0 represents the slices with the total absence of defects, while class 1 identifies highly defective slices. With the term *highly defective,* we indicate slices with a number of defects that exceeds a threshold. The threshold is automatically computed and fixed to the 95th quantile of the empirical cumulative distribution of the percentage area of defects.

Dataset is randomly shuffled and a training and a validation set are defined preserving the initial proportion among the two classes. The training include the 75% of the available samples while the validation set is composed by the remaining 25%. For both case studies (air and Nitrogen blowing), classifier based on Decision Tree (DT) has been carried out and subsequently validated on the respective validation set [19].

The performance has been evaluated computing the Balanced Classification Rate (BCR) as defined in Eq. 5.

$$BCR = \frac{1}{2}\left[\frac{TP}{TP + FN} + \frac{TN}{TN + FP}\right] \tag{5}$$

where TP is the number of unitary values correctly classified, FN the number of unitary values incorrectly classified, TN the number of null values correctly classified, and FP the number of null values incorrectly classified. BCR is more appropriate for imbalanced datasets than the classical accuracy index, as in both available datasets the null class is far more frequent [20–22].

Each node of the trained DT represents the associated process variable, each branch corresponds to a range of values it can assume and finally the leafs correspond to the two defined classes. Through a path from the root to a leaf the procedure detects a process window leading to a specific result, taking into account if the leaf value is unitary or null.

Decision tree classifier can be translated in a simple chain of IF-THEN-ELSE rules becoming easily interpretable by no-skilled operators. This method provides an actual way to support decisions and to extract good process windows to be adopted during production to avoid defects; moreover, it can be adopted to provide a degree (namely *importance*) of how much a process variable affects the analysed target, so that the quality experts can further investigate on it [23].

Table 3 illustrates the very satisfactory performances of the classifier, while Table 4 shows the selected variables that mostly affect the target for the two

Table 3 Classification performances in terms of BCR evaluated on the validation set

Case study	BCR	Accuracy class 0	Accuracy class 1
Air-blowing (%)	99.34	99.78	98.90
N2-blowing (%)	97.65	98.92	96.37

Table 4 Most affecting process variables and the associated normalized relevance

Air blowing		Nitrogen blowing	
Air blade distance	1	Water bath temperature	1
Tunnel zone temperature	0.69	Air blade distance	0.46
Line speed	0.29	Hot briddle zone temperature	0.39
Air blade pression	0.25	Line speed	0.14
Fans speed	0.15	Air blade pression	0.14
Top-roll zone temperature	0.08	Fans speed	0.11
Water bath temperature	0.07	Air blade height	0.05

sub-problems, which is representive of the 95% of the information content. The proposed method is generic and do not require any a priori assumptions, for this reason it can be employed in other applications [24].

6 Comparison to Common Concepts

The described HR data model natively provides a solution for the problems of flat steel production data synchronization and material tracking, which need to be solved before a through-process data analysis can be performed in this environment.

In the present implementation, the model has been realized by means of a common RDBMS as the available amount of data allows to realize the full model data during ETL procedures and to store it completely on the server. However, this model could also be realized using Big Data Management Technologies and MapReduce to improve its scalabilty.

Figure 9 shows the median query response and finishing times, already mentioned in Sect. 5.1, depending on the number of queried coils. According to the linear approximation (dotted lines), it can be stated that the response time seems to be almost independent on the number of queried coils, whereas for the finishing time there is a slight increase. This behaviour is mainly achieved due to the shift of the tracking consideration and the data synchronization to the ETL procedures and the use of an adequate index structure on the RDBMS.

In contrast, the situation for a common data-warehouse concept, which is optimized for a fast per-coil data access, is shown in Fig. 10. It shows the finishing times for a visualisation of coil-sets of 10, 100 and 1000 coils and indicates the distribution of the finishing times measured in 5 runs for each coil-set size. Before the visualisation result can be presented, each coil has to be tracked individually and the data has to be synchronized. This leads to a clear dependency of the finishing times on the number of queried coils and a query time of approx. 10 min for a visualisation of 1000 coils.

In other words, when 1000 coils are queried, the presented data model is approx. 300 times faster than the common data-warehouse concept. This relation is even

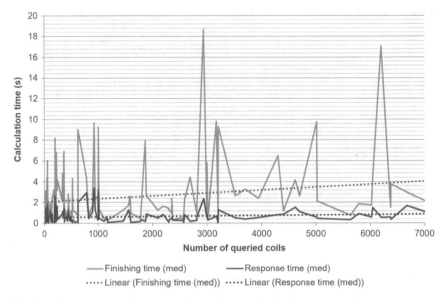

Fig. 9 Performance of the implemented visualisation solution (event-based data)

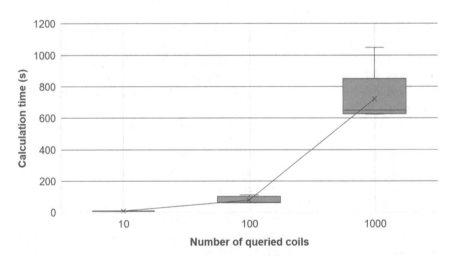

Fig. 10 Query performance of a common data-warehouse concept (event-based data)

increasing when more coils are queried making our concept outstanding fast. Thus, for the first time ever the HR query response times allow an instant interactive analysis of HR data as soon as a quality problem occurs, enabling a new dimension of quality data assessment for flat steel production [14].

7   Conclusion

For an effective application of Big-Data technologies in manufacturing industries, it is not sufficient to store a massive amount of raw data. Instead, a full production model is mandatory to enable through process synchronization of all available measuring data.

Therefore, in this chapter we describe a suitable production model for flat steel production, able to realize fast, flexible and focused access to industrial Big-Data, due to a new multi-scale data representation across production steps. Using a three-tier architecture, we could successfully implement this approach at two industrial sites and proof its usability. Moreover, as well for a fast data visualisation supporting the interactive investigation of quality problems as for providing source data for HR cause-and-effect analysis using more than one aggregated value per coil, we could show that this new concept performs far better than any state-of-the-art data model in terms of query response time.

Concluding it can be stated that standard data-warehouse concepts are not appropriate to utilise the full potential of modern measuring equipment in flat steel production, as an efficient statistical evaluation of multi-coil HR data is not adequately supported. On the other hand, new technologies combined with a suitable production model can provide valuable input to quality engineers and plant operators already from the very beginning.

Acknowledgements   The work described in the present paper was developed within the project entitled "Refinement of flat steel quality assessment by evaluation of high-resolution process and product data—EvalHD" (Contract No RFSR-CR-2012-00040) that has received funding from the Research Fund for Coal and Steel of the European Union. The sole responsibility of the issues treated in the present chapter lies with the authors; the Commission is not responsible for any use that may be made of the information contained therein.

References

1. Bartel, J., Decker, B., Falkenberg, G., Guzek, R., Janata, S., Keil, T. et al. (2012). Big Data im Praxiseinsatz: Szenarien, Beispiele, Effekte. Bundesverband Informationswirtschaft, Telekommunikation und neue Medien e.V. (BITKOM).
2. Freytag, J.-C. (2014). Grundlagen und Visionen großer Forschungsfragen im Bereich Big Data. *Informatik-Spektrum, 37,* 97–104.
3. Katibah, E., & Stojic, M. (2011). New Spatial Features in SQL Server Code-Named 'Denali'. SQL Server Technical Article. https://msdn.microsoft.com/en-us/library/hh377580.aspx.
4. PostGIS. http://postgis.net/.
5. GeoServer. http://www.geoserver.org.
6. Apache JMeter. https://jmeter.apache.org/.
7. Brandenburger, J., Schirm, C., Melcher, J., Ferro, F., Colla, V., Ucci, A., et al. (2016). Refinement of Flat Steel Quality Assessment by Evaluation of High-Resolution Process and Product Data (EvalHD). European Commission, Directorate-General for Research and Innovation.
8. thyssenkrupp Rasselstein. (2015). Wege der Produktion. Brochure.

9. Brandenburger, J., Piancaldini, R., Talamini, D., Ferro, F., Schirm, C., Nörtersheuser, M., et al. (2014). *Improved Monitoring and Control of Flat Steel Surface Quality and Production Performance by Utilisation of Results from Automatic Surface Inspection Systems (SISCON)*. European Commission, Directorate-General for Research and Innovation.
10. Brandenburger, J., Schirm, C., & Melcher J. (2016). Instant interactive analysis—how visualisation can help to improve product quality. In *Surface Inspection Summit SIS*. Europe, Aachen.
11. Tanner, C. C., Migdal, C. J., & Jones, M. T. (1998). The Clipmap: A virtual Mipmap. In: *Proceedings of the 25th Annual Conference on Computer Graphics and Interactive Techniques* (pp. 151–158). ACM.
12. Nielsen, J. (1993). *Usability Engineering*. Morgan Kaufmann Publishers Inc.
13. OGC OpenGIS. (2010). *Web Map Tile Service Implementation Standard*. Open Geospatial Consortium Inc.
14. Brandenburger, J., Colla, V., Nastasi, G., Ferro, F., Schirm, C., & Melcher, J. (2016). Big data solution for quality monitoring and improvement on flat steel production. In *7th IFAC Symposium on Control, Optimization and Automation in Mining, Mineral and Metal Processing MMM*, Vienna.
15. Brandenburger, J., Stolzenberg, M., Ferro, F., Alvarez, J. D.; Pratolongo, G., & Piancaldini, R. (2012) *Improved Utilisation of the Results from Automatic Surface Inspection Systems (IRSIS)*. European Commission, Directorate-General for Research and Innovation.
16. Borselli, A., Colla, V., Vannucci, M., & Veroli, M. (2010). A fuzzy inference system applied to defect detection in flat steel production. In *IEEE World Congress on Computational Intelligence, WCCI*.
17. Cateni, S., Colla, V., & Nastasi, G. (2013). A multivariate fuzzy system applied for outliers detection. *Journal of Intelligent and Fuzzy Systems, 24*, 889–903.
18. Cateni, S., Colla, V., & Vannucci, M. (2007) A fuzzy logic-based method for outliers detection. In *Proceedings of the IASTED International Conference on Artificial Intelligence and Applications* (pp. 561–566).
19. Cateni, S., Colla, V., & Vannucci, M. (2010). Variable selection through genetic algorithms for classification purposes. In: *Proceedings of the 10th IASTED International Conference on Artificial Intelligence and Applications, AIA* (pp. 6–11).
20. Vannucci, M., Colla, V., Cateni, S., & Sgarbi, M. (2011). Artificial intelligence techniques for unbalanced datasets in real world classification tasks. In: *Computational Modeling and Simulation of Intellect: Current State and Future Perspectives* (pp. 551–565).
21. Vannucci, M., & Colla, V. (2011). Novel classification method for sensitive problems and uneven datasets based on neural networks and fuzzy logic. *Applied Soft Computing Journal, 11*, 2383–2390.
22. Vannucci, M., & Colla, V. (2015). Artificial intelligence based techniques for rare patterns detection in the industrial field Smart Innovation. *Systems and Technologies, 39*, 627–636.
23. Cateni, S., Colla, V., & Vannucci, M. (2009). General purpose input variables extraction: A genetic algorithm based procedure GIVE a GAP. In *ISDA 2009—9th International Conference on Intelligent Systems Design and Applications* (pp. 1278–1283).
24. Cateni, S., Colla, V., Vignali, A., & Brandenburger, J. (2017). Cause and effect analysis in a real industrial context: study of a particular application devoted to quality improvement. In *WIRN 2017, 27th ItalianWorkshop on Neural Networks June 14–16*, Vietri sul Mare, Salerno, Italy.

Parallel Generation of Very High Resolution Digital Elevation Models: High-Performance Computing for Big Spatial Data Analysis

Minrui Zheng, Wenwu Tang, Yu Lan, Xiang Zhao, Meijuan Jia, Craig Allan and Carl Trettin

Abstract Very high resolution digital elevation models (DEM) provide the opportunity to represent the micro-level detail of topographic surfaces, thus increasing the accuracy of the applications that are depending on the topographic data. The analyses of micro-level topographic surfaces are particularly important for a series of geospatially related engineering applications. However, the generation of very high resolution DEM using, for example, LiDAR data is often extremely computationally demanding because of the large volume of data involved. Thus, we use a high-performance and parallel computing approach to resolve this big data-related computational challenge facing the generation of very high resolution DEMs from LiDAR data. This parallel computing approach allows us to generate a fine-resolution DEM from LiDAR data efficiently. We applied this parallel computing approach to derive the DEM in our study area, a bottomland hardwood wetland located in the USDA Forest Service Santee Experimental Forest. Our study demonstrated the feasibility and acceleration performance of the parallel interpolation approach for tackling the big data challenge associated with the generation of very high resolution DEM.

Keywords High-performance and parallel computing · Spatial domain decomposition · Very high resolution DEM · LiDAR data

M. Zheng · W. Tang (✉) · Y. Lan · M. Jia · C. Allan
Department of Geography and Earth Sciences, University of North Carolina at Charlotte, 28223 Charlotte, USA
e-mail: wtang4@uncc.edu

M. Zheng · W. Tang · Y. Lan · M. Jia
Center for Applied GIScience, University of North Carolina at Charlotte, 28223 Charlotte, USA

C. Trettin
U.S. Forest Service, Center for Forested Wetlands Research, 29434 Cordesville, USA

X. Zhao
School of Resources and Environmental Science, Wuhan University, Wuhan, China

© Springer Nature Singapore Pte Ltd. 2018
S. S. Roy et al. (eds.), *Big Data in Engineering Applications*,
Studies in Big Data 44, https://doi.org/10.1007/978-981-10-8476-8_2

1   Introduction

Digital elevation models (DEM) allow for representing the topographic surface of the Earth by providing spatial location and the elevation information over a geospatial area [26]. As a common data source for topographic analysis, DEM data can be produced from a series of technologies, exemplified by Light Detection and Ranging (LiDAR) technologies. Over the past few years, LiDAR data that provides details of geographic features have been increasingly collected to generate DEMs for the delineation and analysis of topographic surfaces, which are essential in a suite of science and engineering domains, such as hydrologic engineering [8, 40], geographic information science and surveying engineering [27, 38], and environmental engineering [13, 18, 41]. Since 1990s, a series of studies have been reported in terms of using LiDAR-derived DEM to support, for example, microtopography analysis [4, 10, 16, 19, 20], plant species distribution [21], and landslide detection [23].

The generation of very high resolution DEMs from LiDAR is well established [6]. However, the computational demand of generating a DEM at very high resolutions (e.g., 0.5, 0.05 m, or even 0.01 m) from LiDAR data is often problematic. The generation of high or very high resolution DEM requires longer computing times together with large storage space requirements as compared to low resolution applications. In other words, the generation of very high resolution DEM is usually accompanied with a big data issue [31, 42] because the volume of the data increases exponentially as spatial resolution becomes finer. To resolve this big data issue facing the generation of very high resolution DEM, high-performance and parallel computing (HPC) represents one possible solution [11, 17, 22].

HPC employs multiple processors (e.g., CPUs) instead of a single one for accelerating a computational problem of interest [39]. Typically, multiple processors used by HPC form into a computing cluster, each of which includes a head node (or master node) and multiple computing nodes connected through network switch [39]. The basic algorithm of parallel computing is to split the entire task of the computational problem into sub-tasks, and then deploy these sub-tasks to multiple processors on the computing cluster for concurrent computation. Once all sub-tasks are completed, the computing nodes return the results to the head node for aggregation. HPC have witnessed an increasing number of applications in scientific fields, such as bioinformatics, molecular dynamics and environmental applications [24, 28, 32, 33, 37].

Thus, to tackle the big data issue, in this study we will generate very high resolution LiDAR-derived DEM using a HPC approach. Our results demonstrate that the HPC is an efficient and effective approach for developing very high resolution DEM that provide representations of topographic features. The parallel computing approach to accelerate the generation of the very high resolution DEM via spatial interpolation is applicable to the widely available LiDAR data thereby expanding the potential application of this high resolution data. For this study we utilized a landscape that is representative of the lower coastal plain in the Southeastern U.S. where small difference in topographic features may have significant ramifications to a wide array of considerations from water management to ecological processes.

2 Study Area and Data

Our study area is the USDA (United States Department of Agriculture) Forest Service Santee Experimental Forest (https://www.srs.fs.usda.gov/charleston/santee). The USDA Forest Service Santee Experimental Forest (SEF) was established in 1937 with the Francis Marion National Forest, South Carolina, with a total area of 2,468 ha (latitudes are from $33.12165°$ to $33.192979°$, and longitude from $-79.752968°$ to $-79.839113°$; see Fig. 1). The purpose of the SEF is to provide a basis for experiments, demonstration trials and long-term monitoring of a variety of field-scale ecological, hydrological, and climatic properties. The forest is representative of the lower coastal plain landscape which is being rapidly developed. It is characterized by very low relief, and gauged watersheds on the SEF are used for monitoring hydrologic responses within 1st, 2nd and 3rd order watersheds that are connected to the East Branch of the Cooper River that flows into Charleston estuary and subsequently the Atlantic Ocean. The study area is characterized by mixed pine-hardwood forests in the uplands and bottomland hardwood forested wetlands. The wetlands are influenced by freshwater tidal cycles and non-tidal systems. An important characteristic of the upland and wetland forests is the spatially distributed microtopographic features (hummocks and hollows), which are impacted differentially by fluctuating water levels in this low relief landscape. The microtopography influences they hydrologic storage properties of a watershed [1] as well as biogeochemical processes, especially related to the carbon cycle in forested wetlands [2, 35].

This LiDAR contains 20 tiles covering our study area (in total 31,561,291 points). The averaged point density is 1 point/m^2 of the LiDAR data managed in point cloud form (see Table 1). Other geospatial data including road networks, streams, and the boundary of the SEF were also available.

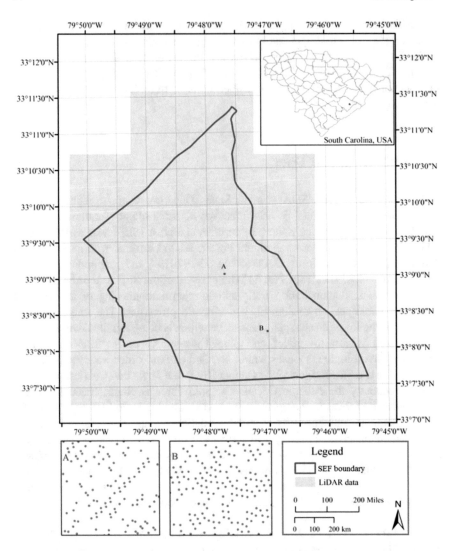

Fig. 1 Study area: santee experiment forest (USDA forest service) within the lower coastal plain, South Carolina

Table 1 Summary of LiDAR data of the study area

Published year	2007
Total tiles of dataset	20
Total size of dataset	3.5G
Geometry type	Point cloud
Unit	Meter
Point density	1 point/m^2

3 Methodology

In this study, we designed a parallel spatial interpolation approach to generate a very high resolution DEM from the LiDAR data in our study region. We performed a spatial domain decomposition on the LiDAR data. Based on the decomposition results, we developed a parallel computing approach for accelerating the generation of high-resolution DEM by applying spatial interpolation of the LiDAR points.

3.1 Spatial Interpolation of LiDAR Data

Spatial interpolation is an approach that predicts the value of an unknown region (point here) based on a number of its surrounding points which values are known. Spatial interpolation can be applied for generating a continuous surface of any geographic variables (e.g., elevation, rainfall, temperature) from sampled locations (as control points). Alternative spatial interpolation algorithms exist, including inverse distance weighted (IDW), Kriging and Spline [25]. All interpolation algorithm can be classified into two basic types: global and local interpolation. The major difference between global and local interpolation is the scope of data used for estimating values of points of interest. Global interpolation uses the entire dataset to estimate points with an unknown value. Local interpolation only considers the points located in a neighborhood distance from the point of interest [36]. IDW is a form of local interpolation algorithms. For massive spatial data, local interpolation has significant advantages because local interpolation can be partitioned into subdomains based on the location of the neighborhood.

In this study, we focus on using the IDW interpolation method for the generation of very high resolution DEM from LiDAR data. As Zimmerman et al. [43] illustrated, IDW predicts the values of unknown locations using a weighted average of points with known values within a certain distance or a given number of nearest points (e.g., 10–30). The weight is inversely proportional to distance between points. The formula for IDW is as follows:

$$r = \frac{\sum_{i=1}^{m} d_i^{-p} v_i}{\sum_{i=1}^{m} d_i^{-p}} \tag{1}$$

where r is the value of a point to be estimated. v_i is the value of a sampling point. m is the number of nearest neighbors. p is the coefficient of the power function and d_i is the distance from m known nearest sampling points to the estimated point r. The power p controls the influence of neighboring points on determining the estimated value of the unknown location of interest.

Cross validation is often needed to select optimal parameters of spatial interpolation from a number of candidates. In this study, we use a Jackknife method (also known as leave-one-out approach) for cross validation. Jackknife is based on removing one sample point of the dataset at a time, and repeatedly estimating value using the remaining points in the dataset [34]. The cross validation performance can be evaluated by the root-mean-square error (RMSE):

$$RMSE = \sqrt{\frac{\sum_{i=1}^{n}(V_{it} - V_i)^2}{n}} \tag{2}$$

where V_{it} is the interpolated value of sample i using remaining $n - 1$ records, V_i is the observed value of sample i, and n is the number of samples in the dataset of interest.

3.2 Parallel Interpolation for the Generation of DEM

The generation of DEM data from massive LiDAR data faces computational challenges. In this study, we developed a parallel spatial interpolation approach for the generation of DEM based on LiDAR data. The past two decades have witnessed a variety of studies on the use of parallel spatial interpolation to solve the computational challenge facing massive spatial data. Armstrong and Mariciano [3] used an IDW interpolation method with a MIMD (multiple instruction, multiple data) parallel processing environment. A few years later, Cramer and Armstrong [5] evaluated static and dynamic domain decomposition strategies for parallel interpolation on using IDW algorithm. Guan and Wu [11] investigated the power of multicore-based parallel computing platforms for generating DEM from massive LiDAR data, in which the IDW interpolation algorithm was used. Huang and Yang [17] proposed a grid computing solution based on the Condor platform (aka, Condor, see https://research.cs.wisc.edu/htcondor/) for the spatial interpolation of DEM using IDW from a large spatial dataset. Li et al. [22] developed a general framework for parallel processing of large-scale LiDAR data, and used DEM generation for Colleton County in South Carolina as a case study to demonstrate the utility of the framework implemented based on a map-reduce mechanism.

There are typically four steps to design a parallel computing algorithm, including partitioning, communication, agglomeration and mapping [9, 29]. With these four steps, Fig. 2 shows the framework of our parallel computing approach for the generation of DEM based on spatial interpolation.

Partitioning is the first step of the parallel spatial interpolation for the generation of DEM. In this study, we used spatial domain decomposition for the partitioning of a large spatial interpolation problem into sub-problems for acceleration. Spatial domain decomposition is one of the spatial strategies that is popular in parallel spatial modelling particularly with big spatial data. The decomposition strategy divides a dataset into several subtasks based on different task requirements, and then schedule these tasks on multiple processors (i.e., computing node). All computing nodes' results are returned and aggregated on the head node. Over the past

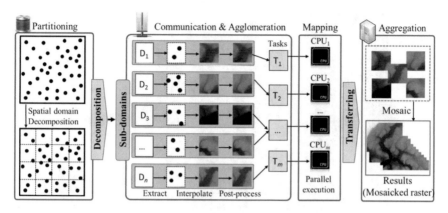

Fig. 2 Framework of the parallel computing approach of spatial interpolation for the generation of very high resolution DEM

several years, domain decomposition is already proved its benefits in accelerating spatial modelling tasks when using parallel computing, and a number of publications on parallel processing based on spatial domain decomposition have been reported. For example, Ding and Densham [7] stressed that one- or two-dimensional decomposition can be used in many types of data: single data type (i.e., binary or categorical) with regular (e.g., square, rectangular, triangle) or irregular shape (e.g., LiDAR data in this study), mixed data types (i.e., binary and categorical) with regular or irregular shape. Because two-dimensional decomposition needs less communication, they also pointed out two-dimensional decomposition is more efficient than one-dimensional decomposition in regular shape data. Besides the one- or two-dimensional decomposition method, there exists other decomposition methods, such as quadtree domain decomposition [36]. We chose to use two-dimensional regular spatial domain decomposition strategies in our study for the parallel spatial interpolation for the generation of DEM data based on LiDAR data. The spatial domain covering our study area is split into a matrix of subdomains in rectangular shapes (see Figs. 2 and 3). Originally, these rectangular sub-domains are non-overlapping.

Handing of communication among tasks associated with subdomains is the second step of parallel computing algorithms. Overlapping spatial domain decomposition is often associated with spatial analysis algorithms in need of information from neighborhood scope [5, 7]. Usually, a non-overlapping domain decomposition is the most efficient way for the parallelization of spatial analysis algorithms. However, non-overlapping decomposition may lead to incorrect results when spatial analysis algorithms depend on neighborhood information, such as IDW algorithm in this study. Therefore, overlapping spatial domain decomposition is often preferred for parallelizing spatial analysis algorithms with neighborhood scope rather than the non-overlapping solution [7]. The overlapping regions of subdomain depends on the neighborhood scope of the focal geospatial features for

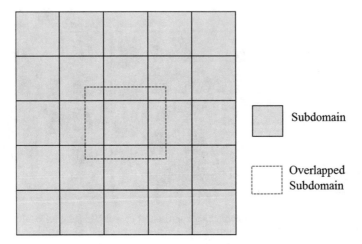

Fig. 3 Illustration on the spatial domain decomposition of parallel spatial interpolation

the spatial analysis algorithms. For example, when using IDW algorithm, the radius of overlapping subdomains is defined by the longest distance from neighboring points to the focal point of interest. A series of meaningful extensions and applications of overlapping domain decomposition have been conducted through the years. For example, Shepard [30] involved overlapping subdomains for parallelizing nearest neighbor search operations. Hohl et al. [14, 15] implemented overlapped subdomains to include neighborhood points within a threshold distance (bandwidth here) to correct edge effects for parallel kernel density estimation.

IDW algorithm used in spatial interpolation in this study relies on the scope of neighborhood because a number of neighboring points are used to estimate the value of a focal point of interest. If the original subdomains that are non-overlapping are directly used, the communication among computing nodes for those neighboring points that may be located in different computing nodes needs to be addressed (e.g., using a message-passing mechanism; see [39]). In this study, instead of message-passing parallelism, we chose to use an alternative approach to address this situation: each subdomain is extended based on a buffer analysis operation (see Fig. 3). As a result, a series of overlapping subdomains are generated that can be used to extract the sub-datasets for spatial interpolation. The radius of the buffer analysis should be larger than the longest distance from neighboring points to the focal point of interest. Thus, it is not necessary to directly handle communication among computing nodes for the parallel spatial interpolation on multiple processors. In other words, there is no overhead for inter-processor communication.

The computing performance of our parallel approach is evaluated using speedup and efficiency. Speedup and efficiency are the two indexes to evaluate the computing performance of a HPC [39]. Speedup (s) is a measure of relative performance between execution time using parallel computing (t_p) with n CPUs and execution time using sequential computing (t_s; one CPU here), defined as:

$$s = \frac{t_s}{t_p} \tag{3}$$

The theoretically maximum speedup is n with n CPUs (say, linear speedup). Efficiency (e) is a standardized metric that is the ratio of speed up over the number of CPUs used for the computation:

$$e = \frac{s}{n} \tag{4}$$

Studies in the literature demonstrated parallel implementation of DEM interpolation can substantially accelerate the overall computation. For example, Guan and Wu [11] used a parallel solution with pipelining algorithm in their spatial interpolation for DEM generation. The computing time of their algorithm was reduced from 50 to 12 min, with a speedup of 4.2 based on 8 processors. In Huang and Yang [17]'s study, the best speedup for interpolating DEM interpolation using Condor is 16, on the basis of 20 CPUs. Li et al. [22] applied their general-purpose computing framework for parallel processing of DEM interpolation on a Hadoop cluster, and the speedup with 10 computing nodes reached 5.38.

3.3 Implementation

The generation of very high resolution LiDAR-derived DEM in this study uses a set of software packages. We used ESRI ArcMap (version 10.3; http://desktop.arcgis.com/en/arcmap/) for IDW spatial interpolation and the mosaic function for the aggregation of all results from different subdomains into a single dataset. Python script is used to implement spatial domain decomposition. We developed a Python script that combines a series of ArcGIS functionalities to automate our parallel domain decomposition and spatial interpolation process.

4 Experiment and Results

4.1 Setting up of High Performance Computing Environments

The high-performance computing resources used for this study are a Windows-based cluster (Sapphire at the Center for Applied Geographic Information Science, University of North Carolina at Charlotte). The Windows-based cluster consists of

M. Zheng et al.

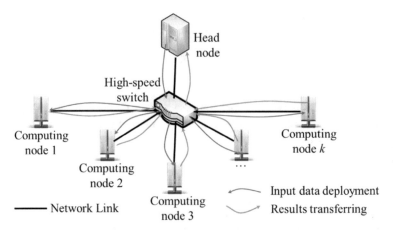

Fig. 4 Illustration on the architecture of a windows-based cluster

20 computing nodes, each of which has 2 CPUs (Intel Core 2 Duo CPU with 3.00 GHz) and 4 GB of memory. The computing nodes are connected through a gigabit band network switch. ArcGIS 10.3 and Python are installed on the head and computing nodes of the cluster (see Fig. 4). We use Microsoft HPC Cluster Manager [12] for job scheduling.

4.2 Result of Very High Resolution LiDAR-Derived DEM

We designed an experiment including five treatments to investigate the impact of granularity level on parallel computing performance. The spatial domain decomposition strategies for the five treatments are 10×10, 15×15, 20×20, 25×25, and 30×30. As a result, the number of decomposed tiles with LiDAR data is 100, 225, 400, 625, and 900 subdomains for these five treatments (see Fig. 5). Each decomposed tile corresponds to a sub-task that can be further aggregated into a task deployed on a processor.

The distance of the buffer used for creating overlapping regions for each subdomain was set to 1 m. The parameter of power coefficient, p (Eq. 1) for IDW-based spatial interpolation is 3.34 with 0.06 RMSE (Eq. 2) for leave-one-out cross validation. Based on the 0.05 m spatial resolution, the number of rows and columns of spatial interpolation results (as in raster data) is $148,217 \times 140,105$. All the treatments are computed on the Window cluster discussed in Sect. 4.1.

Figure 6 depicts the microtopographic detail of this low relief bottomland wetland environment derived from the high resolution DEM developed from our computing procedure. Such data are required to apply spatially distributed hydrologic and biogeochemical models in this and similar wetland environments.

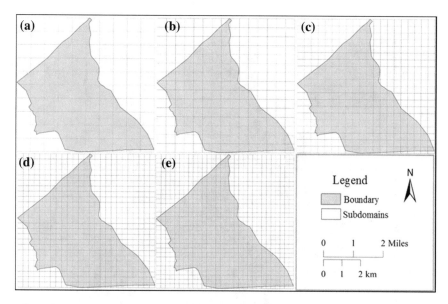

Fig. 5 Map of spatial domain decompositions with different granularities (**a** 10 × 10 subdomains, **b** 15 × 15 subdomains, **c** 20 × 20 subdomains, **d** 25 × 25 subdomains, **e** 30 × 30 subdomains)

4.3 Computing Performance

Table 2 summarized the computing time of these five treatments in response to the number of CPUs used for parallel acceleration. The number of CPUs increases from 2 to 28 with an increment of 2. Usually, the sequential time is computed using the entire dataset on a single CPU. However, the spatial interpolation cannot be directly applied to the entire dataset because the size of the matrix (148,217 × 140,105) used to host the entire spatial interpolation results are too large (which consumes huge amounts of computer memory). Thus, in this study, we used the summation of computing time for each sub-task as an alternative of sequential computing time. The sequential time of the five treatments are 74,866.51 s (about 20.8 h), 70,853.72 s (about 19.7 h), 64,543.72 s (about 17.9 h), 64,343.94 s (about 17.9 h), and 66,869.5 s (about 18.6 h), respectively. As we could observe from Table 2, the execution time for each of the five treatment decreases as more CPUs are added. From Fig. 7, we could see that the execution time rapidly declines when 4 CPUs are used. When the range of used CPUs from 4 to 18, computing time shows a slowly decreasing trend. After 18 CPUs are involved, the parallel computing time is relatively constant (most runs are within 1 h). In particular, when the number of CPUs used for parallel spatial interpolation is 28, the lowest range of computing time is from 3,734 to 2,512 s with increasing the number of decompositions.

From Fig. 7 and Table 2, we see that with the use of 25 × 25 or 30 × 30 decompositions results in limited reductions in parallel computing time. Instead,

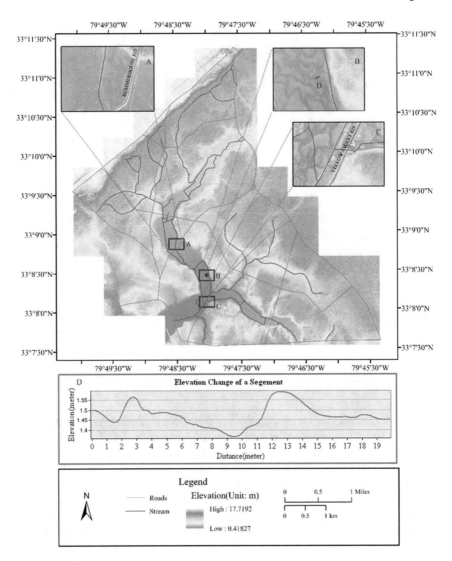

Fig. 6 Map of very high resolution LiDAR-derived DEM (spatial resolution: 0.05 m; landscape size: 148,217 × 140,105)

very fine partitioning will likely lead to increase in computing time because the time spent on spatial domain decomposition time is rapidly increasing. Figure 8 shows computing time for each step of the parallel spatial interpolation (including decomposition, spatial interpolation, and post-processing) over five treatments. Figure 8a illustrates the scenario of sequential computing time and Fig. 8b parallel computing time using 28 CPUs. We observe that spatial interpolation dominates the entire process both for sequential and parallel computing. The computing time spent

Table 2 Computing performance of the experiment of five decomposition granularity levels over number of CPUs (unit: seconds)

CPU	10 × 10	15 × 15	20 × 20	25 × 25	30 × 30
2	37,560.69	35,984.50	32,394.33	32,247.05	33,449.69
4	19,422.50	18,258.56	16,298.61	16,225.25	16,756.99
6	12,981.45	12,129.59	10,892.55	10,894.40	11,240.28
8	10,257.82	9,248.81	8,492.82	8,164.69	8,465.50
10	8,528.61	7,437.02	6,813.05	6,732.07	6,802.18
12	7,250.45	6,351.11	5,605.52	5,512.69	5,705.71
14	6,285.16	5,693.98	4,913.88	4,733.35	4,871.34
16	5,158.56	5,084.68	4,913.88	4,194.72	4,306.56
18	5,329.11	4,494.84	3,893.03	3,863.67	3,784.10
20	4,592.07	3,819.58	3,525.73	3,463.27	3,486.67
22	4,083.29	3,816.80	3,209.88	3,064.32	3,115.70
24	3,950.40	3,360.74	2,961.28	2,869.90	2,937.03
26	3,889.81	3,255.98	2,735.48	2,679.55	2,690.08
28	3,733.85	2,982.76	2,644.97	2,492.91	2,511.66
Sequential	74,866.51	70,853.72	64,543.72	64,343.94	66,869.50

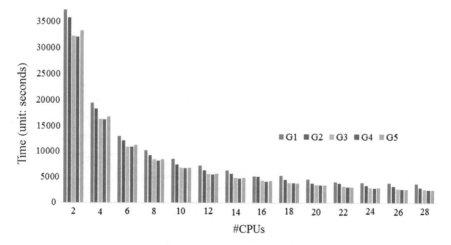

Fig. 7 Parallel computing time of the experiment of spatial domain decomposition granularities over different numbers of CPUs (G1: 10 × 10 subdomains; G2: 15 × 15 subdomains; G3: 20 × 20 subdomains; G4: 25 × 25 subdomains; G5: 30 × 30 subdomains)

on post-processing generally tends to be the shortest among the three steps. For sequential computing, the spatial interpolation and post-processing times decrease slightly when spatial domain decomposition becomes finer, but decomposition time tends to increase. Further, the total execution time tends to decrease since computing time spent on spatial interpolation the step that dominates the entire

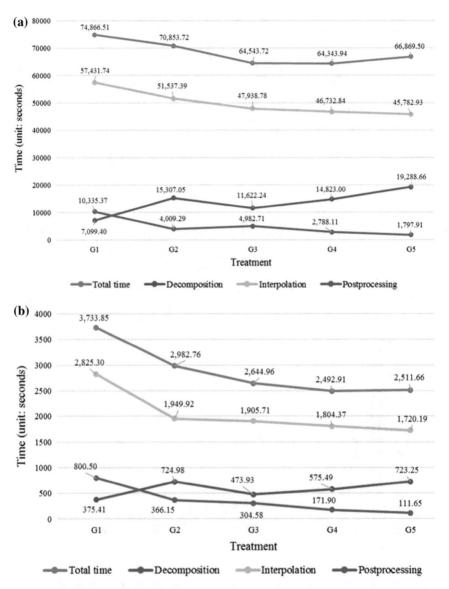

Fig. 8 Computing time for each step of parallel spatial interpolation over five treatments (**a** sequential computing time, **b** parallel computing time; time unit: seconds; #CPUs: 28; G1: 10 × 10 subdomains, G2: 15 × 15 subdomains; G3: 20 × 20 subdomains; G4: 25 × 25 subdomains; G5: 30 × 30 subdomains)

computing time exhibits a decreasing pattern in response to finer decomposition granularity. Likewise, the parallel computing time (Fig. 8b) shows similar patterns in response to increase in the granularity of spatial domain decomposition,

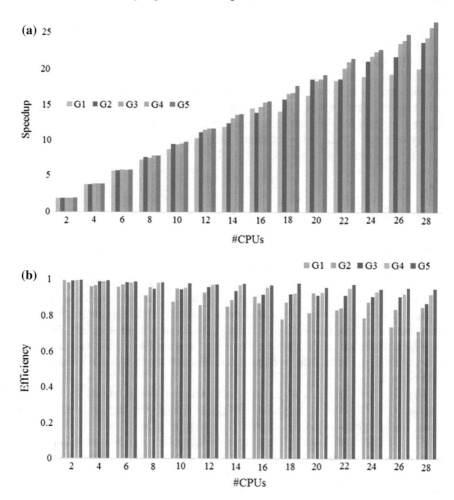

Fig. 9 Speedup and efficiency results in response to the number of CPUs (**a** speedup; **b** efficiency; G1: 10 × 10 subdomains, G2: 15 × 15 subdomains; G3: 20 × 20 subdomains; G4: 25 × 25 subdomains; G5: 30 × 30 subdomains)

but change in the total parallel computing time from 25 × 25 to 30 × 30 decompositions is marginal (about 20 s).

Figure 9 demonstrates speedup and efficiency results over the number of CPUs for the five treatments. Generally, speedup tends to increase as the number of CPUs increases (Fig. 9a). But, increase in speedup when more CPUs are used tend to be slow for coarse spatial domain decomposition. Efficiency results (Fig. 9b) show a decreasing pattern. When the number of CPUs employed is less than 6, the efficiencies of five treatments are close to 1 because the computing performance for each CPU is similar. But, except for the finest decomposition (30×30 tiles), efficiencies of all decomposition treatments tend to decrease as the number of CPUs becomes larger. For 30×30 tiles, the efficiencies stay close to the highest efficiency (i.e., 1) under most circumstances. When spatial domain decomposition becomes finer, the computing time associated with each decomposed subdomains (tiles) tend to be smaller. Thus, the computing time for those tasks that are aggregated from multiple subdomains tends to be balanced. As a result, the efficiency of parallel computing for fine spatial domain decomposition tends to be higher than coarse decomposition.

5 Conclusion

In this study, we demonstrated that a high performance and parallel computing approach to interpolate LiDAR data for generating very high resolution DEM. Those DEMs play an essential role in topographic analyses with a focus on micro-level features, which are often of particular importance for a suite of geospatially related science and engineering applications. In this study, the spatial resolution of 0.05 m was used. Our parallel spatial interpolation results show that very high resolution DEMs provide substantial support for delineating micro-level detail of topographic surfaces such as hummocks or hollows in a low relief wetland environment.

The high-performance and parallel computing solution proposed in this study demonstrated its ability to accelerate the generation of very high resolution DEM using spatial interpolation. As more CPUs were introduced, the execution time tends to decrease substantially (e.g., from 20.8 h to 41 min when using 28 CPUs). Our results suggest that when spatial domain decomposition becomes finer, the efficiency of parallel computing tends to be lowered (though the generation of DEM still gains acceleration benefits). In other words, spatial domain decomposition strategies are pivotal in reaping the high-performance computing power for big spatial data analytics.

Future work will concentrate on the following aspects. First, we will extend our approach to a more in depth microtopography analysis. More DEM-derived metrics (e.g., slope and surface roughness) will be introduced to explain and differentiate the microtopography features in our study area. Second, we will further examine other spatial domain decomposition strategies for the acceleration of the generation of very high resolution DEM. Third, but not last, we will apply the proposed parallel computing approach to other study regions.

Acknowledgements We thank support from US NSF XSEDE Supercomputing Resource Allocation (SES170007), and USDA Forest Service grant "Development and Operation of a Web GIS-enabled Data Management System for the Santee Experiment Forest".

References

1. Amoah, J. K. O., Amatya, D., & Nnaji, S. (2013). Quantifying watershed surface depression storage: determination and application in a hydrologic model. *Hydrological Processes, 27*(17), 2401–2413.
2. Anderson, C. J, & Lockaby, B. G. (2011). Forested wetland communities as indicators of tidal influence along the Apalachicola River, Florida, USA. *Wetlands, 31*(5), 895.
3. Armstrong, M. P., & Marciano R. (1993). Parallel spatial interpolation. In *Autocarto-Conference.*
4. Brubaker, K. M., Myers, W. L., Drohan, P. J., Miller, D. A., & Boyer, E. W. (2013). The use of LiDAR terrain data in characterizing surface roughness and microtopography. *Applied and Environmental Soil Science*, 13. https://doi.org/10.1155/2013/891534
5. Cramer, B. E., & Armstrong, M. P. (1999). An evaluation of domain decomposition strategies for parallel spatial interpolation of surfaces. *Geographical Analysis, 31*(2), 148–168.
6. Deilami, K., & Hashim, M. (2011). Very high resolution optical satellites for DEM generation: A review. *European Journal of Scientific Research, 49*(4), 542–554.
7. Ding, Y., & Densham, P. J. (1996). Spatial strategies for parallel spatial modelling. *International Journal of Geographical Information Systems, 10*(6), 669–698.
8. Emerson, C. H., Welty, C., & Traver, R. G. (2005). Watershed-scale evaluation of a system of storm water detention basins. *Journal of Hydrologic Engineering, 10*(3), 237–242.
9. Foster, I. (1995). *Designing and building parallel programs* (Vol. 78). Boston: Addison Wesley Publishing Company.
10. Griffin, L. F., Knight, J. M., & Dale, P. E. R. (2010). Identifying mosquito habitat microtopography in an Australian mangrove forest using LiDAR derived elevation data. *Wetlands, 30*(5), 929–937. https://doi.org/10.1007/s13157-010-0089-8.
11. Guan, X., & Huayi, W. (2010). Leveraging the power of multi-core platforms for large-scale geospatial data processing: Exemplified by generating DEM from massive LiDAR point clouds. *Computers and Geosciences, 36*(10), 1276–1282.
12. HPC. (2016). Windows HPC Cluster Manager. https://technet.microsoft.com/en-us/library/ff919397.aspx.
13. Hickey, R., Smith, A., & Jankowski, P. (1994). Slope length calculations from a DEM within ARC/INFO GRID. *Computers, Environment and Urban Systems, 18*(5), 365–380.
14. Hohl, A., Delmelle, E. M., & Tang, W. (2015). Spatiotemporal domain decomposition for massive parallel computation of space-time kernel density. *ISPRS Annals of the Photogrammetry, Remote Sensing and Spatial Information Sciences, 2*(4), 7.
15. Hohl, A., Delmelle, E., Tang, W., & Casas, I. (2016). Accelerating the discovery of space-time patterns of infectious diseases using parallel computing. *Spatial and Spatio-Temporal Epidemiology, 19,* 10–20.
16. Huang, C.-H., & Bradford, J. M. (1992). Applications of a laser scanner to quantify soil microtopography. *Soil Science Society of America Journal, 56*(1), 14–21.
17. Huang, Q., & Yang, C. (2011). Optimizing grid computing configuration and scheduling for geospatial analysis: An example with interpolating DEM. *Computers and Geosciences, 37*(2), 165–176.
18. Jensen, R. P., Bosscher, P. J., Plesha, M. E., & Edil, T. B. (1999). DEM simulation of granular media—structure interface: Effects of surface roughness and particle shape. *International Journal for Numerical and Analytical Methods in Geomechanics, 23*(6), 531–547.
19. Knight, J. M., Dale, P. E. R., Spencer, J., & Griffin, L. (2009). Exploring LiDAR data for mapping the micro-topography and tidal hydro-dynamics of mangrove systems: An example from southeast Queensland, Australia. *Estuarine, Coastal and Shelf Science, 85*(4), 593–600.
20. Komiyama, A., Santiean, T., Higo, M., Patanaponpaiboon, P., Kongsangchai, J., & Ogino, K. (1996). Microtopography, soil hardness and survival of mangrove (Rhizophora apiculata BL.) seedlings planted in an abandoned tin-mining area. *Forest Ecology and Management, 81*(1), 243–248.

21. Lassueur, T., Joost, S., & Randin, C. F. (2006). Very high resolution digital elevation models: Do they improve models of plant species distribution? *Ecological Modelling, 198*(1), 139–153.
22. Li, Z., Hodgson, M. E., & Li, W. (2016). A general-purpose framework for parallel processing of large-scale LiDAR data. *International Journal of Digital Earth*, 1–22.
23. McKean, J., & Roering, J. (2004). Objective landslide detection and surface morphology mapping using high-resolution airborne laser altimetry. *Geomorphology, 57*(3), 331–351.
24. Milne, L., Lindner, D., Bayer, M., Husmeier, D., McGuire, G., Marshall, D. F., et al. (2008). TOPALi v2: A rich graphical interface for evolutionary analyses of multiple alignments on HPC clusters and multi-core desktops. *Bioinformatics, 25*(1), 126–127.
25. Mitas, L., & Mitasova, H. (1999). Spatial interpolation. *Geographical Information Systems: Principles, Techniques, Management and Applications, 1*, 481–492.
26. Moore, I. D., Grayson, R. B., & Ladson, A. R. (1991). Digital terrain modelling: A review of hydrological, geomorphological, and biological applications. *Hydrological Processes, 5*(1), 3–30.
27. Naoum, S., & Tsanis, I. K. (2003). Hydroinformatics in evapotranspiration estimation. *Environmental Modelling and Software, 18*(3), 261–271.
28. Prasannakumar, V., Vijith, H., & Geetha, N. (2013). Terrain evaluation through the assessment of geomorphometric parameters using DEM and GIS: Case study of two major sub-watersheds in Attapady, South India. *Arabian Journal of Geosciences, 6*(4), 1141–1151.
29. Rauber, T., & Rünger, G. (2013). *Parallel programming: For multicore and cluster systems*, Springer Science & Business Media.
30. Shepard, W. E. (2000). *A parallel approach to searching for nearest neighbors with minimal interprocess communication*. uga.
31. Tang, W., & Feng, W. (2017). Parallel map projection of vector-based big spatial data: Coupling cloud computing with graphics processing units. *Computers, Environment and Urban Systems, 61*, 187–197.
32. Tang, W., & Wang, S. (2009). HPABM: A hierarchical parallel simulation framework for spatially-explicit agent-based models. *Transactions in GIS, 13*(3), 315–333.
33. Tang, W., Feng, W., Zheng, M., & Shi, J. (2017). Land cover classification of fine-resolution remote sensing data. In *Reference module in earth systems and environmental sciences*. Elsevier.
34. Tomczak, M. (1998). Spatial interpolation and its uncertainty using automated anisotropic inverse distance weighting (IDW)-cross-validation/jackknife approach. *Journal of Geographic Information and Decision Analysis, 2*(2), 18–30.
35. Trettin, C. C., Czwartacki, B. J., Allan, C. J., & Amatya, D. M. (2016). Linking freshwater tidal hydrology to carbon cycling in bottomland hardwood wetlands. In Stringer, C. E., Krauss, K. W., Latimer, J. S. (Eds.), *Headwaters to estuaries: Advances in watershed science and management-proceedings of the fifth interagency conference on research in the watersheds* (p. 302). March 2–5, 2015, North Charleston, South Carolina. e-General Technical Report SRS-211. Asheville, NC: US Department of Agriculture Forest Service, Southern Research Station.
36. Wang, S., & Armstrong, M. P. (2003). A quadtree approach to domain decomposition for spatial interpolation in grid computing environments. *Parallel Computing, 29*(10), 1481–1504.
37. Wang, S., & Armstrong, M. P. (2009). A theoretical approach to the use of cyberinfrastructure in geographical analysis. *International Journal of Geographical Information Science, 23*(2), 169–193.
38. Werner, M. G. F. (2001). Impact of grid size in GIS based flood extent mapping using a 1D flow model. *Physics and Chemistry of the Earth, Part B: Hydrology, Oceans and Atmosphere, 26*(7–8), 517–522.
39. Wilkinson, B., & Allen, M. (1999). *Parallel programming: Techniques and applications using networked workstations and parallel computers*. Prentice-Hall.

40. Wise, S. (2000). Assessing the quality for hydrological applications of digital elevation models derived from contours. *Hydrological Processes, 14*(11–12), 1909–1929.
41. Wu, S., Li, J., & Huang, G. H. (2008). A study on DEM-derived primary topographic attributes for hydrologic applications: Sensitivity to elevation data resolution. *Applied Geography, 28*(3), 210–223.
42. Zikopoulos, P., & Eaton, C. (2011). *Understanding big data: Analytics for enterprise class hadoop and streaming data*, McGraw-Hill Osborne Media.
43. Zimmerman, D., Pavlik, C., Ruggles, A., & Armstrong, M. P. (1999). An experimental comparison of ordinary and universal kriging and inverse distance weighting. *Mathematical Geology, 31*(4), 375–390.

Big-Data Analysis of Process Performance: A Case Study of Smart Cities

Alejandro Vera-Baquero and Ricardo Colomo-Palacios

Abstract This chapter presents a data-centric software architecture that provides timely data access to key performance indicators (KPIs) about process performance. This architecture comes in the form of an analytical framework that lies on big-data and cloud-computing technologies aimed to cope with the demands of the crowd-sourced data analysis in terms of latency and data volume. This framework is proposed to be applied to the Smart Cities and the Internet of Things (IoT) arenas to monitor, analyse and improve the business processes and smart services of the city. Once the framework is presented from the technical standpoint, a case study is rolled out to leverage this process-centric framework and apply its fundamentals to the smart cities realm with the aim of analysing live smart data and improve the efficiency of smart cities. More specifically, this case study is focussed on the improvement of the service delivery process of the Open311 smart services deployed in the city of Chicago. The outcomes of the test show the ability of the systems to generate metrics in nearly real-time for high volumes of data.

Keywords Smart cities · Internet of Things · Big data · Cloud computing
Business process analytics

1 Introduction

Smart cities is an emerging discipline that is gaining ground in recent years with the aim at offering advanced and innovative services to citizens. The smart city concept encompasses a wide range of technology and ubiquitous ICT (Information and Communications Technology) solutions that are applied to a city in multiple

A. Vera-Baquero (✉)
Universidad Carlos III de Madrid, Getafe, Spain
e-mail: averabaq@gmail.com

R. Colomo-Palacios
Østfold University College, Halden, Spain
e-mail: ricardo.colomo-palacios@hiof.no

© Springer Nature Singapore Pte Ltd. 2018   41
S. S. Roy et al. (eds.), *Big Data in Engineering Applications*,
Studies in Big Data 44, https://doi.org/10.1007/978-981-10-8476-8_3

domains, such as e-government and public administration, intelligent transportation systems, traffic management, public safety, social, health-care, educational, e-commerce, building and urban planning, environmental, energy, etc. [1]. A myriad of innovative services can be put in citizens' hands in order to improve their quality of lives. At nowadays, a large number of problems of big cities can be avoided, or mitigated to some extent, by integrating smart cities technology into service management [2]. Pervasive ICT systems and cloud technology can surround urban citizens by an ubiquitous digital eco-system where a great variety of internet-connected devices interact to each other [1] to provide a powerful environment of smart objects with the capabilities to digitally manage, monitor and track physical objects [3], thereby making valuable information of individuals accessible within an *Internet of Things* (IoT) context.

IoT represents intelligent end-to-end systems that enable smart solutions to arise by means of a diverse range of technologies, including sensing, communications, networking, computing, information processing, and intelligent control technologies [4]. By extension, IoT aims to connect heterogeneous devices worldwide which entail an enormous density of information connecting billions of objects. These smart objects can be any technological equipment with Internet processing capabilities, such as computers, tablets, smartphones, smart TVs, Global Positioning Systems (GPS), sensors, built-in vehicle devices, and so on. The interaction of these systems throughout the provision of Smart City services may generate a very large amount of data that can be used to analyse and improve the throughput and operative efficiency of those services. Nevertheless, traditional approaches are not suitable for dealing with high volumes of data to such extent. The exponential growth of data within an IoT context demands the introduction of innovative and emerging technologies with the capabilities to deal with high-latency response time on systems that manage high volumes of data. Traditional RDBMS systems and conventional data warehouses platforms are not suitable for managing vast amount of data in the order of terabytes (TB) petabytes (PB) or even exabytes (EB) of information, and drastic improvements are needed "to scale databases to data volumes beyond the storage and/or processing capabilities of a single large computer" [5].

Big-data technology has emerged as response to the existing limitation found on traditional data systems for handling very large datasets [6]. The term "Big-data" is commonly used to refer to a set of technological components that have the capabilities to empower data-intensive analysis on very large and complex datasets whose size spans beyond the ability of traditional software tools to capture, collect, integrate, unify, store and analyse hundreds of millions terabytes of information [7]. According to recent studies [8], big data approaches, when aligned with business and supported by the right people, can make a big impact in business. In the big data field, research efforts are being driven by big-data technology to extract meaning and infer knowledge from very large datasets [9], and the need to enable data-intensive processing over massive datasets has introduced a new form of data analytics called Big Data Analytics [10].

The evolution of big data has been driven by the rapid growth of application demands, cloud computing, and virtualized technologies. Cloud computing provides on-demand computational resources for offering data-intensive processing over big data [11], and to a certain extent, the advances of cloud computing foster the adoption and development of big data solutions [12]. Thereby, Big-data, cloud-computing and IoT are terms that are closely related to each other, and these are especially important on IT solutions when trying to analyse the operations management involved in the services provided by IoT applications.

In a nutshell, ICT plays a key role when adopting smart city technology and when supporting the provision of seamless ubiquitous services [1]. Big data techniques are conceived as the powerful tool to exploit all the potential of the Internet of Things and the smart cities [13]. In this regard, research effort must be conducted to provide innovative ICT solutions that can deal with the demands aforementioned. Herein, we presents a big-data and cloud-based analytic framework that can be applied to the Smart Cities and IoT arenas to monitor, analyse and improve the business processes and smart services of the city.

2 Smart Cities, IoT and Big-Data Analysis

Investing on smart city technology can become a business-competitive and attractive environment [1], especially on a slowing-down economy such as the present one. Paroutis et al. [14] study the positive impact of applying a smart city strategy on urban areas in conditions of economic recession. The right choice of a Smart City strategy is crucial for contributing to the sustainability and acceleration of the economic growth of the city.

In this regard, business process improvement can help smart cities to make them, not only even smarter but more efficient and profitable. For instance, health care monitoring systems could help managers to reduce costs and improve the efficiency of the service by alleviating the short-age of resources and personnel. Likewise, intelligent transportation systems could assist traffic analysts in reducing congestion and improving the roads safety. Additionally, the monitoring and analysis of business processes that flow through smart distribution systems, could aid business users to improve the quality and reduce the cost of goods and services in very large and complex supply chains [4]. This is especially interesting in business scenarios that involve very complex and highly distributed processes like supply chain management [15]. For instance, manufacturing processes produce massive amounts of data due to the continuous execution of process operation and control [16]. The execution of business processes is lineal by nature, whereby their event data is originated on a sequential manner. This leads analytical systems to process data streams with a known start but an uncertain end. On very high transactional environments, the adoption of big data technology become a must in order to analyse event streams in real-time [17].

The use cases aforementioned are only a mere illustration of the myriad of potential opportunities to apply business process initiatives on smart cities services. This converges to the adoption of smart cities, internet of things and big-data by the industry as a means to apply business-process expertise in the improvement of operation and production management.

Business process intelligence initiatives must address diverse technical challenges. One of the most challenging is the heterogeneity, also known in this context as big data chain [18], where business process improvement (BPI) solutions have to deal with the heterogeneous systems landscape of large enterprises and complex administrative procedures that stands within and across several distinct units across the city. This becomes even more complex when business processes run beyond the software boundaries of a single organization. At nowadays, most of the business transaction data involving stakeholders, processes, products and services, are increasingly available beyond the corporate boundaries, "including significant data from social networks and the Internet-of-things" [19]. In the case of smart cities, these complexities are even intensified when the processes require the citizens' interaction, and where the use of smart objects can become the main source to interact with the public sector and governments affairs. The internet of things is a clear example of how smart devices can be a powerful data source and key-enabler to improve cities' efficiency. By having a forward thinking, in a not so distant future, we can forecast how home inter-connected devices could provide valuable event data about processes such as energy consumption over time, or tracking vehicles journeys as processes with in-vehicles smart devices. This would help local authorities bringing sustainability to the city by reducing congestion, improving the road safety and controlling CO_2 emission [20].

Classic business intelligence (BI) platforms have become a powerful tool to business users for decision making. These systems have been traditionally used for discovering trends and relationships in large, complex business data sets. However, the use of traditional BI systems is not sufficient to meet today's business needs. They normally are business domain specific and have not been sufficiently process-aware to support the needs of process improvement type activities, especially on large and complex distributed processes, where it entails integrating, monitoring and analysing a vast amount of dispersed event logs, with no structure, and produced on a variety of heterogeneous environments. In addition, the continuous generation of event data in IoT contexts over time cannot be efficiently managed by means of traditional storage systems, which are not adequate to manage event data in the order of hundreds of millions of linked records. In turn, the monitoring and analysis of high volumes of data is a data-intensive process that usually exceeds the processing capabilities of a single large computer, thereby requiring robust and complex supporting systems that can easily scale over time to meet the processing demands in terms of latency and data volume. Hence, emerging data-centric systems aimed to improve cities' efficiency must be big-data ready for enabling elastic-scalable data analysis.

In this chapter we leverage a former work in the area of big-data analytics with the aim of bringing business-process expertise to the smart cities and internet of things arenas. This will help cities to make them, not only even smarter but more efficient and profitable. We propose the adoption of its fundamentals to improve the processes of urban services and we extend some of the core architectural components to make the framework suitable to be applied on such domains, thus driving city managers to achieving a well-running, smart and efficient city.

3 Technological Solution

The technological solution proposed is based on the analytical framework widely discussed in [21–23], which is focused on the improvement of process performance through the monitoring and analysis of the execution outcomes of business processes. This framework has the ability to provide cloud computing services to third-party applications in a timely fashion. These services rely on a big-data infrastructure that enables the system to perform data-intensive computing on processes whose executions produce a vast amount of event data that cannot be efficiently processed by means of traditional systems. Furthermore, this data usually comes from a variety of heterogeneous platforms that are continuously producing enterprise business events which must be unified across disparate platforms.

Real-time, low latency monitoring and analysing of business events for decision making is key, but difficult to achieve. The difficulties are intensified on processes whose enterprise data cross organizational boundaries. Distributed processes usually flow across heterogeneous systems such as business process execution language (BPEL) engines, ERP systems, CRMs, workflows, document management systems, etc. The heterogeneity of these supporting systems makes the collection, integration and analysis of high volume business event data extremely difficult [24].

The software architecture discussed herein overcomes this pitfall by introducing a generic event model that is an extension of the BPAF (Business Process Analytics Format) [25], hereinafter exBPAF [22]. This format enables heterogeneous systems to produce interchangeable event information regardless of the underlying concerns of the source systems, and it hosts the information required to enable the system to perform analytical processes over them, as well as representing any derived measurement produced during the execution of any process flow. This is essential to provide the framework of a concrete understanding and representation of what needs to be monitored, measured and analyzed, and it can contribute to the continuous improvement of processes by giving insights into process performance.

One of the major contributions of this approach, among others, relies on the ability to offer to business users the opportunity of availing process execution path information at very low latency rates. The monitoring of cross-organizational business processes is achieved by listening (in or nearly real-time) state changes and business actions from operational systems. This is attained by collecting, unifying and storing the execution data outcomes across a collaborative network, where each node represents a participant organization.

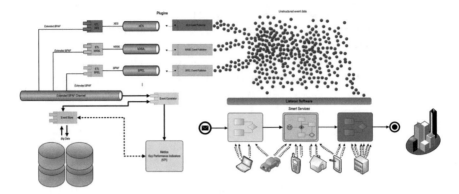

Fig. 1 Collection, unification and storage of live smart event data

Event Data Gathering

The data collection is done by bespoke listening software which is responsible for capturing the events from crowd-sourced heterogeneous data and publishing them to the network throughout the ActiveMQ message broker (see Fig. 1). The legacy listener software emits the event information to different endpoints depending on the message format provided. Currently, the platform supports a variety of widely adopted formats for representing event logs such as XES, MXML or even raw BPAF. Consequently, a different set of plug-ins are available per supported event format, and in turn, each plug-in incorporates specific ETL (Extract, Transform and Load) functions to convert source event streams into the proposed extended BPAF. Once the events are transformed, then they are forwarded to a specific channel for processing. The event correlation module is subscribed to this channel listening continuously for new incoming events.

Event Data Correlation

The data unification is achieved by correlating hundred millions of unstructured events across heterogeneous systems and devices. Event correlation refers to the determination of the sequence of events produced by the execution of inter-related and consecutive process instances or activities. Event correlation is an essential part of the proposed framework for achieving the correct identification of process execution sequences. Without the ability to correlate events it would not be possible to generate metrics per process instance or activity [23]. The event correlation algorithm is out of scope at the present paper, but it basically relies on finding the associated process instance or activity in the event repository based on the exBPAF definition [22].

Event Data Storage

Once the events are correlated, they are stored in the event repository in order it can be found by its successors. This entails the event correlation mechanism to deal with very large volumes of data in a global and distributed business process execution repository, whereby its events must be readily accessible at minimum latency. This pitfall is addressed by designing the event repository as big-data cloud storages that can be directly mined by the event correlation module. This repository allows the system to scale out easily on readily available hardware, which is essential for dealing with the data-intensive processing demands of the event correlation process. Herein, the event repository is implemented using the HBase product as big-data storage. HBase is a NoSQL, versioned, column-oriented data storage system that provides random real-time read/write access to big data tables and runs on top of the Hadoop Distributed Filesystem. HBase features powerful scaling capabilities. HBase clusters expand on commodity of HRegionServers, thus linearly increasing the storage and processing capacity. The technical documentation of this product reveals extraordinary clustering capabilities for providing data-intensive processing on large data tables. The distributed event repository is implemented as big-data tables over HBase, thereby exploiting its outstanding features for providing timely access to key data.

The overall architect solution (see Fig. 2) is a cloud-based platform that comprises a set of cloud-computing nodes that provides analytical services to third party applications, namely BASU (Business analytics service unit) and GBAS (Global Business Analytical Service) modules. These service components have the capabilities for collecting data originating from distributed heterogeneous systems, storing the enterprise data and inferring knowledge from the gathered information.

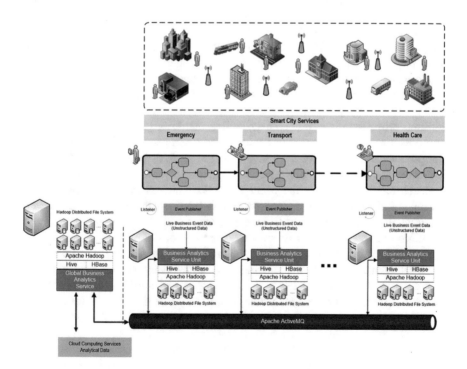

Fig. 2 Architecture overview

These modules monitor and analyse operational activities within both, local and global contexts. In a local context, the processes reside within an organization (inter-departmental) on where they can be analysed and optimized by the means of BASU units. These units are attached to individual organizations for performing the managerial activities of their internal processes. In contrast, the GBAS component supports the monitoring, analysis and optimization of large and complex supply-chain processes like manufacturing and retail distribution.

The successful integration of those components through enterprise service bus adapters completes the high-level architecture depicted on Fig. 2. Each of these components relies on the provision of a number of clustering nodes that provide big data support. In turn, each clustered configuration is deployed on a cloud-based environment, thereby empowering the scalability and performance features of the IT solution proposed at the easily commodity of hardware.

Within the context of this work, any smart object such as tablets, smartphones, smart TVs, sensors, built-in vehicle and home devices, etc. can be a potential source of information for the analytical framework to monitor and track a myriad of processes wherein those devices are involved across the city. Moreover, if that live smart data is gathered and put together on a large enough scale, it could be used to detect undesirable patterns in the process or even anticipate future actions by incorporating predictive analysis or machine learning methods. The software architecture discussed herein does no support those advanced analytic methods yet, but it is part of on-going work. Nevertheless, and as it will be shown later in the next section, the architecture has the capability to correlate large amounts of events and generate metrics and KPIs in the order of few milliseconds with a small cluster size, thereby making the system a big-data ready platform for enabling elastic-scalable data analysis in the context of smart cities and IoT.

It is important to mention that the software architecture has evolved with respect to the previous work in order to meet the actual IoT demands in terms of dealing with the crowdsourcing heterogeneity and guaranteeing high performances with zero downtimes. In the regard of dealing with heterogeneous devices and sensors that produce event data, we have leveraged the work carried out by [26, 27] that introduces the concept of Thingsonomy to tackle the variety of events in IoT. This approach aims to decouple the event semantics between the smart objects (producers) and the event listeners (consumers). In this way, there is no longer need to construct the events in exBPAF format at source, which otherwise it might be infeasible to achieve on an IoT scenario due to incompatibility issues among the myriad of multiple devices and manufacturers. This new layer abstracts the IoT event data generation from the data ingestion at the event gathering module.

With respect to guaranteeing high performance and reliability on the solution, we have introduced a consistent hashing approach in front of the cluster of BASU nodes in order to empower the concept of data locality and speeding up the event sequence lookups. The generation of metrics and KPIs makes a heavily use of a distributed cache which is key for featuring monitoring capabilities in nearly real-time. In this regard, every time an event is correlated in the stream chain, that event is cached and associated to the event stream. Upon process completion, the entire event stream is read from the cache for that particular instance and forwarded to the subsequent modules for processing. In a crowd-sourced environment such as IoT, it is crucial the event correlator module to scale out without sacrificing performance. A consistent hashing solution on the event gathering module helps in this matter by evenly forwarding the events of a particular instance to the exact same BASU node, which has the predecessor events in-cache already, and hence it prevents the KPI generator from performing repeated remote calls to other cache nodes, thereby improving performance and reducing inter-communication costs.

4 Case Study: Smart Cities

We present a case study aimed to monitor and analyse the processes involved on smart services offered by the city of Chicago. The city of Chicago adopted in 2012 a common standard for 311 reporting known as "Open311" [28]. This open standard is being adopted worldwide in multiple urban areas, and brings governments the ability to build uniform interoperable systems that allow citizens to interact with their cities in the form of a broad range of information and services.

These services allow people to submit a service request, track its progress, and receiving a notification feedback when the issue is resolved. These Open311-based systems also have the ability to provide residents with customized alerts based on their localized area [29].

The main goal of using Open311 technology is to increase the efficiency of these services while "providing the public with data on the city's service delivery". Open311 aims to improve the efficiency of 311 call centres, i.e. reduce the number of calls and abandon rates, remove the duplicate calls, reduce the caller wait times or even improve distribution and management of agents. The Open311 standard stands on the openness, innovation, and accountability in city services by offering citizens multiple methods for requesting, track and acknowledges response of a wide range of 311 services [30]. These services can be invoked from smart applications that are built conformed to the Open311 standard and which are accessible from multiple smart devices [31]. Whilst an individual can report a non-emergency issue (ideally with a photo) at a given location, such as potholes, broken streetlights, garbage, vandalism, and so on, these issues are recorded and routed to the relevant authority by means of a Customer Relationship Management (CRM) system that automatically forwards the service requests to the appropriate City department, governmental or non-governmental agency [32].

This standard has been rolled out in the city of Chicago, and in order to leverage the massive amount of data reported by their residents, we have connected our framework to the real operational systems that register citizens' reporting issues through their public Open311 interface. This interface comes in the form of a set of APIs that gives third party applications the ability to collect real residents' request data in real-time. As it will be covered later, we have implemented event listeners that consume those services, thereby getting straight access to such data.

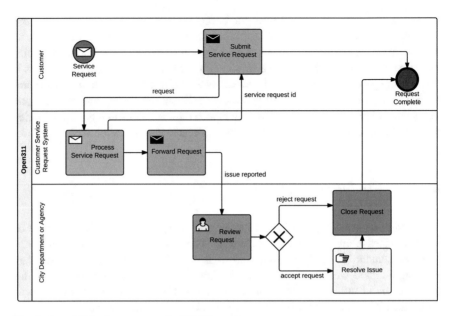

Fig. 3 Open311 business process in BPMN notation

In addition, those listeners are responsible for the transformation of the incoming data streams into such a way that they can be processed by the analytical input channels of the framework. In a nutshell, we aim to highlight that this case study has been conducted on real use cases harnessed by a real dataset. These interfaces are extensively documented and publicly available through the development site of the city of Chicago (http://dev.cityofchicago.org/docs/open311).

The case study presented herein focuses on the improvement of the service delivery process of the Open311 smart services deployed in the city of Chicago. The Fig. 3 depicts the business process in BPMN notation that we aim to monitor, analyse and improve. As stated in [32], the goals of the Open311 department of the city of Chicago, and by analogy to the purposes of this case study, are:

(1) Record efficiently all requests for non-emergency City services and forward them to the proper governmental and non-governmental agencies.
(2) Assist City departments, governmental and non-governmental agencies deliver improved customer service and manage resources more efficiently.
(3) Monitor and provide consistent, essential performance management reports and analysis of City services delivery.

We have leveraged the analytical framework proposed in this paper and applied its fundamentals to this case study in conjunction with a methodology proposed as part of the works [21, 22], so-called CBI4PI methodology (Cloud-based Infrastructure for process intelligence). This methodology aims to put real BPI technology in hands of business users by following a methodical process that aims guide the deployment of the analytical framework in the smart cities domain. The implementation methodology is rolled out in the following sections and applied to this case study.

Methodology to Apply the Analytical Framework in the Smart Cities Domain (CBI4PI)

4.1 PHASE 1. Definition

4.1.1 Identification of Scope and Boundaries

In this phase we identify the business nodes that are globally involved on the business process aforementioned. In this case study, the business entities determination would correspond to specify the different jurisdictions that the service request goes through during processing. The Open311 API used in this case study only provides one jurisdiction, so only one node is identified in this step.

4.1.2 Definition of Sub-processes, Activities and Sub-activities

In this case study we aim to analyse the performance of the service delivery process of the Open311 smart services. For this purpose, we define a global process that represents the service request that may flow through diverse business nodes (in this case there is only one), whilst we define a sub-process as the set of activities and tasks defined in the business process depicted in Fig. 3.

4.1.3 Determination of Level of Detail Within Business Processes

The level of detail is fixed by the information scope supplied by the Open311 API. The data structure of the API is able to provide data according to the activities (with some minor considerations) of the process thereof. Therefore, the data gathering and analysis must include the activity level of those tasks specified in the business process (see Fig. 3).

4.1.4 Development of Model Tables

For constructing the model table we have identified the list of 311 service requests (process) and its activities (tasks). The following table outlines the process model developed and highlights those activities that are discarded. These tasks are rejected mainly because they are either irrelevant, supply useless information for decision making or the operational systems (Open311 API) are unable to provide visibility on those activities (Tables 1 and 2).

Table 1 Service request process definition

Process definition ID	Process name
4ffa4c69601827691b000018	Abandoned vehicle
4ffa9cad6018277d4000007b	Alley light out
4fd3bd72e750846c530000cd	Building violation
4fd3b167e750846744000005	Graffiti removal
4ffa971e6018277d4000000b	Pavement cave-in survey
4fd3b656e750846c53000004	Pothole in street
4fd6e4ece750840569000019	Restaurant complaint
4fd3b9bce750846c5300004a	Rodent baiting/rat complaint
4fd3b750e750846c5300001d	Sanitation code violation
4ffa995a6018277d4000003c	Street cut complaints
4ffa9f2d6018277d400000c8	Street light out
4ffa9db16018277d400000a2	Traffic signal out
4fd3bbf8e750846c53000069	Tree debris

Table 2 Service request process model

Process	Activity	Act. Parent	Properties
[Service Request]	Submit.Service. Request		ServiceRequestId, Address, Location, Channel
	Forward.Request		ServiceRequestId, Address, Location, Channel
	Review.Request		ServiceRequestId, Address, Location, Channel
	Resolve.Issue		ServiceRequestId, Address, Location, Channel, Dupl
	Close.Request		ServiceRequestId, Address, Location, Channel, Dupl

5 PHASE 2. Configuration

5.1 Business Nodes Provisioning and Software Boundaries Identification

We deployed one BASU node in a test environment for evaluating the approach and loaded the process model developed in the previous step. This phase is crucial to identify the specific software requirements and how to interact with the Open311 API interface. Since there are no jurisdictions defined in the dataset, all process execution flows cross just one business node. This is the simplest scenario possible, notwithstanding this step is still useful to determine source software elements that can affect to the proper identification of process instances (service requests) upon the event sequence arrival.

5.1.1 Selection of Event Data Format

We selected exBPAF as the event format since we do not require integration with other process mining tools.

5.1.2 Event Correlation Data Determination

This phase is critical to recreate successfully the inbound service request paths at destination. For the purpose of this case study, the correlation data to be used is the service request identification number that is managed internally by the Open311 systems. This information will uniquely identify a specific request, and thus the process instance along the sequence of events.

5.1.3 Listeners Implementation

For the implementation of the listener we developed an Open311 API client with ETL (Extract, Transform and Load) support. The listener invokes the Open311 services to retrieve the status information of all requests available to date. The Open311 API endpoint acts as data source (extract phase), the incoming data in JSON format is analysed and converted into events in exBPAF format (transformation phase), and those events are then forwarded to the framework endpoint (load phase).

Processes may have different activities for representing the "Resolve.Issue" task. These activities could be merged into a single activity ("Resolve.Issue") as modelled, however, from an analytical perspective, it is worthwhile to analyse the performance of its sub-tasks with the aim of identifying bottlenecks, perform root cause analysis or diagnosis intents. For this purpose, the listener implements an inner process interpreter that identifies different business patterns per process, and it sets out its expected activities along with its events. This is possible to achieve as long as the Open311 interface provides this information in the data payload. In such a case the interpreter determines the sub-activities and infers their execution time from the body response. Below it is outlined an event sample that is generated by the listener along with the payload received from the API (Table 3).

5.1.4 Selection of Metrics and KPIs

The set of metrics and KPIs selected for the purpose of this case study are specified below. The standard metrics are outlined in the Table 4 for representing behavioural measures, and Table 5 for the structural ones.

Table 3 Sample of data conversion of an exBPAF event

Open311 API response payload	exBPAF Data Event
"notes": [*{* *"datetime": "2014-12-15T16:39:54-06:00",* *"summary": "Request opened",* *"type": "opened"* *},* *{* *"datetime": "2014-12-16T12:42:00-06:00",* *"summary": "Inspect for Violation",* *"description": "Completed",* *"type": "activity"* *},* *{* *"datetime": "2014-12-16T12:42:04-06:00",* *"summary": "Request closed",* *"type": "closed"* *}* *]*	*<?xml version="1.0" encoding="UTF-8" standalone="yes"?>* *<ns2:Event Timestamp="2014-12-16T12:42:00-06:00"* *ActivityDefinitionID="Inspect for Violation"* *ProcessName="Sanitation Code Violation"* *ProcessDefinitionID="4fd3b750e750846c5300001d"* *EventID="37ffca0a-c8b2-4e46-be23-5cd0ce6c8408"* *xmlns:ns2="http://www.uc3m.es/softlab/basu/event">* *<EventDetails PreviousState="Open.Running.InProgress"* *CurrentState="Closed.Completed"/>* *<Correlation>* *<CorrelationData>* *<CorrelationElement value="14-02142035"* *key="ServiceRequestId"/>* *</CorrelationData>* *</Correlation>* *<DataElement value="open" key="status"/>* *<DataElement value="Completed" key="statusNotes"/>* *<DataElement key="duplicate"/>* *<Payload value="4344 N HAMLIN AVE, CHICAGO, IL"* *key="address"/>* *<Payload value="41.95999687807092" key="lat"/>* *<Payload value="-87.72289882814333" key="long"/>* *<Payload value="phone" key="channel"/>* *<Payload value="60618" key="zip"/>* *<Payload value="17" key="police"/>* *<Payload value="39" key="ward"/>* *</ns2:Event>*

Table 4 Standard behavioural measures

DSS-standard measure	Description
Throughput time	Total amount of time for a process or activity
Change-over time	Time elapsed for the process or activity to be forwarded to a specific department or agency (process), or assigned to a specific operator (activity)
Processing time	Time elapsed since the process or activity is forwarded or assigned until the task starts to be processed
Waiting time	Effective amount of time taken for a process or activity to complete
Suspended time	This metric is unused in this case study since this state is unknown by the Open311 API

The metrics outlined above are deduced by querying and filtering the event data gathered from the listeners. The details of how this calculation is performed are out of scope in this paper but further information can be found at [22, 23]. Regarding to the KPI selection, and only for illustration purposes, we outline some KPIs that the framework can deal with, but not limited to.

Table 5 Standard structural measures

DSS-standard measure	Description
Running cases	Number of service requests processed
Successful cases	Number of service requests that were resolved successfully
Failed cases	Number of service requests that were not solved (by any reason but duplication)
Aborted cases	Number of service requests that were cancelled due to duplication

Behavioural KPIs

Performance: This KPI measures the average time of request resolution per request submitted. A threshold value may be agreed at design time during the KPIs definition stage or let the framework to infer this value after a given period of time.

$$Pf(x) = \frac{\sum_i i_{throughput}}{\sum_{i \in} 1}$$

$Pf(x)$ Performance of the 311 service "x".

Efficiency: This KPI measures the relation between the net processing time taken for request resolution and the gross processing time. This measure gives an insight into the efficiency of the resolution per process (service request type).

$$Ef(x) = \frac{\sum_{i \in x} i_{proce\sin g}}{\sum_{i \in x} i_{troughoput}}$$

$Ef(x)$ Efficiency of the 311 service "x".

Structural KPIs

Productivity: This KPI measures the production efficiency of the smart services by calculating the ratio of request resolutions per request submissions.

$$Pd(x) = \frac{SC(x)}{RC(x)}$$

$Pd(x)$ Productivity of the 311 service "x"
$SC(x)$ Number of successful cases for all instances of process "x"
$RC(x)$ Number of running cases for all instances of process "x"

Duplication rate: This KPI measures the rate of duplicate requests. This is calculated by identifying the number of duplicated requests per submission. A high rate indicates waste of resources that are assigned to review existing issues that are already open, solved or in progress.

$$Dup(x) = \frac{AC(x)}{RC(x)}$$

$Dup(x)$ Duplication rate of the 311 service "x"
$AC(x)$ Number of aborted cases for all instances of process "x"
$RC(x)$ Number of running cases for all instances of process "x"

Whilst the correlation of event data must be perform at very low latency rates, the KPI's generation can be generated in batch for historical or predictive analysis. Notwithstanding, decision makers normally aim to react quickly to undesired situations, thus a quick identification of non-compliant scenarios is desirable. Therefore, the BAM-like component of the framework plays an important role in the KPIs configuration. This feature is very useful to managers as they may decide whether or not establish thresholds per process or activity. This depends on whether there already exists historical information, as this will allow the system to infer the expected execution time. In such case, the thresholds might be set in the BAM component to generate alerts for detecting bottlenecks and non-compliant situations in nearly real-time.

5.2   PHASE 3. Execution

The evaluation has been accomplished successfully in a test environment that follows the infrastructure depicted on Fig. 2. A large amount of event data was generated by the Open311 API, and the analytical framework was fed with up-to-date service requests status information.

5.3   PHASE 4. Control

Functional Perspective
Throughout the monitoring and analysis of the outcomes we detected a high rate of duplicate requests. This high rate may indicate a significant waste of resources that are assigned to review issues that are already open, solved or in progress. Due to multiple citizens can potentially report simultaneously the same issue upon a time at the same area, the process performance can be improved significantly by avoiding the creation of false positive requests on the operating systems. This overhead could be greatly alleviated by introducing a verification step during the request submission activity. Whereby the smart application might check or estimate whether the same request has already been notified by another citizen according to the given date, location, and estimated area. Likewise, the smart reporting service could response back with the current status of the issue if duplicated, namely if it is already in progress, resolved or rejected.

Table 6 Performance analysis

	IO operation	Average (ms)	Standard deviation (ms)	Throughput (events/s)
Event correlation	Read	0.21791576	0.444654234	4589
	Write	1.721771801	1.422962964	580
Metric generation	1 read + 1 write	1.939687562	1.639354638	515

Performance Perspective

For the evaluation of the architecture from a performance perspective, the framework's infrastructure was deployed on a Cloudera (CDH 4.7.1) instance using a 4-node configuration, thereby exploiting the clustering capabilities of the proposed solution. The volume of data rose to nearly 500 GB of raw data, and the correlation algorithm performed reading operations in the order of few milliseconds for such volume of data. It is important to highlight that this is a prospective study and further efforts are ongoing to progressively increase the volume of data to the levels of TBs of information. As already sated, the most significant finding is that the read operations performed in the order of few milliseconds (see Table 6) and read execution time remained stable over time and did not increase as the number of events grew. As discussed in previous sections, the framework correlates the events as they arrive by finding their predecessors in the big data storage, whereby this entails one read operation plus a write for storing the event in the repository in order that it can be found by its successors. The table bellows summarizes the results of the experiment and shows that the IT solution features a high performance and is able to produce timely metrics by monitoring around 500 events per second using a small cluster size.

5.4 PHASE 5. Diagnosis

According to the methodology description in [33], this steps aims to find weaknesses on the model, and once they are detected and identified they must be eliminated from the operational systems. In such a case, the business process is re-designed and re-deployed on the operational environment, and the improvement lifecycle starts over again on a continuous refinement basis. Since we are using a dataset as an input, and we have no control over the operational systems, we cannot go further on this case study. Hence, this phase is out of scope in this paper.

6 Discussion: Frameworks Comparison

At this point, we have given an architectural overview of the analytical framework and its associated BPI methodology. This has been applied to a real-world use case in the area of smart cities and IoT by covering important aspects such as integrating heterogeneous crowed-sourced data and timely data analysis. Despite of having come up with a comprehensive vision of the solution, we consider that it is worthwhile to compare it with other relevant existing frameworks in the research community. Unfortunately, and to the best of authors' knowledge, there does not exist any solution that can be used for a like-for-like comparison with the one presented in this chapter. The arguments for this assertion are manifold. First, our approach consists of a BPI methodology that relies on the specific analytical platform proposed, which in turn is based on a bespoke correlation algorithm [22]. Second, such algorithm has been implemented using a determined big-data technology based on secondary indexes in HBase for speeding up event lookups. And third, the experiment was run on a specific environment with very specific settings, and the outcomes may strongly vary depending on the infrastructure and hardware configurations, i.e. cluster size and overall processing power. In addition, this is intensified when the use case runs on big-data contexts, such as the present one, whereby the computing capabilities become key for measuring framework's throughput. Hence, we cannot treat the solution as a whole for comparison purposes, and thereby we have broken down the comparison analysis into two different categories: Methodology and Frameworks for business process data analysis. In order to simplify the analysis comparison, we have chosen the most relevant work from the literature on each category, to the best authors' criteria, and we have undertaken a qualitative comparison between them from each category. As we will see below, research efforts are still needed to contribute with big-data analytics solutions that can serve as a link between existing BPI methodologies and data analytics platforms, especially on big-data ready frameworks, which are key to efficiently cope with the processing demands of heterogeneous data on IoT.

Methodology
The following table outlines relevant business process improvement methodologies in the literature. They are all based on the concept of continuous process improvement but with different lines of action. As we can see, there is a noticeable absence of key-driving systems and tools that can support their process improvement initiatives, whereby there is a clear disconnection between the improvement program definition itself and the supporting tools that can carry them out. The methodology used in the approach (CB4PI) tries to close the loop by focusing on the performance dimension, which is the measure we are trying to improve on the smart services (Table 7).

Table 7 Evaluation of different BPI methodologies

Main oriented goal	BPI methodology	Lifecycle	Supporting tools	Authors
Process quality	TQM (Total Quality Management)	PDCA (Plan Do Check Act)	N	[34, 35]
	Six Sigma	DMAIC (Define, Measure, Analyse, Improve and Control)	N	[36–38]
Process optimization	BPMM (Business Process Maturity Model)	CMM (Capability Maturity Model)	N	[39]
	BPR (Business Process Reengineering)	Standard BPM	N	[40, 41]
Process visibility	TAD (Tabular Application Development)	IMIDDI (Identification, Modeling, Improvement, Development, Design and Implementation)	P	[42]
Process performance	CBI4PI	Standard BPM	F	[43]

F Fully accomplished or mentioned; *P* Partially accomplished or implicitly mentioned; *N* Not accomplished or not mentioned

Data Analytics Frameworks

The following table compares three well-known frameworks about inferring knowledge from business processes' execution. The proposed framework does intend to complement the existing approaches by filling the existing gap in terms of real-time monitoring on highly distributed big-data environments. As per previous sections, the system has demonstrated to have great capabilities to provide monitoring activities in real-time as well as gathering disperse event logs regardless of operational system technology and ubiquitous location. This could close the loop between event generation and post-execution analysis by contributing with the provision of real-time monitoring activity services. In this regard, the system could complement existing tools and serve as event collector for supporting process mining functionality in real-time. Likewise, it could also consume process mining services to provide functionality beyond the monitoring and performance analysis features. Whereby, crowed-sourced data collected from smart services could be used in real-time for applying process mining techniques such as process model discovering, conformance checking, and so forth (Table 8).

Table 8 Evaluation of process-oriented analytics frameworks

Framework	BPI Methodology	Conformance Checking	Model Auto-discovery	Monitoring	Control	Post-execution analysis	Heterogeneous data	Domain Agnostic	Big Data enabled	Authors
Process mining tools (ProM, Disco)	–	F	F	N	F	F	F	F	N	[44]
iWISE	Six Sigma	N	N	F	P	P	P	F	N	[45]
CBI4PI	CBI4PI	N	N	F	P	S	F	F	F	[22]

F Fully accomplished or mentioned; *P* Partially accomplished or implicitly mentioned; *S* Scheduled, planned and/or on-going work but not there yet; *N* Not accomplished or not mentioned

7 Conclusions and Future Work

We have presented a data-centric architecture that supports business process analysis on highly distributed environments. The big-data based analytical features of the framework were exploited in order to enable the system to deal with very large amount of data that can be generated by a myriad of smart objects during service execution. Thereafter, we tested the applicability of the work's fundamentals to the smart cities arena and we successful monitored and analysed the vast amount of service requests submitted by citizens of Chicago. The analysis was performed in a very short response time basis, and we detected potential areas of improvement in the smart service processes. Likewise, big-data technology has been used to effectively manage large volumes of event data and providing analysis in nearly real time using low hardware costs. Likewise, the system relies on an event-driven architecture to conduct the data integration and enabling a platform-independent solution for collecting and unifying data from external systems and regardless of their underlying technology which is ideal for IoT based constructions.

Further research may include the gradual incorporation of services for supporting advanced functionality that can be supported by emerging technologies and optimization techniques. The provision of simulation techniques would highly empower the cloud-based functionality since the structured data may serve as an input to simulation engines. This will enable business users to anticipate actions by reproducing what-if scenarios, as well as performing predictive analysis over augmented data that constitutes a base of hypothetical information. Likewise, this would enable analysts to reproduce live process instances and re-run event streams in simulation mode for diagnosis purposes and root cause analysis on non-compliant situations along large and complex distributed processes.

References

1. Piro, G., Cianci, I., Grieco, L. A., et al. (2014). Information centric services in smart Cities. *Journal of Systems and Software, 88,* 169–188. https://doi.org/10.1016/j.jss.2013.10.029.
2. da Silva, W. M., Alvaro, A., Tomas, G. H. R. P., et al. (2013). *Smart cities software architectures: A survey* (pp. 1722–1727). New York, NY, USA: ACM.
3. Skiba, D. J. (2013). The Internet of Things (IoT). *Nursing Education Perspectives, 34,* 63–64.
4. Zheng, J., Simplot-Ryl, D., Bisdikian, C., & Mouftah, H. T. (2011). The Internet of Things [Guest Editorial]. *IEEE Communications Magazine, 49,* 30–31. https://doi.org/10.1109/MCOM.2011.6069706.
5. Borkar, V. R., Carey, M. J., & Li, C. (2012). Big data platforms: What's next? *XRDS, 19,* 44–49. https://doi.org/10.1145/2331042.2331057.
6. Fosso Wamba, S., Akter, S., Edwards, A., et al. (2015). How 'big data' can make big impact: Findings from a systematic review and a longitudinal case study. *International Journal of Production Economics, 165,* 234–246. https://doi.org/10.1016/j.ijpe.2014.12.031.

7. Patel, A. B., Birla, M., Nair, U. (2012). Addressing big data problem using Hadoop and Map Reduce. In *2012 Nirma University International Conference on Engineering (NUiCONE)* (pp. 1–5).
8. Wamba, S. F., Gunasekaran, A., Akter, S., et al. (2017). Big data analytics and firm performance: Effects of dynamic capabilities. *Journal of Business Research, 70,* 356–365. https://doi.org/10.1016/j.jbusres.2016.08.009.
9. Park, H. W., & Leydesdorff, L. (2013). Decomposing social and semantic networks in emerging "big data" research. *J Informetr, 7,* 756–765.
10. Mutschler, B., Reichert, M. U., & Bumiller, J. (2005). *Towards an evaluation framework for business process integration and management.* Los Alamitos: IEEE Computer Society Press.
11. Talia, D. (2013). Clouds for scalable big data analytics. *Computer, 46,* 98–101. https://doi.org/10.1109/MC.2013.162.
12. Chen, M., Mao, S., & Liu, Y. (2014). Big data: A survey. *Mobile Networks and Applications, 19,* 171–209. https://doi.org/10.1007/s11036-013-0489-0.
13. Jara, A. J., Genoud, D., & Bocchi, Y. (2014). Big data for smart cities with KNIME a real experience in the SmartSantander testbed. *Software: Practice and Experience n/a-n/a.* https://doi.org/10.1002/spe.2274.
14. Paroutis, S., Bennett, M., & Heracleous, L. (2014). A strategic view on smart city technology: The case of IBM Smarter Cities during a recession. *Technological Forecasting and Social Change, 89,* 262–272. https://doi.org/10.1016/j.techfore.2013.08.041
15. Kache, F., & Seuring, S. (2017). Challenges and opportunities of digital information at the intersection of big data analytics and supply chain management. *International Journal of Operations & Production Management, 37,* 10–36. https://doi.org/10.1108/IJOPM-02-2015-0078.
16. Qin, S. J. (2014). Process data analytics in the era of big data. *AIChE Journal, 60,* 3092–3100. https://doi.org/10.1002/aic.14523.
17. Alippi, C., Ntalampiras, S., & Roveri, M. (2017). Designing HMMs in the age of big data. In P. Angelov, Y. Manolopoulos, L. Iliadis, et al. (Eds.), *Advances in big data* (pp. 120–130). Cham: Springer International Publishing.
18. Janssen, M., van der Voort, H., & Wahyudi, A. (2017). Factors influencing big data decision-making quality. *Journal of Business Research, 70,* 338–345. https://doi.org/10.1016/j.jbusres.2016.08.007.
19. Baesens, B., Bapna, R., Marsden, J., & Vanthienen, J. (2016). Transformational issues of big data and analytics in networked business. *Management Information Systems Quarterly, 40,* 807–818.
20. Hamilton, A., Waterson, B., Cherrett, T., et al. (2013). The evolution of urban traffic control: Changing policy and technology. *Transportation Planning and Technology, 36,* 24–43. https://doi.org/10.1080/03081060.2012.745318.
21. Vera-Baquero, A., Colomo-Palacios, R., & Molloy, O. (2013). Business process analytics using a big data approach. *IT Professional, 15,* 29–35. https://doi.org/10.1109/MITP.2013.60.
22. Vera-Baquero, A., Colomo Palacios, R., Stantchev, V., & Molloy, O. (2015). Leveraging big-data for business process analytics. *The Learning Organization, 22,* 215–228. https://doi.org/10.1108/TLO-05-2014-0023.
23. Vera-Baquero, A., Colomo-Palacios, R., & Molloy, O. (2015). Measuring and querying process performance in supply chains: an approach for mining big-data cloud storages. *Procedia Computer Science, 64,* 1026–1034. https://doi.org/10.1016/j.procs.2015.08.623.
24. Vera-Baquero, A., & Molloy, O. (2013). Integration of event data from heterogeneous systems to support business process analysis. In A. Fred, J. L. G. Dietz, K. Liu, & J. Filipe (Eds.), *Knowledge discovery, knowledge engineering and knowledge management* (pp. 440–454). Berlin, Heidelberg: Springer.
25. Müehlen, M. zur & Swenson, K. D. (2011) BPAF: A standard for the interchange of process analytics data. In: M zur Muehlen & J Su (Eds.), *Business process management workshops* (pp. 170–181). Berlin Heidelberg: Springer.

26. Hasan, S., & Curry, E. (2015). Thingsonomy: Tackling variety in Internet of Things events. *IEEE Internet Computing, 19,* 10–18. https://doi.org/10.1109/MIC.2015.26.
27. Hasan, S. & Curry, E. (2014). Approximate semantic matching of events for the Internet of Things. *ACM Transactions on Internet Technology, 14,* 2:1–2:23. https://doi.org/10.1145/2633684.
28. (2012). Code for America innovation team arrives in Chicago to develop new open 311 system. US Fed News Service. US State News.
29. (2012). Text messaging capabilities are added to Chicago's 311 and city alerts system. US Fed News Service. US State News.
30. Nuttall, R. (2014). *Bring Open311 to Pittsburgh.* Pittsburgh City Pap.
31. (2014). What is Open311? Open311. Retrieved November 21, 2014 from http://www.open311.org/learn/.
32. (2014). City of Chicago—311 City Services. Retrieved November 7, 2014 from http://www.cityofchicago.org/city/en/depts/311.html.
33. Vera-Baquero, A., Colomo-Palacios, R., & Molloy, O. (2014). Towards a process to guide big data based decision support systems for business processes. *Procedia Technology, 16,* 11–21. https://doi.org/10.1016/j.protcy.2014.10.063.
34. Dale, B. G. & Cooper, C. L. (1994). Introducing TQM: The role of senior management. *Management Decision, 32,* 20–26. https://doi.org/10.1108/00251749410050660.
35. Kiran, D. R. (2016). *Total quality management: Key concepts and case studies* (1st ed.). India: Butterworth-Heinemann.
36. Pande, P. S., Neuman, R. P., & Cavanagh, R. R. (2000). *The six sigma way: How GE, Motorola, and other top companies are honing their performance.* McGraw Hill Professional.
37. Breyfogle, F. W. (2003). *Implementing six sigma, second edition: Smarter solutions using statistical methods* (2nd ed.). Hoboken, NJ: Wiley.
38. Harry, M., & Schroeder, R. (2006). *Six sigma: The breakthrough management strategy revolutionizing the world's top corporations* (51634th ed.). New York: Crown Business.
39. De Bruin, T., & Rosemann, M. (2005). Towards a business process management maturity model. In D. Bartmann, F. Rajola, J. Kallinikos, et al. (Eds.), *Faculty of science and technology* (pp. 1–12). CD Rom: Verlag and the London School of Economics.
40. Harrington, H. J. (1991). *Business process improvement: The breakthrough strategy for total quality, productivity, and competitiveness.* McGraw-Hill Education.
41. James Harrington, H. (1995). Continuous versus breakthrough improvement: Finding the right answer. *Business Process Re-engineering & Management Journal, 1,* 31–49. https://doi.org/10.1108/14637159510103211.
42. Damij, N., Damij, T., Grad, J., & Jelenc, F. (2008). A methodology for business process improvement and IS development. *Information and Software Technology, 50,* 1127–1141. https://doi.org/10.1016/j.infsof.2007.11.004.
43. Vera-Baquero, A., Colomo-Palacios, R., & Molloy, O. (2014). *towards a process to guide big data based decision support systems for business processes.* Toria, Portugal: SciTePress—Science and and Technology Publications.
44. van der Aalst, W. (2016). *Process mining.* Berlin Heidelberg, Berlin, Heidelberg: Springer.
45. Molloy, O. & Sheridan, C. (2010). A framework for the use of business activity monitoring in process improvement. *E-Strategies for Resource Management Systems: Planning and Implementation,* 21–46.

Implementing Scalable Machine Learning Algorithms for Mining Big Data: A State-of-the-Art Survey

Marjana Prifti Skënduli, Marenglen Biba and Michelangelo Ceci

Abstract The growing trend of Big Data drives additional demand for novel solutions and specifically-designed algorithms that will perform efficient Big Data filtering and processing, recently even in a real-time fashion. Thus, the necessity to scale up Machine Learning algorithms to larger datasets and more complex methods should be addressed by distributed parallelism. This book chapter conducts a thorough literature review on distributed parallel data-intensive Machine Learning algorithms applied on Big Data so far. The selected algorithms fall into various Machine Learning categories, including (i) unsupervised learning, (ii) supervised learning, (iii) semi-supervised learning and (iv) deep learning. The most popular programming frameworks like MapReduce, PLANET, DryadLINQ, IBM Parallel Machine Learning Toolbox (PML), Compute Unified Device Architecture (CUDA) etc., well suited for parallelizing Machine Learning algorithms, will be cited throughout the review. However, this review is mainly focused on the performance and implementation traits of scalable Machine Learning algorithms, rather than on framework wide-ranging choices and their trade-offs.

Keywords Big Data · Data mining · Machine learning · Unsupervised machine learning · Supervised machine learning · Semi-supervised machine learning

M. P. Skënduli (✉) · M. Biba
Department of Computer Science, University of New York in Tirana, Tirana, Albania
e-mail: marjanaprifti@unyt.edu.al

M. Biba
e-mail: marenglenbiba@unyt.edu.al

M. Ceci
Department of Computer Science, University of Bari, Bari, Italy

© Springer Nature Singapore Pte Ltd. 2018
S. S. Roy et al. (eds.), *Big Data in Engineering Applications*,
Studies in Big Data 44, https://doi.org/10.1007/978-981-10-8476-8_4

1   Introduction

During the recent years researchers have been witness to a dramatic, persistent increase of the sheer size of Big Data, often referred to as the data deluge. This ever growing surge of data continues to double every 2 years [1] as different format data pours in from sensors, devices, smartphones, digital platforms, independent or connected applications. As the amount of data expands exponentially, nearly all of it carries everyone's digital footprint, unleashing a new era of opportunities for businesses and people around the world. According to the Digital Universe Study, the projected size of our digital universe by 2020 will reach 44 zettabytes, or 44 trillion gigabytes [1]. This estimation opens up a whole new set of opportunities, but also challenges that need tackled.

The term "Big Data" coupled with the relevant awareness of Big Data phenomenon, was coined in 1998 from John Mashey, a SGI (Silicon Graphics) scientist who wrote a slide deck entitled "Big Data and the Next Wave of InfraStress" [2]. While this prominent term has been initially spotted from the industry representatives of the computer science community, it was right after Mashey's notes, when the term was first used in an academic source, the book under the authorship of Weiss and Indurkhya [3]. The concept of Big Data is continuing to evolve and to be reassessed, as it becomes the driving force behind many ongoing waves of the digital transformation, including artificial intelligence, data science and the Internet of Things (IoT). It is already true that Big Data has drawn the attention of many researchers, promising to revolutionize many fields like commerce, business, scientific research, government, national security, global economy, society administration, and so on [4].

There is huge potential and highly useful information hidden in the increasing volume of data, which can only be harnessed through powerful and accurate applications, tools and technologies that we still lag behind. Thus, a new scientific paradigm has emerged, the data-intensive scientific discovery (DISD), often associated to the evolving range of Big Data problems. Jim Gray's definition of this Fourth Paradigm, as a new method of pushing forward the frontiers of knowledge, has enabled new data-intensive computing technologies for gathering, manipulating, analyzing and displaying data according to Big Data requirements [5].

Big Data's nature is not simply defined by its volume, but far more from its complexity. According to a research report carried from Doug Laney under the META group in 2001 [6], it is described as being three dimensional. Subsequently, it was widely accepted as the "3 V-s" model that stands for data Volume (quantity/ amount of data), Velocity (speed of data generation) and Variety (type, nature, and format of data). In 2012, Gartner (former META) updated the definition grounded on this model, and consequently two more dimensions were added; Value (insights and impact) and Veracity (trustworthiness/quality of captured data), which contributed in achieving a mature definition of Big Data. The best way to describe Big Data according to Zhou et al. [7] is by organizing these five dimensions into a stack,

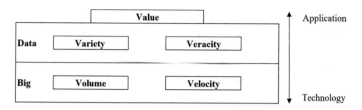

Fig. 1 Big Data Stack. *Source* [7]. Retrieved from http://www.sciencedirect.com/science/article/pii/S0925231217300577

comprising three layers respectively Big, Data and Value in a bottom—up fashion (see Fig. 1).

The upper layer represented from the Value dimension is more concerned with applications that exploit the strategic power buried in Big Data, while the bottom layer consisting of Volume and Velocity relies mainly on technological advances [7]. So far, we can envision Big Data as a messy compilation of large complex sets of data, incredibly massive, "dirty", noisy, vague, partial and moreover heterogeneous. Yet it offers myriad opportunities and valuable insights buried in the data, worthy to be leveraged through altered thinking, tools, and techniques. It is fitting therefore that we investigate some of the techniques that lie behind Big Data mining. Current research confirms that, state-of-the-art data processing approaches and traditional analytical platforms appear unsuitable to capture the full value residing in Big Data. However, considerable effort has been dedicated to the improvement of current approaches, aiming to give birth to novel multidisciplinary ones [8].

The latest techniques applied for the Big Data analytics comprise several cutting edge analytics, involving statistics, data mining, Machine Learning, neural networks, social network analysis, pattern recognition, signal processing, optimization methods and visualization approaches [9]. Among them, Machine Learning (ML) has received considerable attention from both industry and academia researchers. ML is an important subfield of Artificial Intelligence (AI), whose core principle is to design algorithms that permit machines to evolve behaviors (learn) based on empirical data. ML tools are currently deemed ideal in learning to recognize patterns on their own and make predictions, which have resulted pretty promising in the AI kit for business applications and human behaviors, such as computer vision, speech recognition, facial recognition, object recognition, linguistic translation, neuroscience, health, Internet of Things etc. The interest spurred in ML paradigms and algorithms is tightly coupled with ML's ability to exploit complex large volume of data within limited run times, and uncover more fine-grained patterns that are ultimately used to make more timely and accurate predictions than ever before [7]. The Big Data era provides remarkable rich information sources for ML algorithms to exploit hidden patterns and construct predictive models. The authors Zhou et al. [7] have discussed about opportunities and challenges that Big Data present to ML, thoroughly investigating most of them.

Among the key opportunities they highlight the quest that Big Data poses on ML to introduce a new way of thinking and novel algorithms to manage the technical challenges [10]: (i) the key role that Deep Learning has come to play, (ii) the unique opportunities for co-design of system and ML that have enabled hardware accelerations; the promised opportunity for research on workflow management and task scheduling, (iii) the importance of privacy-preserving ML, (iv) the extraordinary opportunity for learning with humans in the loop and (v) the high potential to create real-world impact. Yet, Big Data present critical challenges for traditional ML in terms of model scalability, adaptability, usability, high data dimensionality, streaming data and distributed computing.

The exponential increase of data volume, drives additional demand for developing mechanisms that will perform efficient data filtering and processing, recently even in real-time. Thus, the necessity to scale up Machine Learning to larger datasets and more complex methods can only be addressed by distributed parallelism, as suggested from current research. Distributed parallelism is the key to delivering fast parallel execution of tasks combined with the optimal usage of computational resources.

The increased attention to large-scale Machine Learning applications is attributed both to the evolution of hardware and programming frameworks and to the proliferation of large datasets across many modern applications. The vast choice of parallel and distributed platforms that implement efficient Machine Learning algorithms on large-scale data, include customizable integrated circuits (e.g., Field-Programmable Gate Arrays-FPGAs), custom processing units (e.g., general purpose Graphics Processing Units-GPUs), multiprocessor and multicore parallelism, High-Performance Computing (HPC) clusters connected via fast local networks, and datacenter-scale virtual clusters which are lately offered on a rent base from cloud computing providers [11].

Thus, the aim of this paper is to conduct a thorough literature review on distributed parallel data-intensive Machine Learning algorithms applied on Big Data so far. The algorithms that will be in the focus of this review fall into various Machine Learning categories, including unsupervised learning, supervised learning, semi-supervised learning and deep learning. The most popular programming frameworks like MapReduce, PLANET, DryadLINQ, IBM Parallel Machine Learning Toolbox (PML), Compute Unified Device Architecture (CUDA) etc., well suited for parallelizing Machine Learning algorithms, will be cited throughout the review. According to Bekkerman [11] the key segregating aspects between these frameworks relate to the parallelism granularity, algorithm customization degree, flexibility to bring together various programming paradigms, dataset scaling and online or offline execution fashion. However, this review will be mainly focused on the performance and implementation traits of scalable Machine Learning algorithms, rather than on platforms and framework wide-ranging choices and their trade-offs.

The chapter is structured as follows: Sect. 2 discusses unsupervised Machine Learning for Big Data, Sect. 3 presents recent work for supervised approaches,

Sect. 4 describes research for semi-supervised methods, Sect. 5 presents deep learning methods and Sect. 6 concludes with future work and challenges.

2 Unsupervised Machine Learning Algorithms on Big Data

Unsupervised learning represents a canonical setting of Machine Learning. Moreover, unsupervised learning algorithms make use of training data to attain a prediction function f, which is later applied to test instances. Given the fact that training data is provided in the form of unlabeled data, it is exactly from the later that we extract hidden value and insights. Data clustering is one of the most salient forms of unsupervised learning. The clustering process aims to partition an unlabeled dataset into a predetermined amount of disjoint sets (k), called clusters. Data instances assigned to the same cluster (intra-cluster) should presumably share a high degree of similarity, whereas the inter-cluster similarity is low. Intra-cluster similarity is defined in various ways including Euclidean distance, cosine similarity (especially applicable for vector data) etc. Indeed, clustering algorithms have been successfully employed in frameworks that support applications such as document retrieval, pattern classification, image segmentation and customer segmentation.

To begin with, one of the most popular yet simple clustering algorithm is K-means clustering, which dates back in 1955. Nevertheless, K-means is still in use and according to Wu et al. [12] it has been classified as one of the top 10 algorithms in data mining. Given a dataset $X = \{x_1, x_2, ..., x_n\}$, the basic principle of K-means is to assign n points to k clusters, while all points should belong to a center with the nearest distance for each cluster. Particularly, in K-means, the total distance between each data point and a representative point (centroid) of the cluster to which it is assigned is minimized. Conventionally, the K-means clustering algorithm adopts Euclidean distance to calculate the distances between data points and centroids. It may be obvious that the most intensive task that occurs in K-means algorithms is the calculation of distances, which is greatly tackled by introducing parallel execution of distance calculations. Researchers [13] proposed a parallel K-means clustering algorithm based on MapReduce, to which they namely refer as Parallel K-means (PKmeans) algorithm. They proved that their algorithm achieves reasonably good performance improvements in terms of speedup, size up and scale up, by simply employing MapReduce iterative jobs for parallelism. This approach is further improved from Anchalia et al. [14], that have successfully implemented the K-Means Clustering Algorithm over a distributed environment using Apache^{TM} Hadoop, stressing the key role of Mapper and Reducer routines introduced by them.

Another aspect considered for improvement in K-means algorithms, concerns the selection of initial centroids. For a proper initialization of k-means, Arthur and Vassilvitskii [15] proposed a modification of k-means namely k-means++, which improves both speed and accuracy of k-means by augmenting k-means with a simple, randomized technique. As a result, k-means++ results O (log k)-competitive with the optimal clustering [15]. Five years later, Bahmani et al. [16], proposed a scalable k-means++ (K-means‖) which samples O(k) points per iteration, repeatedly for O(log n) iterations, thus resulting in O(k log n) points as candidates [16]. Authors have proven that K-means‖ outperforms k-means ++ in both sequential and parallel settings, under experimental settings. Improvements of k-means clustering algorithm in terms of reliability and efficiency, have been introduced from [17, 18], that proposes a MapReduce k-means++ algorithm which can drastically reduce the number of MapReduce jobs by using only one MapReduce job to obtain k centers. In turn, this method addresses both issues concerning proper initialization and communication costs. To a broad extent, the K-means is often seen as a serialization algorithm, encompassing many iterative processes. Recent research [19, 20] signals that exposure to increased data volume, points out some challenges to traditional K-means clustering translated as exponentially increased time and space complexity. The performance of the framework employed in such cases, will worsen due to the increased number of iterations carried from the computations and the vast amount of redundant distance calculations, resulting in waste of computing resources. A distinctive, recent proposal of [19, 20] presents a twofold strategy behind a significant increase of efficiency and performance of parallel k-means clustering algorithm based on MapReduce. Authors' novel proposal consists on eliminating redundant distance calculations and replacing the intra-cluster Euclidean distance with Manhattan distance calculation model. The results they have gathered while conducting extensive experiments on real medical data, show a significant outperformance of their algorithm named IMR-KCA.

Density-based clustering algorithms are deemed popular given their relatively easy implementation and straightforward nature. In principle, density-based clustering algorithms can exploit clusters of different shapes and sizes even in noisy datasets, without necessitating prior information on the number of clusters (unlike K-means). With respect to the rationale behind them, clusters are dense regions in the data space, separated by regions of lower object density, thus a cluster is defined as a maximal set of density-connected points. The most representative clustering algorithm of this family is DBSCAN (Density-Based Spatial Clustering of Applications with Noise) proposed back in 1996 from Martin Ester, Hans-Peter Kriegel, Jörg Sander and Xiaowei Xu. Eventual drawbacks discovered in the classical DBSCAN algorithm, have inspired researchers to come up with various improved extensions of the initial one. The main drawback of DBSCAN is the high computational complexity in querying nearest neighbors. DBSCAN can suffer poor performance when dealing with high-dimensional and large-scale datasets. In order to respond to the ultimate challenges posed from the data deluge, there can be employed three main strategies to retain the computation efficiency: data indexing structures, parallel computing and dividing data sets [21]. With respect to the first strategy, new indexing structures are being embodied into DBSCAN with the purpose to reduce time complexity from $O(n^2)$ to $O(n \log n)$ and address large-scale data requirements. The quest to deal with the second strategy, parallel computing, has put forward many extensions of the DBSCAN algorithm. P-DBSCAN is a novel version of parallel DBSCAN in distributed environment, which was introduced from [22]. An important contribution has been made from [23], who proposed the so called DBSCAN-MR algorithm, tuned for parallel processing on the Hadoop platform, to solve the scalability problem. Patwary et al. [24] proposed a new parallel DBSCAN algorithm (PDSDBSCAN) relying on graph algorithmic concepts mainly. An interesting density-based clustering algorithm for Big Data using Hadoop, is Cludoop proposed from [25]. This algorithm is established on top of the serial clustering algorithm CluC, acting as a plugged-in clustering on parallel mapper together with cell descriptions (rather than the whole cell) during transmission, significantly reducing the number of distance calculations. The third strategy, consisting of data set division, aims to reduce time consumption by breaking down large datasets into smaller ones, followed by the application of DBSCAN. This approach assumes the employment of K-means and/or similar clustering techniques, during the pre-processing stage.

Hierarchical clustering, as a widely used clustering technique provides richer representation means, due to its native capacity to suggest the potential group structures in large-scale datasets. It can be further classified as agglomerative (a bottom-up approach) and divisive (a top-down approach). One of the most representative algorithms here is the agglomerative clustering algorithm called Single-linkage hierarchical clustering (SLINK). It's still a preferred analysis tool aimed to perform early-stage knowledge discovery, due to its simplicity and quadratic time complexity. Yet enough, SLINK does not scale well for large datasets, due to its high time complexity and inherent data dependencies. A recent publication [26] presented an efficient Spark-based single linkage hierarchical clustering algorithm, referred to as SHAS. SHAS algorithm translates the single-linkage hierarchical clustering problem to the minimum spanning tree (MST) problem. Authors have figured out that SHAS is memory efficient and can be scaled out linearly in Spark based environment, unlike with MapReduce. Additionally, there exists a new novel optimization of SLINK algorithm, Grid-SLINK proposed lately from [27]. Authors managed to achieve the optimization in GridSLINK, reducing the required number of distance calculations. This is attained by manipulating spatial locality of data points along with an adaptive gridding technique. In result, a second version dGridSLINk is also parallelized for distributed memory systems, showing good speedup and scalability.

Spectral clustering and its family methods have received increased attention. Algorithms that fall under this category, cluster data points based on Eigenvalue decompositions of affinity, dissimilarity or kernel matrices [28]. Spectral clustering treats clustering as a graph partitioning problem, without making strong assumptions on the statistics of clusters. Consequently, the good clustering results make spectral clustering algorithms, easy to implement and reasonably fast for sparse datasets of several thousand elements. However, they appear to be sensitive to the choice of parameters and computationally expensive for large-scale datasets. This computational complexity issue has been addressed in several studies. One effective approach to alleviate the time complexity was proposed from [29], who used the

Nysttrom approximation method to the similarity matrix. Power iteration clustering (PIC), is also deemed approriate to tackle the computational complexity of spectral clustering. This is achieved by replacing the Eigen decomposition of the similarity matrix with a small number of iterative matrix-vector multiplications. The original PIC approach was pushed forward from [28], who implemented a parallel power iteration clustering (p-PIC) based on the MPI message passing library interface, to handle large-scale data. Further, Jayalatchumy and Thambidurai implemented p-PIC in MapReduce achieving fast, scalable and accurate results for Big Data handling [30]. A very recent study carried from Priscilla and Chilambuchelvan, extended the original Power Iteration Clustering approach to the introduction of a Parallel Deflated Power Iteration Clustering (P-DPIC) MapReduce framework [31]. The P-DPIC algorithm is proposed with the promise to overcome inter-class collision problem in PIC. Experimental data revealed that by increasing the number of nodes in the mapper, the overall performance of P-DPIC also increased. The authors claim that their P-DPIC algorithm offers better accuracy, has smaller complexity and it is suitable for large, sparse, and multiclass datasets [31].

Subspace clustering, often represented from the classical SUB-CLU algorithm, aims to discover all lower-dimensional clusters hidden in subspaces of high dimensional data. According to Parsons et al. [32] subspace clustering algorithms fall under to main categories, top-down and bottom-up. The top-down algorithms find an initial clustering in the full set of dimension and evaluate subspaces iteratively refining clusters with smaller subspaces. While, bottom-up algorithms seek to find dense regions in low dimensional spaces and combine them to form clusters. They are further divided into grid-based and density-based methods [32]. Zhu et al. [33] proposed CLUS, a novel parallel algorithm rooted on SUBCLU algorithm and implemented in Spark. CLUS makes full use of Spark's in-memory primitives, while applying a new dynamic data partitioning method and executing multiple tasks density-based clustering (DBSCAN) tasks in parallel. Obviously, CLUS outperforms SUBCLU in terms of execution time, as it succeeds to minimize communication costs between nodes, to maximize overall CPU usage and to balance the load among them time. CLU is very promising with regards to Big Data exploitation, since it scales well with respect to the number of dimensions and the size of datasets. Bo Zhu and Alberto Mozo have further extended their work, with the presentation of a very recent parallel subspace clustering algorithm which was implemented on top of Spark, called Spark2Fires. It reduces time complexity (from exponential to quadratic) of both base and final cluster subspace generation, via approximation. Spark2Fires has been tested synthetic large datasets, demonstrating good scalability, accuracy and efficiency [34].

Bi-clustering, also known as co-clustering, aims to cluster both rows and columns of a matrix into groups, in a simultaneous fashion. Bi-clustering examines inter-related submatrices of rows and columns, which is pretty different compared to simple clustering. Additionally, it appears powerful in discovering hidden local patterns that remain unapparent to basic unsupervised algorithms such as K-means. Research shows that bi-clustering has high applicability in various application including text mining, bioinformatics, collaborative filtering, and graph mining.

There is evidence on several co-clustering algorithms proposed so far. Bongjune Kwon and Hyuk Cho have conducted an extensive review of co-clustering algorithms, focusing on the unified view of co-clustering algorithms, also known as Bregman co-clustering (BCC) framework, which includes six Euclidean distance and six I-divergence co-clustering algorithms. Subsequently, they face the scalability challenge by successfully parallelizing the twelve co-clustering algorithms in the BCC framework using message passing interface (MPI) [35]. A later, yet very interesting proposal came from Papadimitriou and Sun, who designed a distributed Co-clustering (DisCo) framework, developed on MapReduce. DisCo introduces practical approaches in the context of distributed data pre-processing and co-clustering [36].

3 Scaling of Supervised Machine Learning on Big Data

Supervised learning is a Machine Learning task, which infers a function from supervised data standing for labeled training data. Ultimately, supervised learning aims to identify and construct a function f that produces accurate predictions on previously unseen data or just test data, which are further subjected to verification and classification. Two most classical supervised tasks are classification and regression. Classification is a technique which is widely applied to determine the class of variables, equal to future trends prediction in real world scenarios. A classification algorithm is a supervised learning algorithm that analyzes the training data and produces an inferred function, which is called a classifier if the output is discrete. Whereas, a regression algorithm produces an inferred function which is called a regression function because the output is continuous. Supervised learning is widely applied in applications such as spam filtering, image recognition, speech recognition, text categorization, fraud detection, information extraction and retrieval, bioinformatics etc. A brief summary of noted classification and regression algorithms follows below.

Decision tree classification is one of the most popular classification methods in the rich context of data mining applications. A decision tree is a directed tree consisting of a root node, internal node or decision node and leaf nodes. The root has no incoming edges, while the rest of nodes with exactly one incoming edge are noted as decision nodes. During the training phase, each decision node splits the instance space so that the tree is constructed in a recursive fashion and the performance of the classifier is optimized. The decision rule constitutes the path from the root node to the leaf node, and is used to determine which class a new instance belongs to. A popular decision tree algorithm is C4.5. It has gained the attention of several researchers, which have continuously introduced improved to C4.5. However, the challenges posed from Big Data cannot be afforded from traditional decision tree algorithms including the classical C4.5 algorithm. First of all, building a decision tree on large-scale data appear time consuming, and secondly even though decision trees classification algorithms can accommodate parallel computing, data distribution should be optimized. Several C4.5 algorithm draw-backs are addressed very recently from [37]. In their paper the authors presented a parallelized version of the C4.5 decision tree learning algorithm called MR-C4.5-Tree based on MapReduce. The construction of MR-C4.5-Tree nodes is achieved via two different parallel methods; first ensuring the best splitting attribute via the information entropy-based parallelized attribute selection method (MR-A-S) and then by partitioning training data into subsets via a data splitting method called (MR-D-S). The authors tackled the over-partitioning issue using three termination conditions respectively, the depth of tree, the minimum of samples, and the minimum accuracy rate. During the experimental stage, MR-C4.5-Tree exhibits feasibility and the good performance [37].

Random forest is an ensemble classifier consisting on many (an arbitrary number) decision trees, that can be applied on both classification and regression tasks. Random forest grows an arbitrary number of classification trees. For classification problems, the ensemble of decision trees vote for the most popular class. While, in the regression problems, the ensemble uses averaging to achieve an estimate of the dependent variable. Using tree ensembles can lead to significant improvement in prediction accuracy and control over-fitting. With respect to this category, [38] have presented SMRF, a scalable Random Forest algorithm that performs data classification based on MapReduce. SMRF algorithm has three stages: initializing, generating and voting. The authors draw the conclusion that SMRF algorithm is much more suitable for Big Data classification as compared to any traditional Random Forest algorithm [38].

Fuzzy Rule-Based Classification Systems often referred to as FRCB-s, represent classification systems that can handle uncertainty and ambiguity effectively [39]. A promising algorithm has been proposed from [40], who extends the existing linguistic fuzzy rule-based classification system called Chi-FRBCS to a new version suited to Big Data, called Chi-FRBCS-BigData. This algorithm used MapReduce framework to learn and fuse rule bases. The interpretable model behind this algorithm is able to exploit large-scale data with fast response times and good accuracy. Authors have developed two versions of the algorithm, respectively Chi-FRBCS-BigData-Max and Chi-FRBCS-BigData-Ave. The first one achieves better accuracy, while the second one achieves faster results. According to authors the best version of the algorithm is the one that fits best with user's needs [40]. A subsequent recent study carried from almost the same researchers [41], proposed the very first Evolutionary Fuzzy system suited for Big Data problems. The initial knowledge base is built from the Chi-FRBCS-BigData algorithm and by means of a genetic tuning of the 2-tuples Data base. As a result, the fuzzy labels will be perfectly contextualized within every subset of the problem, and the coverage of the Rule Base will be enhanced. In the final step, the knowledge Bases from each Map process are joined to build an ensemble classifier [41].

Naïve Bayes is a supervised learning method as well as a statistical method for classification, grounded on the statistical Bayes' theorem (named after Thomas Bayes who proposed this theorem) with independence assumptions among variables [39]. This learning method assumes an underpinning probabilistic model, allowing us to capture uncertainty about a model by defining probabilities of the outcomes. In summary, Bayesian classification offers a convenient perspective for assessing and evaluating many learning algorithms. It has been extensively employed in applications such as text classification, spam filtering, collaborative filtering, sentiment analysis, etc. Among many implementations and a wide range of developed algorithms, a recent version of Naive Bayes classifier has been implemented on MapReduce for sentiment analysis. The authors' goal has been to evaluate how well scales up the Naïve Bayes classifier (NBC) in large-scale datasets. The results provided from this study are very promising, in that the NBC accuracy improved and approached 82% when the dataset size increases [42].

Support vector machines (SVM) represent a core Machine Learning technology that reveal robust theoretical foundations and excellent empirical successes in many pattern recognition applications such as isolated handwritten digit recognition, text categorization and information retrieval according to Bekkerman [11]. Support Vector Machines are key classification and regression tools, based on which have been developed several SVM software models, such as lightSVM, ls-SVM, libSVM and so on. LibSVM is often referred to as the most proficient and applicable SVM. Needless to say but, SVM-s also suffer from a widely recognized scalability problem in both terms of memory use and computational time. The improvement of SVM based algorithms has increasingly attracted many researchers of the field. Sun and Fox [43], developed a parallel SVM model (parallel LibSVM) based on the existing research on SVM-s and on Twister MapReduce framework. The model proposed from the researchers, divides the training samples into subsections. Then, each subsection goes through a training session provided from the SVM model. More specifically, mappers using LibSVM accomplish the task to train in parallel every subSVM. The non-support vectors are filtered via subSVMs. The support vectors of each subSVM are used as the input of the next layer subSVM. As a result, the global SVM model will be obtained through iteration [43]. Variations of the parallelized SVM algorithm on MapReduce but not limited to it, have been successfully implemented in email classification [17], sentiment analysis [44], biotechnology for protein to protein interaction prediction [45] and so on.

Artificial neural networks (ANNs) are also efficient and attractive to be employed in classification and regression tasks. The most common ANN is the back-propagation neural network knowns as PBNN due to its sensational function approximation and generalization properties. Initially it was developed as a solution to the problem of training multi-layer perceptrons. The paper [46] proposed a parallelized back propagation neural network algorithm PBPNN. The challenging goal to improve the algorithm accuracy is met by including bootstrapping and majority voting. Substantially, bootstrapping assures original data information are maintained in sub-dataset, while majority voting provides the means to generate strong classifiers. The experimental work carried from the authors, underlines that PBPNN can significantly outperform the classical BPNN in terms of accuracy and stability. Additionally, given the iterative nature of the jobs to be carried, the evaluations of the authors point out that amongs Spark, HaLoop and MapReduce, the first one fits best with the proposed PBPNN algorithm [46].

Multiple linear regression is a widely adopted regression task, which is characterized from a very high training time and may be prone to failure when applied on large datasets. An audacious work in this subtle area, has been completed from Rehab and Boufares. The authors have proposed a multiple linear regression model on MapReduce (MLR-MR) to enhance speedup and scalability over large-scale datasets. The distributed training method behind MLR-MR, combines the QR decomposition and the ordinary least squares method adapted to MapReduce environment. Technically speaking, the MLR-MR algorithm comprises three steps. Initially, the matrix is divided in smaller blocks, then mappers compute a local QR for each block, and finally the results of all the mappers are combined in order to provide the final results. The results contemplated from the authors present a very promising algorithm, whose parallelized version can handle large datasets and solves the out-of-memory problem also [47].

4 Semi-supervised Machine Learning on Big Data

Semisupervised learning is another Machine Learning task that makes use of data collections containing small amounts of labeled data and large amount of unlabeled data. This mixture of labeled and labeled data is used as training data. Generally speaking, semisupervised learning is quite appropriate when dealing with large amounts of unlabeled Big Data, whose labelling cost is both high and time consuming. The two most important techniques in semisupervised learning are: co-training and active learning [39].

The co-training learning process requires two views of the data. It assumes that each example is partitioned into two distinct views and both views are independently sufficient. Initially, it learns to separate classifiers for each view, and further the most confident predictions of each classifier on unlabeled data is used to iteratively create labeled trained data [39]. An interesting co-training Machine Learning algorithm is the one proposed from Hariharan and Subramanian [48]. This is one of the few (still limited number) research project focused explicitly on large scale multi-view learning on programming models such as MapReduce. The researchers came up with a co-training Multiview learning algorithm, using both consensus and complementary principles. They proposed a computational design based on data structures such as mapping table, label file and bi-directional Reducers. In turn, this design reduces the overall I/O requirements, and computational needs, since it is not necessary to broadcast all the labels.

Active learning is another semisupervised learning technique in which the learner actively chooses which examples to label and the final goal is to reduce the number of labeled examples needed for learning. The Back-Propagation BP (PCAL-BP) algorithm, is a decent neural network algorithm adaptable to Big Data requirements [49]. The PCAL-BP algorithm choses samples and punishments based on the absolute value of the prediction error, aiming to improve the efficiency of learning large data. This approach induces reduction of the learning effort and provides high precision. Compared to 16 varied classical classification algorithms, the PCAL-BP algorithm clearly outperformed 14 of them [49].

5 Deep Machine Learning on Big Data

Deep learning is a growing field of interest, a subset of Machine Learning (ML), which aims to solve problems that have resisted the best attempts of the artificial intelligence community so far. Deep Learning can be definitely seen as ML's bleeding edge. It encompasses a set of supervised and unsupervised ML techniques that are based on learning hierarchical representations of data in deep architectures for classification [50]. Early representations were inspired by the advances in neuroscience, and specifically on the human brain mechanisms and neural coding. Subsequently, other representations turned out to be better at simplifying the learning task. One of the most applicable frameworks for deep learning is the combination of a Deep Belief Network (DBN) with Restricted Boltzman Machine (RBM). RBMs are successfully used to construct a training model under unsupervised conditions. Afterwards, the DBN uses it for supervised classification [39]. Deep learning has dramatically improved the state-of-the-art in many applications. In terms of deep learning strengths, impressive results have been achieved in audio classification, image recognition, natural language processing, fraud detection, threat detection, biomedical informatics, sentiment analysis, and log analysis. Deep neural network architectures can be adapted to many types of problems and domains and apparently their hidden layers reduce the need for feature engineering. In terms of limitations, deep learning algorithms are usually not suitable as general-purpose algorithms because they require a very large amount of data. In fact, they are usually outperformed by tree ensembles for classical machine learning problems. In addition, they are computationally intensive to train, and they require much more expertise to tune. In spite of all the compelling achievements in large-scale deep learning, this field is still in its early stages. A lot of research efforts should be paid to address many significant challenges posed from the Big Data realm.

6 Challenges and Future Work

The Big Data era is already underway, and we are invited to accommodate large-scale data and make use of the gold mine insights we can get out of them. According to estimations [1], this year we are going to cross into the zettabyte regime in terms of data volume. Handling these high volume, heterogeneous and dynamic data is becoming a must. Hence, next to the problem of collecting and storing the data sea, the subject of efficient learning and adaption in the Big Data context becomes paramount. Amounts of data that we store should be coupled with novel algorithms, customized applications and scalable tools. The emergence of new sources of data, the quick release of Big Data platforms, advanced business analytics, the existence of too many engines and so on, have contributed in creating a messy and large Big Data ecosystem. We need to gain control over this data deluge, standardize the ecosystem so that it can deliver high quality information on the promise to make our life healthier, our tasks effortless, and our businesses more agile and successful.

With respect to this big picture the aim of this review is to acknowledge the challenges along with the opportunities in the Big Data era. An extensive review of seminal research has been conducted to ensure that the audience will be informed on the most distinctive and/or promising Machine Learning algorithms augmented so far in scalable Big Data platforms. While conducting this technical review on scalable Machine Learning algorithms, the authors include an appreciation of their performance against computational complexity. Categorizing Machine Learning algorithms is an intricate goal. Scientific literature suggests several reasonable approaches; they can be grouped into generative/discriminative, parametric/ non-parametric, supervised/unsupervised, and so on. However, the authors here introduce another approach to categorizing algorithms by machine learning task. Despite advantages and disadvantages, no one algorithm works best for every problem. Of course, the algorithms you select must be appropriate for your problem/domain, which is exactly about picking the right machine learning task.

Machine Learning is showing the most promise at providing tools that will benefit science, industry, and society. The authors guide the reader through the difficult process of assessing and deploying the best-fit Machine Learning algorithms to the Big Data context. Closed attention has been given to the identification of challenges imposed from Big Data on current Machine Learning tools. These challenges are summarized below:

1. **Scalability**: The data deluge is all about those voluminous data, pouring in from sensors, devices, smartphones, digital platforms, independent or connected applications that we need to distill and harness. Algorithms that mine Big Data should be designed with care, taking into account the growing pace and the increasing complexity of data. They should be flexible enough to adapt with the evolving nature of Big Data, while sticking to the accuracy and time efficiency requirements.

2. **Distributed data**: Conventional data analytic tools relied on centralized/ standalone tools that process quantifiable datasets. As opposed to that, Big Data are generated from diverse, multiple, autonomous and heterogeneous sources, often distributed across multiple physical and technical sites. Therefore, algorithms that mine Big Data need to be distributed in order to genuinely handle the data sea.

3. **Evolving data**: The shift from static data to dynamic data generated on the go, in real time pace, brings up the necessity to come up with sophisticated mining algorithms. They should be well-suited to fast and accurate real-time analysis performed on dynamic datasets.

4. **Heterogeneous data**: Big Data can be envisioned as a messy compilation of large complex sets of data, incredibly massive, unstructured, "dirty", noisy, vague, partial and moreover heterogeneous due to their provenience. Yet, it offers myriad opportunities and valuable insights buried in the data, questing to be handled from new algorithms, altered tools and techniques.

The above challenges divulge into future goals to be meet from research and development.

References

1. IDC/EMC. (2014, April). The digital universe of opportunities: Rich data and the increasing value of the internet of things. Retrieved from https://www.emc.com/leadership/digital-universe/2014iview/executive-summary.htm.
2. Mashey, J. R. (1999). Retrieved from http://static.usenix.org/event/usenix99/invited_talks/mashey.pdf.
3. Weiss, S., & Indurkhya, N. (1998). *Predictive data mining: A practical guide*. Morgan.
4. Jin, X., W. Wah, B., Cheng, X., & Wang, Y. (2015). Significance and challenges of Big Data research. *Big Data Research*, 59–64.
5. Hey, A. J., Tansley, S., & Tolle, K. M. (2009). *The fourth paradigm: Data-intensive scientific discovery*. WA: Microsoft Research Redmon.
6. Laney, D. (2001). 3-D data management: Controlling data volume, velocity and variety. META Group Research Note.
7. Zhou, L., Pan, S., Wang, J., & V. Vasilakos, A. (2017, May 10). Machine learning on Big Data: Opportunities and challenges. *Neurocomputing, 237*, 350–361. http://doi.org/10.1016/j.neucom.2017.01.026.
8. Alippi, C., Ntalampiras, S., & Roveri, M. (2016). Designing HMMs in the age of big data. In *Advances in Big Data: Proceedings of the 2nd INNS Conference on Big Data* (pp. 120–130). Springer International Publishing.
9. Chen, P., & Zhang, C.-Y. (2014). Data-intensive applications, challenges, techniques and technologies: A survey on Big Data. *Information Sciences*, Elsevier.
10. Chen, X.-W., & Lin, X. (2014). Big data deep learning: challenges and perspectives. In *IEEE Access, 2*, 514–525. https://doi.org/10.1109/access.2014.2325029.
11. Bekkerman, R. A. (2011). Scaling up machine learning: parallel and distributed approaches. In *Proceedings of the 17th ACM SIGKDD International Conference Tutorials* (pp. Article 4, 1). San Diego, California. http://dx.doi.org/10.1145/2107736.2107740.
12. Wu, X., Kumar, V., Ross Quinlan, J., Ghosh, J., Yang, Q., Motoda, H., ... Steinberg, D. (2008, January). *Knowledge and Information Systems, 14*(1), 1–37.
13. Zhao, W., Ma, H., & He, Q. (2009). Parallel k-means clustering based on MapReduce. In *IEEE International Conference on Cloud Computing. CloudCom 2009* (Vol. 5931, pp. 674–679). Berlin, Heidelberg: Springer.
14. Anchalia, P. P., Koundinya, A. K., & Srinath, N. K. (2013). MapReduce design of k-means clustering algorithm. In *013 International Conference on Information Science and Applications (ICISA)* (pp. 1–5). Suwon. https://doi.org/10.1109/icisa.2013.6579448.
15. Arthur, D., & Vassilvitskii, S. (2007). k-means++: the advantages of careful seeding. In *Proceedings of the Eighteenth Annual ACM-SIAM Symposium on Discrete Algorithms* (pp. 1027–1035). New Orleans, Louisiana: Society for Industrial and Applied Mathematics.
16. Bahmani, B., Moseley, B., Vattani, A. K., & Vassilvitskii, S. (2012). Scalable k-means++. In *Proceedings of the VLDB Endowment* (pp. 622–633). VLDB Endowment. http://dx.doi.org/10.14778/2180912.2180915.

17. Xu, K., Wen, C., Yuan, Q., He, X., & Tie, J. (2014). A MapReduce based parallel SVM for email classification. *Journal of Networks, 9*(6), 1640–1647.
18. Xu, Y., Qu, W., Li, Z., Min, G., Li, K., & Liu, Z. (2014). Efficient k-means++ approximation with MapReduce. *IEEE Transactions on Parallel and Distributed Systems*, 3135–3144.
19. Tang, Z., Liu, K., Xiao, J., Yang, L., & Xiao, Z. (2017, March 23). A parallel k-means clustering algorithm based on redundance elimination and extreme points optimization employing MapReduce. *Concurrency and Computation: Practice and Experience.* http://dx.doi.org/10.1002/cpe.4109.
20. Tang, Z., Liu, K., Xiao, J., Yang, L., & Xiao, Z. (2017). A parallel k-means clustering algorithm based on redundance elimination and extreme points optimization employing MapReduce. *Concurrency and Computation: Practice and Experience.* https://doi.org/10.1002/cpe.4109.
21. Lv, Y., Ma, T., Tang, M., Cao, J., Tian, Y., Al-Dhelaan, A., & Al-Rodhaan, M. (2016, January 1). An efficient and scalable density-based clustering algorithm for datasets with complex structures. *Neurocomputing*, 9–22. http://doi.org/10.1016/j.neucom.2015.05.109.
22. Chen, M., Gao, X., & Li, H. (2010). Parallel DBSCAN with priority r-tree. In *2010 2nd IEEE International Conference on Information Management and Engineering* (pp. 508–511). Chengdu. https://doi.org/10.1109/icime.2010.5477926.
23. Dai, B.-R., & Lin, I.-C. (2012). Efficient map/reduce-based DBSCAN algorithm with optimized data partition. In *2012 IEEE Fifth International Conference on Cloud Computing* (pp. 59–66). Honolulu. https://doi.org/10.1109/cloud.2012.42.
24. Patwary, M. A., Palsetia, D., Agrawal, A., Liao, W.-K., Manne, F., & Choudhary, A. (2012). A new scalable parallel DBSCAN algorithm using the disjoint-set data structure. In *Proceedings of the International Conference on High Performance Computing, Networking, Storage and Analysis* (pp. 1–11). Salt Lake City, Utah: IEEE Computer Society Press.
25. Yu, Y., Zhao, J., Wang, X., Wang, Q., & Zhang, Y. (2015). Cludoop: An efficient distributed density-based clustering for Big Data using Hadoop. *International Journal of Distributed Sensor Networks, 11*(6).
26. Jin, C., Liu, R., Chen, Z., Hendrix, W., Agrawal, A., & Alok, C. (2015). A scalable hierarchical clustering algorithm using spark. In *Proceedings of the 2015 IEEE First International Conference on Big Data Computing Service and Applications* (pp. 418–426). Washington, DC, USA: IEEE Computer Society.
27. Goyal, P., Kumari, S., Sharma, S., Kuma, D. R., Kishore, V., Balasubramaniam, S., & Goyal, N. (2016). A fast, scalable SLINK algorithm for commodity cluster computing exploiting spatial locality. In *2016 IEEE 18th International Conference on High Performance Computing and Communications; IEEE 14th International Conference on Smart City; IEEE 2nd International Conference on Data Science and Systems (HPCC/SmartCity/DSS)* (pp. 268–275). IEEE. https://doi.org/10.1109/hpcc-smartcity-dss.2016.0047.
28. Yan, W., Brahmakshatriya, U., Xue, Y., Gilder, M., & Wise, B. (2013, March). p-PIC: Parallel power iteration clustering for Big Data. *Journal of Parallel and Distributed Computing, 73*(3), 352–359. http://doi.org/10.1016/j.jpdc.2012.06.009.
29. Fowlkes, C., Belongie, S., Chung, F., & Malik, J. (2004). Spectral grouping using the Nystrom method. *IEEE Transactions on Pattern Analysis and Machine Intelligence*, 214–225.
30. Jayalatchumy, D., & Thambidurai, P. (2014). Implementation of P-Pic algorithm in Map Reduce to handle Big Data. *IJRET: International Journal of Research in Engineering and Technology*, 113–118.
31. Priscilla, G. A., & Chilambuchelvan, A. (2016). A fast and parallel implementation of reduction based power iterative clustering algorithm in cloud. *International Journal of Computer Science and Information Security*, 81–87.
32. Parsons, L., Haque, E., & Liu, H. (2004). Subspace clustering for high dimensional data: A review. *SIGKDD Explorations Newsletter*, 90–105.
33. Zhu, B., Mara, A., & Mozo, M. (2015). CLUS: Parallel subspace clustering algorithm on Spark* (pp. 175–185). Poitiers, France.

34. Zhu, B., & Mozo, A. (2016). Spark2Fires: A new parallel approximate subspace clustering algorithm. In I. et al. (Ed.), *Communications in computer and information science* (pp. 147–154). Prague, Czech Republic.
35. Kwon, B., & Cho, H. (2010). Scalable co-clustering algorithms. In *Algorithms and Architectures for Parallel Processing: 10th International Conference* (pp. 32–43). Busan, Korea: Springer Berlin Heidelberg.
36. Papadimitriou, S., & Sun, J. (2008). DisCo: Distributed co-clustering with Map-Reduce: A case study towards petabyte-scale end-to-end mining. In *Proceedings of the 2008 Eighth IEEE International Conference on Data Mining* (pp. 512–521). Washington, DC, USA: IEEE Computer Society. https://doi.org/10.1109/icdm.2008.142.
37. Mu, Y., Liu, X., Yang, Z., & Liu, X. (2017). A parallel C4.5 decision tree algorithm based on MapReduce. *Concurrency and Computation: Practice and Experience.* https://doi.org/10.1002/cpe.4015.
38. Han, H., Liu, Y., & Sun, X. (2013). A scalable random forest algorithm based on MapReduce. In *2013 IEEE 4th International Conference on Software Engineering and Service Science* (pp. 849–852). IEEE. https://doi.org/10.1109/icsess.2013.6615438.
39. Gupta, P., Sharma, A., & Jindal, R. (2016). Scalable machine-learning algorithms for Big Data analytics: A comprehensive review. *Wiley Interdisciplinary Reviews: Data Mining and Knowledge Discovery*, 194–214.
40. del Río, S., López, V., Benítez, M., & Herrera, F. (2015). A MapReduce approach to address big data classification problems based on the fusion of linguistic fuzzy rules. *International Journal of Computational Intelligence Systems*, 422–427.
41. Fernandez, A., del Río, S., & Herrera, F. (2016). A first approach in evolutionary fuzzy systems based on the lateral tuning of the linguistic labels for Big Data classification. In *2016 IEEE International Conference on Fuzzy Systems (FUZZ-IEEE)* (pp. 1437–1444). IEEE. https://doi.org/10.1109/fuzz-ieee.2016.7737858.
42. Liu, B., Blasch, E., Chen, Y., Shen, D., & Chen, G. (2013). Scalable sentiment classification for Big Data analysis using Naïve Bayes classifier. In *2013 IEEE International Conference on Big Data* (pp. 99–104). IEEE.
43. Sun, Z. S., & Fox, G. (2012). Study on parallel SVM based on MapReduce. In *Proceedings of the International Conference on Parallel and Distributed Processing Techniques and Applications (PDPTA)* (p. 1).
44. Khairnar, J., & Kinikar, M. (2014). Sentiment analysis based mining and summarizing using SVM-MapReduce. *International Journal of Computer Science & Information Technology, 5* (3), 4081.
45. You, Z.-H., Yu, J.-Z., Zhu, L., Li, S., & Wen, Z.-K. (2014). A MapReduce based parallel SVM for large-scale predicting protein–protein interactions. *Neurocomputing, 145*, 37–43. http://doi.org/10.1016/j.neucom.2014.05.072.
46. Liu, Y., Xu, L., & Li, M. (2016). The parallelization of back propagation neural network in MapReduce and spark. *International Journal of Parallel Programming*, 1–20.
47. Rehab, M. A., & Boufares, F. (2015). Scalable massively parallel learning of multiple linear regression algorithm with MapReduce. In *2015 IEEE Trustcom/BigDataSE/ISPA* (Vol. 2, pp. 41–47).
48. Hariharan, C., & Subramanian, S. (2013). Large scale multi-view learning on MapReduce. In *Proceedings of 19th International Conference on Advanced Computing and Communications.*
49. Wang, S., Zhao, Q., & Ye, F. (2013). A new back-propagation neural network algorithm for a Big Data environment based on punishing characterized active learning strategy. *International Journal of Knowledge and Systems Science*, 32–45.
50. Ranzato, M. A., Boureau, Y.-L., & LeCun, Y. (2007). Sparse feature learning for deep belief networks. In *Proceedings of the 20th International Conference on Neural Information Processing Systems* (pp. 1185–1192). USA: Curran Associates Inc.

Concepts of HBase Archetypes in Big Data Engineering

Ankur Saxena, Shivani Singh and Chetna Shakya

Abstract All the technology that has been used for the big data handling is inspired by technology that was explain in the Google paper back in 2003. HBase is of the top most used and preferred open source distributed system developed by the Apache including apache zookeeper, apache Hadoop HBase provide random access for the storing and retrieving the data. In HBase we can store any type of data in any format, data can be structured and semi structured. It is very malleable and dynamic in case of data model. It is a No-SQL database i.e. it doesn't let any inter row transactions to occur. Unlike traditional systems HBase run on multiple or a cluster of computers instead of single one, number of computer in a cluster can be increased or decreased as per the requirement. This type of design provide a more powerful and scalable approach for the data handling. This chapter explains about the how efficient HBase architecture and its command, operations are different from traditional systems.

1 Introduction

HBase is a database. You were probably thinking: "why we are studying about it, there are many database out there what so special about it?"

So let's start with its fundamental difference, that it's a non-relational database. For putting it in perspective it don't mind storing integer in one row and string in

A. Saxena (✉) · S. Singh · C. Shakya
Amity University, Noida, UP, India
e-mail: ankursaxena1434@gamil.com; asaxena1@amity.edu

S. Singh
e-mail: ssinghsivani6@gmail.com

C. Shakya
e-mail: chetna12shakya@gmail.com

© Springer Nature Singapore Pte Ltd. 2018
S. S. Roy et al. (eds.), *Big Data in Engineering Applications*,
Studies in Big Data 44, https://doi.org/10.1007/978-981-10-8476-8_5

another for same column. Also it don't worry about the type of data and it can store both structure and semi-structure data as well. It has dynamic and flexible data models to work with, also it stores the data differently as it uses row key, column key and timestamp like parameters for indexing, these parameters are also called the key value store and this method of storing is called the multidimensional sorted maps, in some cases these maps can even hold the image of itself previous copy specially in rapidly changing data like in weather pattern related data, etc.

Now let's talk about other key points of HBase. It supports scalability to another level by allowing it to use distributed computing or cluster computing, in this it uses multiple commodity hardware's instead of a single server hardware, and each of these commodity hardware's will provide a bit of storage, cache and computing also all these nodes (commodity hardware system in this case) are equal in their work so even if one of the nodes will come down or went offline due to some reason still the system will work, also as we are using cheaper hardware this makes the whole system non-expensive to setup.

It set's a benchmark in OLAP (online analytical processing) which is essentially the processing of metadata, thanks to its flexible data models, also although relational databases were doing a fine job with ACID (atomicity, consistency, isolation and durability) insurance but HBase is doing it better.

HBase helps in overcoming one of the biggest problem that relational database had of object-relational mismatch.

There are bunch of interesting facts about HBase. So it starts with the development of nutch which was essentially a web-searching project, so what happened is that Hadoop was developed out of nutch in 2007 and eventually become a top-level product of apache its open source nature let curious people to tingle with its code and so it happened, a man named cafarella released an experimental code based on big table (a part of Hadoop), he called this code the HBase and from here on HBase keeps on developing across communities thanks to its open source nature.

HBase is proven to be a powerful tool to be used in places where Hadoop is already implemented also HBase had become a top-level project of apache now HBase is being used by corporations like stumbleupon, Adobe, Facebook, Twitter and many more.

1.1 Installation Requirements

There are different requirement for HBase system installation, we can start with hardware, servers, software, networking etc.

Hardware
It is difficult to specify the particular type of the server Hardware for running HBase as it is capable of running over a wide variety of hardware configurations.

The commonly used one is commodity hardware, region server requires more space for the internal structure management such as the block cache or the mem store whereas the name node servers require more memory for softly processing, so we can make their hardware configurations accordingly which provides greater flexibility.

Hardware requirements for HBase installation can be categorised into 2: servers and networking

Servers

There are 2 type of machines in Hadoop and HBase: masters and slaves, in masters the name node, job tracker and H masters are included and in slave machines data nodes, task trackers and HBase region servers are present. It is not compulsory to use different configurations for different machines but it can be a little beneficial if there is a slight difference between hardware specifications, if possible but mostly the same hardware are used for both because of convince. The master machine doesn't need much storage so there is no requirement to add many disks, since they are more important than slave machines, we require redundant hardware component in case of failure for backup purpose, Slave machines store data and therefore they require plenty of space it depend on the task weather it is for read/write or processing purpose.

Disk: The number of disk is needed to be balanced with the number of CPU core, commonly we have minimum 1 core per disk, 6 disk is a good number in an 8—core server adding any more will not increase the performance.

Networking

In a data centre, servers are typically mounted into 19 racks with 40U or more in height. Switches usually have 24 or 48 ports, and for the channel bonding, the size of the networking should be large enough to provide enough bandwidth. Installing 40 1U servers would need 80 network ports in practice, you need a different setup where you can use multiple rack switches and then shift to a much larger core aggregation switch (CaS). This gives a two-tier architecture, where the distribution is handled by the ToR (Top-of-Rack) switch and the aggregation by the CaS.

Software

This can range from the operating system itself to files system choices and configuration of various auxiliary services.

- *Operating system*

It work's on all different Operating Systems which supports Java platform.

Java verification: enter java-version on the command line and verify that it works and that it prints out the version number.

- *Java*

Java is required for HBase. At minimum java version 6, or later version of java is required. If HBase fails to start with a warning that it was not able to find java then edit the conf/HBase-env.sh file by commenting out the JAVA_HOME line and changing its path to your Java installation path.

- *Hadoop*

For using HBase to its fullest you need Hadoop.

- *SSH*

To manage remote Hadoop and HBase daemons you require ssh. A commonly used software package providing these commands is OpenSSH. The supplied shell scripts make use of SSH to send commands to each server in the cluster.

1.2 Java Installation

Check java version in your system

$ java–version

The output will show the version configuration, path and running environment. Check the configuration of the operating system to install the java according to it (34 bits, 64 bits). Create the directory for the Java by removing JDK/JRE binary. Download the JDK/JRE for the Linux. Copy the binary to the **usr/local/java** directory, in few cases the binary will be download at the home. After that check your directories then edit the file path for java Fig. 1.

1.3 Installation of HBase

Hear, you will learn about the installation of HBase and its running mode, how to deploy and configure HBase (Fig. 2).

There are two modes to run HBase, Standalone and distribution mode. It is important to understand the configuration requirement for the HBase installation, it also depend on your requirement and the program that need to be run on the system. How much storage is required by your system for configuration matters a lot in the HBase installation and running of different nodes on machines. Not just hardware but software also affects the installation, which platform you are using for the HBase whether it is windows, Linux. It can affect the program run (Fig. 3).

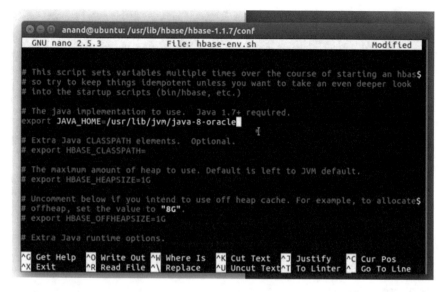

Fig. 1 Export java home path

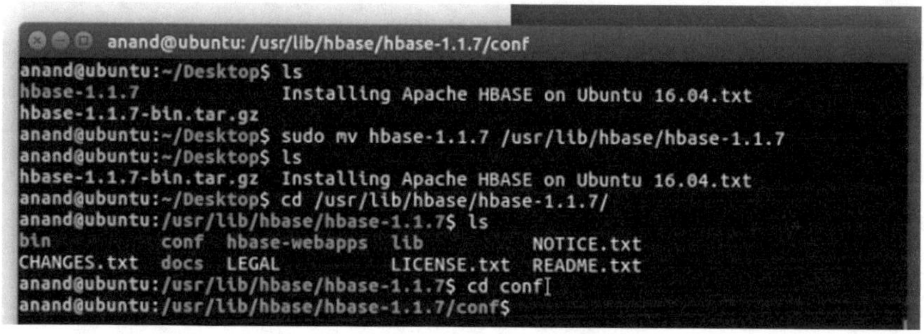

Fig. 2 HBase installation

1.4 Mode

There are three modes to run the program in HBase, Standalone, distributed which is divided into two parts fully distributed and Pseudo-distributed mode. Large-Scale tabular storage format is use in Hadoop (HDFS) Apache HBase. When you have to run HBase in the standalone mode, change the configuration by editing the *conf* directory of the HBase files, for the file distribution deployment. Despite which mode you will use to run HBase, you need to edit the ***conf/HBase-env.sh*** to tell the HBase of the type of java file to use. HBase environment variables like log files location, JMV, heap size etc. will be mention in this file. Set the java point **java_home** at it installation.

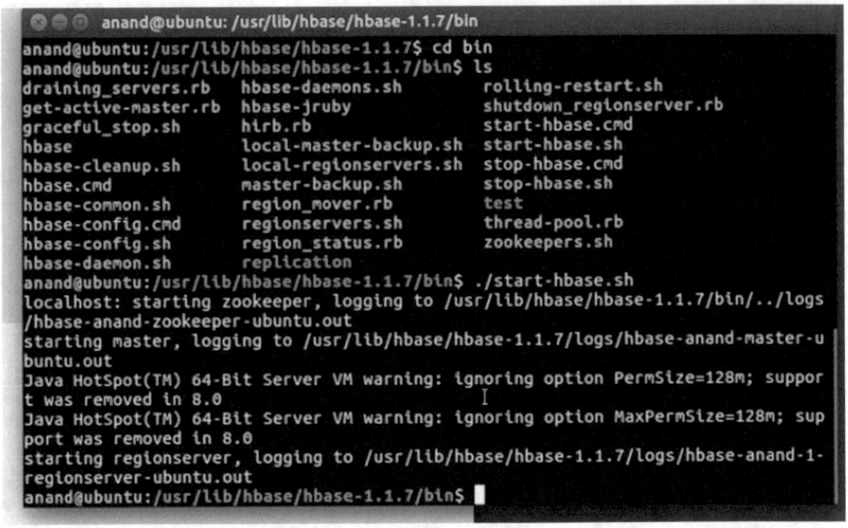

Fig. 3 HBase running

1.4.1 Standalone Mode

Standalone mode is a default mode or a "quick start" to install HBase, rather than HDFS (Hadoop distributed File system) it use local file system instead. Local zookeeper is run in the JMV (Fig. 4).

This mode is used for testing bugs and test run programs. It has no Job Tracker, Task Tracker, Name node or Meta-data to store the information. With no data node on the system it will store the file locally on hard disk.

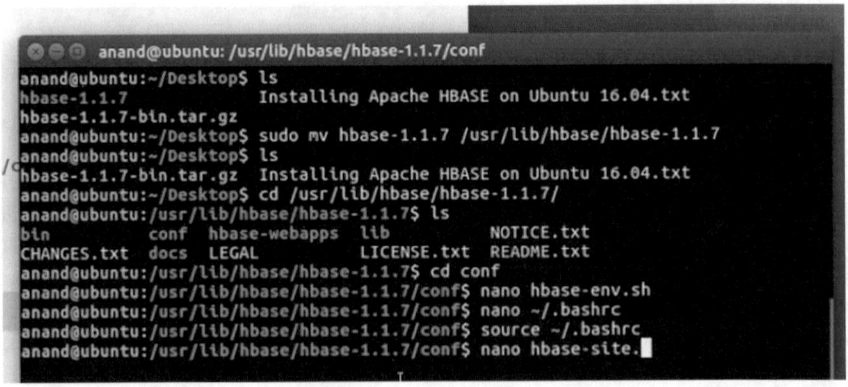

Fig. 4 Changing environmental variable

1.4.2 Distributed Mode

As mention above there are two type of distributed mode pseudo and fully, we will learn about them in detail in this section. Let's start with Pseudo-distributed mode.

Pseudo-distributed Mode

This mode unlike standalone mode have name node, data node, Job tracker, task tracker. It runs on the multiple nodes or the systems. There are two for the distributed mode to work, Pseudo distributed Mode and fully distributed mode. Both work on the multi-server machine.

Pseudo mode run on the multi-server on a single machine, it mimics that process. It stores the file in the HDFS. The number of the cluster you can run 1000 of node on it. We will see how installation is done to for the pseudo-distributed mode. It is important that you have installed the java and Hadoop in your system (Ubuntu), as shown above in the Figs. 1 and 2. Now how to install the HBase on the pseudo-distributed mode.

HBase installation starts with downloading the package from the Apache official site. Extract the tar file

tar –xvf HBase-1.2.2.bin.tar.g2

and copy it in the directory. After the extraction the file will be transfered to the distend file location. After this update the environmental variable with the HBase path.

Create the directory *cd HBase-1.2.2*

(to keep the location of HBase on the home)

>sudo mkdir/usr/lib/HBase

Create storage for the new directory in the file. Then update the configuration file of the changes you made in the file HBase-site.xml and HBase env.sh then run *>cd/usr/lib/HBase/habse-1.1.2/conf* and *>sudo gedit HBase-site.xml*

Copy paste the configuration then exit the geditor. Edit the habse-env.sh on command prompt and export your java home path then exit it. In the next run export, the HBase_home path in the. bashrc file and run it. After all this start Hadoop services by running yar and dfs file. Then start the Hadoop services by *>start-HBase.sh* use *>jps* to check and display the services name on the command prompt. Check whether the Hfile is created on HDFS or not.

Fully Distributed Mode

This is the mode which actually run on the Hadoop cluster. It is much more powerful cluster than the pseudo-distributed mod. With the different number of cluster present, it will be difficult to debug the system which will happen in the stand alone mode because of the working on the only single virtual machine (JMV). Running the fully distributed on the multiple hosts you need to add HBase.cluster. distributed property in the HBase-site.xml. And then need to add the location of the name node, HDFS to write the data in the Hfile. You also need to modify the configuration of the region server file. A fully distributed mode depends upon the zookeeper. Zookeeper runs as the default in the HBase configuration if you need to change or run it independently from the other nodes than go to the zookeeper management use HBASE_MANAGES_ZK variable in conf/habse-env.sh. This will change the management configuration of the zookeeper ensemble. Then go to the zookeeper configuration for any configurationally change, this option will be present in the HBase-site.xml. If you don't change this configuration then it will be store as default single ensemble member in the local host. Which make client to contact it by storing the setting and functioning on the local file system. The other installation process is similar to the pseudo-distribution mode.

1.5 Deployment

When the HBase is properly configured you need to deploy it on the cluster, since Hadoop and HBase is written in the java it is easier to process that. All files can be copied form server to server; each data shared is same. There is some way to do that, Script-based, Apache whirr, Puppet and Chef.

Script-Based, this approach is used for all list type of cluster from small to medium, but it is more considerate for the advance one. It uses the region server configuration of having all files which contain the list of all servers present in the cluster.

Apache Whirr, use of cluster in dynamic environment in now day it is very demanding, like in public or private clouds, Amazon EC2 etc. The main purpose is to quickly provision the servers and then to run the analytical work load. After result extraction simply shut down the cluster, restart it in case of another dynamic workload run. When the cluster is operational, it is easier and useful to abstract provision part because it will not be trivial to program against every one of the API's giving dynamic cluster infrastructure. This is where Whirr show its purpose by supporting the different type of clouds (private and public type mainly) API's and allow the provision of cluster running a different range of services.

Puppet and Chef, it works very much similar to the Apache Whirr, it has central provisioning server which can store all of the configuration, client software, on each server execute that communicate with central sever to keep it update and locally apply them. It supports changing of the running clusters. Master process is used to monitor the updates on the configuration and repository, and then start with appropriate action. This is used as on-the-fly or re-configuring clusters.

Full Operation Configuration

Once you have decided which machines will run which process, revise the configuration such that nodes can be locating by each other. In Order to do that, make sure that all configuration files are synchronized across the cluster. Use a configuration management system to synchronize the configured files, rsync will be a simpler and quick way for doing that. The addition of the zookeeper quorum address in HBase-site.xml is required as well as necessary move for the configuration change of pseudo-distribution to fully distributed mode. XML property to configure the nodes, are given below and also with the address of the node where the zookeeper quorum peer is running:

```
..<name>HBase.zookeeper.quorum..</name>
<Value>mymasternode</value>
```

HBase.zookeeper.quorum property is the comma-separated list of hosts. Zookeeper servers are running on it. If any of the zookeeper servers are down, then another list will be used by HBase. As a default, the zookeeper service will be bound to the port 2181. For changing the port, HBase.zookeeper.property.client Port property will be added to HBase-site.xml and a value will be set for the port that you want for Zookeeper to use.

```
<name> HBase.zookeeper.quorum</name>
<Value>
zk1.example.com:2181,zk2.example.com:20000,zk3.example.com:31111</value
```

2 HBase Architecture

HBase is a system composed of several different systems which are complex enough on their own rights so in order to understand its architecture you need to dive a bit in those systems as well, then we will put those systems together to see the big picture a.k.a the working of HBase.

For the beginners we can start with the types of servers required to set up HBase.

Basically, there are 3 type of servers required for HBase's master and slave con figuration built these are:

- *Region servers—they* provide regions for read and write purpose.
- *HMaster—they* are responsible for the coordination between the region servers and for the administration tasks.
- *ZooKeeper—*they act as the distributed coordination service to maintain server state in the cluster also checks over the availability of the servers and provide notification in case of server failures and many more.

All three of them explained were the basic building blocks of HBase systems we will cover all of them in detail as the chapter progresses (Fig. 5).

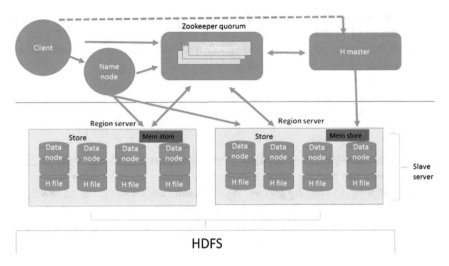

Fig. 5 Architecture of HBase

2.1 HMaster

HMaster is the part of master server of HBase architecture. The main work of HMaster region is to monitor all region servers that are present in the cluster at that time. It serve as an edge for the metadata changes. It runs on name node in a distributed cluster environment (a cluster of computers that are loosely or tightly connected to each other and work together. Figure 2 shows the main functions of HMaster server in HBase [1]. The communication between client and master server is bi-directional, client communicate with HMaster as well as zookeeper for data handling. Communication sometimes depends upon the size of the data or the information that has been provided by the client.

HMaster allot region to region server, to known their working progress and also to know if they are working properly or not. It also monitors the health of region server. DDL operations were oriented by metadata such as remove, create, enable, disable the table, modify the column family by adding column in column family and also moving and assigning of regions.

Some important roles of HMaster in HBase are written below:-

- For the changes in schema and metadata operations as per the client requirement are done by the HMaster.
- It controls the load balancing as well as failure of handling of the load over the other nodes in the distributed cluster.
- Allotting the services to different regions and monitoring their performance.
- It helps in maintaining the performance of nodes in the cluster as well as maintenance of nodes (Fig. 6).

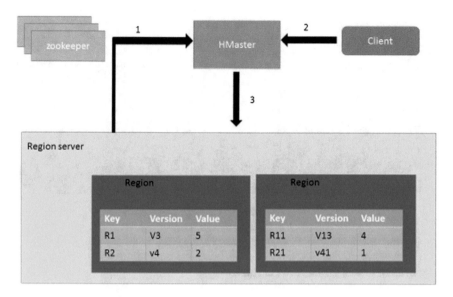

Fig. 6 HMaster in HBase (*1*) monitoring of region servers, (*2*) DDL operation handling (create, delete table), (*3*) assigning of region to region server for recovery and load balancing

2.2 Hregion Server

Region servers are the salve nodes in the HBase. These nodes perform client requests to read, write and delete. It also runs on ever y Hadoop cluster node. Region server runs on HDFS. It has following components:

Block cache: it is the read cache (temporary memory) and frequently stores the memory of read data. Whenever the block cache memory gets full, it deleted/evict the recent used data.

MemStore: it is to write cache. The new data that has yet to be written on disk is stored here. It sorts the data before writing it on the disk. It is present in every region server as per column family.

Hfiles: these are the file which stores the rows which are sorted in key values form on the disk.

WAL: Also known as the write ahead log, it run on the DFS (distributed file system). It stores the new data which is still not persevered over the permanent storage (i.e. disk). In case of failure the stored data is use for recovery of the lost one (Fig. 7).

While splitting the region by region server the master don't participate. The split region will then store in the meta table, which we will talk later. The split region will open on the host region server to report the split to master server.

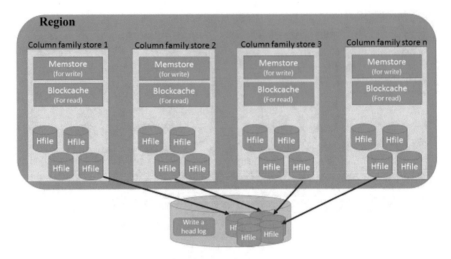

Fig. 7 Inside look of Region server in HBase

2.3 Zookeeper

Zookeeper is the apache open-source distributed application use to coordinate service. It helps in maintain and coordinating the servers. It maintains configuration information of the region server on which assignments or the task is assign to, it also provides synchronization between them, provides recovery if any of the region server crash down and loads the data on other available region server. In most of the cases the client will directly contact the zookeeper to get in touch with HMaster or Region Server. It also helps in repairing the failed nodes in the cluster. It keeps track of all data nodes in the region server and their task. We learned how many things zookeeper manage in HBase and how important it is for the better functioning of the cluster, but it is really important to know how HBase is able to do all this, to understand this we need to learn its technical architecture that has been shown in the Fig. 4. Technical structure of the zookeeper consist of the following nodes (Fig. 8).

Leader Node: It is the node responsible for processing the client write request. Other nodes present in there are follower nodes which simply follow the instruction of leader node to write the data for client.

Not every node present in the zookeeper is the leader node, it is not randomly selected. Leader node is elected by other nodes in the cluster; the node with highest number of vote will be elected as the leader node. For this type of selection the server requires odd number of nodes in the cluster. For example if the cluster is of four nodes and two of them get crash, the zookeeper will be down by half of the number and it would help in deciding a leader node through election. But in odd

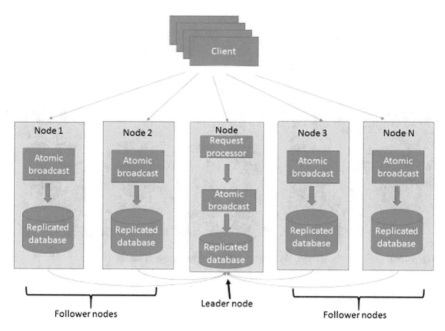

Fig. 8 Zookeeper and client interaction and how client request process through different nodes of the zookeeper quorum

number case even two of the nodes goes offline, there are three of them available to vote and elect a leader with majority number of nodes in services.

Follower Nodes: These nodes are able to process the client read request on their own. But write request need to process through leader node first. They also elect their leader nodes by voting.

Request Processor: This is only present in the leader node for processing the write request of the client to the follower nodes. Once the request is processed through them it will broadcast or transfer the changes to the follower nodes for the data state update.

Atomic Broadcast: This receives the change notification of request from follower nodes and transfer them to the leader node; they are present in both node types.

In-Memory Database/Replicated Database: This component is responsible for storage of the data in zookeeper. Each node has it own database which enable it to process the read request. In file system the data is store in case of recovery during cluster failure. After data is store into the file the only the database will be update for the changes.

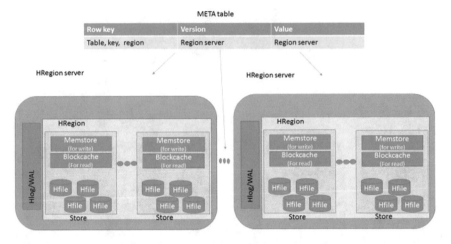

Fig. 9 META Table in HBase, show how data store in HRegion server

2.4 HBase Meta Data File System

HMeta data file system is maintained by the name node. Name node will communicate with client either directly or through the zookeeper. It ensures the availability in the cluster for the data load and auto sharing. Meta tables are the tables which keep list of all regions in the HBase system. It is used to find the region server for the given table key. How data is stored in these table format. It is column-oriented database. The data is stored in the column format table. The Table 1 shows the collection of column families and row id. As shown in the table the column families were present in the key-value schema pairs. Each column family have multiple numbers of columns in the table. These column values will be stored in the disk memory. Row represents the column family collection, column family is the collection of column and collection of row and column is the table (Fig. 9 and Table 1).

When enough amount of data accumulate by the Memstore the whole key-value set pair will written to the new HFile (data is store in key-value pair in HFile) in

Table 1 Shows the storage mechanism in the HBase and column family

Row id		Column family 1			Column family 2			Column family 3		
		Col1	Col2	Col3	Col1	Col2	Col3	Col1	Col2	Col3
R_id 1a		Ca	V1	5	Cf1	V1	8	Cf1:ca	V2	5
R_id 2b		Cb	V2	2	Cf2	V3	6	Cf2:cb	V3	6
R_id 3c		ca	V2	6	Cf3	V2	4	Cf3:ca	V2	7

| Key | | Value | Key | | Value | Key | | Value |

the HDFS. Then the sequential write is done very quickly on the disk. Now the big question is how HBase read and write the data.

2.5 Write and Read in HBase

2.5.1 Write

Client request for write firstly go through the DFS then to the name node for the processing, the request is forward back to the client node. Where after the name node approval the HDFS client forward the request to FSD output stream which put it through the data node i.e. region server. In region server the WAL (write ahead log) store the first important write log, we will study about it in detail later in the chapter. WAL is important to store data in case of any error occurrence while writing the data, that's why it is used. See Fig. 5 (Fig. 10).

After the log entry of the data, it is further forwarded to the memstore where the data is stored on the RAM of data node. Writing process is much faster than the RDBMS. After storing the file it forward to the HFile where data is stored in the HDFS, when the MemCache memory is full, the HFile store the data file directly. After the task is done the acknowledgement is send to the FDS output of the stream of the client node where it transfer the information of task completion to the client, which then revert back the request to close the file after task is done (Fig. 11).

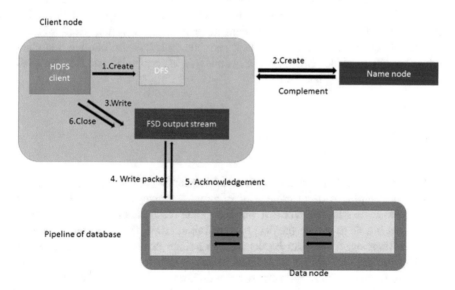

Fig. 10 Write in HBase

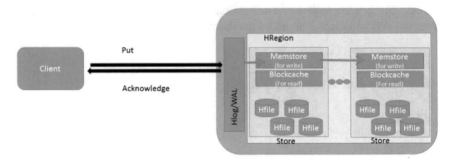

Fig. 11 Write in HBase and how file write in HRegion server

2.5.2 Read

A read request is directly or indirectly send to zookeeper from the client node. In client node the client transfers the request to the DFS from where it transfers to the zookeeper to give the block location as shown in the Fig. 8. Zookeeper is the place where all data status and record of their location is stored on the HRegion server in the META table. It gives the table address to the client as shown in the Fig. 9 step 1. The client goes to the Region where the data table is present which need to be read. Block Cache keeps the record of the previous read on the data. If the client table is found, the block cache returns the result if not then it search the table on the MemStore since data could have been written on the Hfile due to WAL. Even if the file is not present then the client will moved to the HFile because it's where the file could have been stored. After the needed file is located, the required data is taken from here and moves towards the client node FDS input system from where it update the Client of the completed task and acknowledgement is done, the client request to close the file. This is how read and write is done in HBase. Now let's talk about what is WAL (Figs. 12 and 13).

2.6 WAL-Write Ahead Log

WAL is write ahead log which work as life savior in case all data gets lost. It is much more helpful in case for the primary memory damage, in case of server crash WAL restore the data just where it was before the crash happens. Now understand how WAL is able to restore the data from its initial point. The client request for the data modification, which can be delete, put, update or any other command. The modification is done in the key value format which use remote procedure call (RPC) for the wire. These calls are given to different server on the HRegion server. Data written into WAL, and then stored in the MemStore. Which stores on the hard disk. From MemStore the data is written in the HFile after some time period.

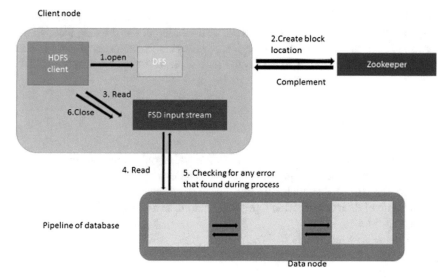

Fig. 12 HBase read operation in client node

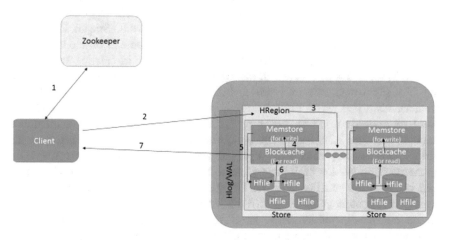

Fig. 13 HBase read operation and how it done step by step in HRegion server

2.7 HLog

Instantiation of every single passing HLog on parameter to HRegion server. Class that implements the WAL is known as the HLog. It keeps tracking of changes in the data. WAL provide amoric, consistency to the data. Sequencing the data makes it easier to keep the track of it, as the region is open to process the request for read or write by the client, it start to read for the high sequence number that stores on the META table in the HFile. Each table store data in key and value range in the

META table, Whenever the data is arrived at a region it is firstly written to the WAL in unpredictable order. HLog work as the searcher for the data location and for the split of the log on different region of HRegion server. This process is done by the help of the HMaster to assign the region server for the task in the split.

2.8 HLog Key

As mention in the Meta table section, the key-value only represent the key type, timestamp, row, column family in it. The write time or the time stamp of the key is noted by the HLogKey which help in keeping the sequential record in the order as per the time of their record. This time stamp help in recording the edit in the written log at that time.

2.8.1 Log Flusher

HRegion Server sort on the basis of Key Value then transfer to WAL. It is then made to a Sequence File. While this has every one of the reserves of being petty, it is unquestionably not. This is one of the base classes in Java IO Stream. Particularly streams structure which is to record a framework from time to time cushioned to redesign and execution as the OS is impressively faster in creating work information in the packs or squares. In the event that you make records independently, IO throughput would be truly unpleasant. In any case, regarding the WAL, this understands a gap where information is the degree that anyone knows written to head yet honestly it is nowhere to be found. To lessen the issue the main stream should be flushed continually. This handiness is given by the LogFlusher class and string. It essentially calls HLog.optionalSync(), which checks if the HBase.regionserver.optional log flush interval, set to 10 s as is ordinarily done, has been beat and if that is the situation conjures HLog.sync(). The other place concerning the synchronizing strategy is HLog.doWrite(). When it has framed the current change in agreement with the stream to check if the HBase.regionserver.flushlogentries parameter, set to 100 according to the typical system, has been beaten and call synchronization additionally. Synchronization itself summons HLog.Writer.sync() and is finished in Sequence File Log Writer. For the present, we expect it flushes the stream to a plate and all is well.

2.8.2 Log Roller

It is important to maintain the persistence of the log on the regular basis and also to restrict the size of the write log, all these things were done bye the Log roller. HBase.regionserver.logroll.period path in the $HBASE_HOME/conf/HBase-site.

xml file. Other parameter that control the role are HBase.regionserver.hlog.block-size and HBase.regionserver.logroll.multiplier. There are different methods which help in maintaining the sequential file by checking the highest to lowest number of file written in the storage file, it is needed to be done to keep the update persistence. HLog.rolwriter() and HLog.cleanoldlogs() are the those methods. The later method is used to un-edit any log that is left in the file, if there is, it is then deleted and leave the other one in the file.

3   HBase Shell

There are multiple ways by which we can interact with the HBase. One is shell, it executes various commands which can perform diverse operations. These operations are performed on the data tables that enhance the data storage efficiency and provide flexible interactions with the clients. Interaction of habse shell with the HBase is for data modeling, table operations and table management.

3.1   HBase versus RDBMS

HBase has many advantages over the traditional RDBMS/Mysql/DB2. Traditional databases has various limitations like larger amount of data (petabytes of data) and diverse variety of data (videos, images, audio etc) cannot be stored in it, it has limited memory and processing capability. Habse has overcome all these limitations. It is the NoSQL, that is the schema less data model. It has non locking concurrency control mechanism so that the real time reads will not conflict with the writes. It can store huge volume and variety of data.

It has one important feature called versioning, this feature is not present in the RDBMS. For example if we consider a customer table having row_id, 2 column families named (personal, order). The personal column family would further have column named (name, location) and order column family would have column (o_id, product). If we enter the record say (row_id = 1, name = robin, location = x, o_id = 01, product = phone). Second record (row_id = 2, name = scot, location = delhi, o_id = 02, product = calculator). Within the first record the location column is empty. In RDBMS this would be stored as NULL and would have internal space. But in case of HBase if particular value is missing it is just absent and won't take extra space. Secondly if we update the second record as row_id = 2, SET order: product = book, this is possible in both RDBMS as well as HBase. But the difference is of versioning. In versioning we need to specify how many versions of a particular column family have to be maintained. Suppose if we set the version value = 2 then last two values would be maintained. In RDBMS if we update the value the previous value would be overwritten and older value would be lost. But in HBase with versioning feature both the values (calculator and book) would be there which means a particular column would have two values and every value would have a timestamp. Timestamp represents the time at which the value was inserted. With the help of timestamp we can compare the older and the fresh value.

The commands of the HBase shell include the general and the table management commands which is categorized as data manipulation language (DMLcommands) and the data definition language (DDL) commands.

3.2 General Commands in HBse Shell

The general HBase shell includes the status, version, table help, whom commands which basically, gives the description about the HBase version that is being used, how we can manipulate the table and the current user of the HBase.

Status—This command will display the details related to the status of the system. The status of the system can be defined in terms of the total number of the server present in the cluster, average load value of the server, the active server count. The different parameters like summary, simple and detailed can be used depending on how detailed status we need.

Syntax:

```
HBase > status
HBase > 'summery'
HBase > 'simple'
HBase > 'detailed'
```

Table help—It includes the commands (gets, puts etc.) that can be used for manipulating the table. It provides help for table reference commands.

Syntax

```
HBase > table_help
```

Whoami—gives the description or the information of the current HBase user.

Syntax

```
HBase > whoami
```

3.3 Data Definitional Language (DDL Commands)

The DDL commands operates on the table. They are responsible mainly for the structural or change in the schema of the table. The data definition language includes create, alter, list, disable, is_disabled, enable, is_enabled, describe, exists, drop, drop_all commands.

Create—used for creating the table which would be specified with the name of the table and the column family name. table configuration can be mentioned optionally.

Syntax

```
Create '< table name >' '<column family>'
```

Alter—In order to alter the schema of the column family we use this command. The table name and the dictionary that would specify the new schema of the column family would be passed.

Syntax

```
HBase > alter < table name > name = 'new column family', version
```

List—it will enlist and display all the tables that are present and created in the HBase.

Syntax

```
HBase > list
```

Disable—whenever we need to delete or drop a table we first need to disable it. This command would disable the named table.

Syntax

```
HBase > disable '<table name>'
```

Is_ disabled—It will verify whether the table that we need to drop is disabled. It will disable all the table that are matching the given regex.

Syntax

```
HBase > is_disable '<table name>'
```

Enable—it will enable a particular named table.
Syntax

```
HBase > enable '<table name>'
```

is_enabled—it checks or verifies the named table is enabled or not.
syntax

```
HBase > is_enabled '<table name>'
```

Describe—it gives the detailed structural information about the named table along with its column families, version, filters etc.
syntax

```
HBase > describe '<table name>'
```

Drop—this command is used when we need to delete or drop the table that is created on the HBase. For this the table must first be disabled by using the disable command.
syntax

```
HBase > drop '<table name>'
```

Drop_all—It will delete all the tables which is matching the given regex.
syntax

```
HBase > drop_all '<matching regex>'
```

Exists—it will check for the particular named table if its existing or not.
syntax

```
HBase > exists '<table name >'
```

3.4 Data Manipulation Language (DML Commands)

The DML commands operate on the data of the table and not on the structure. These commands are applied when we need to manipulate the data present in the table. The data manipulation language includes put, get delete, delete all, count, scan and trunicate commands in it.

Put—when we need to enter the value to the specified table/column/row/cell we use put command.
  syntax

```
HBase > put '<table name>' '<row name>' '<column name>' 'value' time stamp
```

Get—to get the content of the specified row and column we use get command. It fetches the contents from the row or the cell.
  syntax

```
HBase > get '<table name>' '<row>' '<column>' time stamp version
```

Delete—to delete the value of the mentioned cell of specified table/row/column. It deletes the cell value in the table.
  syntax

```
HBase > delete '<table name>' '<row>' '<column>' time stamp
```

delete_all—this command will delete all the values or the cell of the given row. A table name, row is passed. Column name and time stamp is optional.
  syntax

```
HBase > delete_all '<table name>' '<row>'
```

Count—this command will retrieve the rows in the table and count the number of rows present in it.
  syntax

```
HBase > count '<table name>' cache = value
```

Scan—the entire table is scanned and the contents of the table are displayed. It scans and returns the data to the table.
  syntax

```
HBase > scan '<table name>'
```

Trunicate—In this command the specified table is disabled, dropped and recreated.
  syntax

```
HBase > trunicate '<table name>'
```

4 Data Models in HBase

It is responsible for managing the structured as well as semi structured data that is the data with higher level of information in the relation database. The HBase manages such data with high performance, service availability as well as the scalability which is in terms of data size and index size. HBase is an open source, no SQL, column oriented, distributed and sorted data store which is modeled after the google's big table. This HBase is built on the top of the Hadoop distributed file system/Hadoop. Further the data in the HBase can be stored in the form of tables which have multiple rows and fixed number of column families. Variety of data is stored in HBase in the column oriented manner that is data is stored and retrieved in the column. Since all the data is stored together in the column it is easy and quick to retrieve the data.

This column oriented storing technique of HBase is advantageous over the row oriented technique. If we compare in row oriented the data is stored and retrieved in one row at a time and therefore it can read the irrelevant data when only relevant data is required whereas in column oriented the data is stored in columns and it reads and picks up only the data which is required. Also it merges many rows and columns hence is able to perform all its operations over the entire datasets. Thus its very much suited for OLAP (online analytical process).

The data model in the HBase comprises of various logical components in it like tables, column families, columns, rows and cells. This model is basically designed in order to accommodate the structured as well as semi—structured data that could have different field size, data types and columns. The figure mentioned below basically describes the outline of the data model in HBase. It stores semi structured data and having different datatypes data. In HBase Rowkey are used to identify the data in rows. HBase data model having following components. HBase Tables, HBase Row, column, column Family, Column Qualifier and cell (Fig. 14).

HBase Table

Any data that is to be stored in the HBase is stored in the tables. These tables are the collection of rows which are stored in the separate partitions called as regions. Every region is then served by the region server by HBase master. The values that are stored in the region server are directly available to the clients. The number of tables to store the same amount of data in HBase is lower as compared to relational database, the reason is that the HBase is column oriented that allows to store many details in the same table. The normalization rules are not applied here thus reducing the number of tables (Fig. 15).

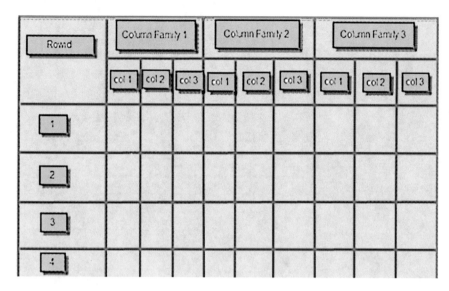

Fig. 14 Schema of HBase storage technique

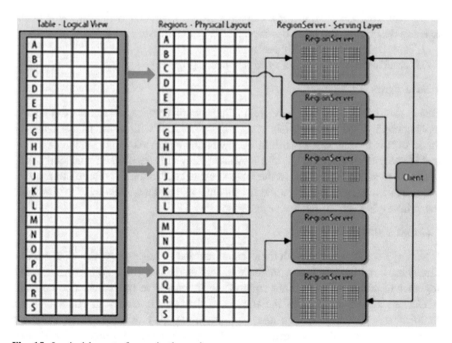

Fig. 15 Logical layout of rows in the regions

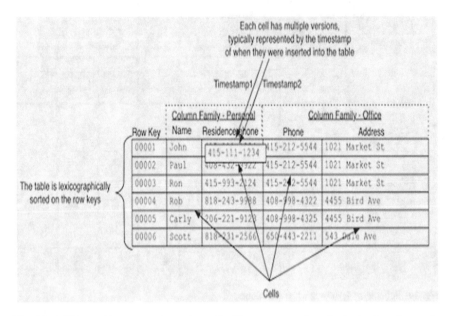

Fig. 16 A HBase table representing column families, rows sorted on the basis of rowkey and a cell that contains the data

To construct the table in the HBase you need to create it with new name or the existing descriptor, the table descriptor in java looks like:

HTableDescriptor (final TableName)

HTableDescriptor (HTableDescriptor desc) (Fig. 16)

HBase Rows

Hbase row consists of row key. Main Purpose of row key is to sort the data at alphabettically and row consists one or more column with value to associate the value so row key is most important. A row is one instance of information in a table and is recognized with the aid of a Rowkeys, they are specific and constantly dealt with as a byte. Each row key in the table is connected to the list of column families which are further connected to the list of time stamp. These row keys is similar to the primary key in the relational database (Fig. 17).

Column Families

It having a set of columns with their values and has storage properties also, column family are common to all rows in hbase table. The data present in the row is clubbed together in column family. One column family can have more than one columns which are stored together in the low level storage file called as Hfile. Column families are basically the strings that are composed of characters. They are responsible for the physical arrangement of the data which is stored in the HBase. Thus they are designed up and can not be easily modified. Column members within the column family would be having the same prefix (Fig. 18).

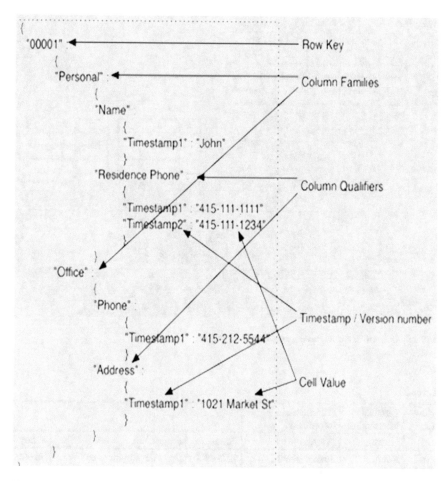

Fig. 17 Representation of row in HBase table

Column Qualifier

It used to index the piece of data and it fixed in table creations. Column is present within the column family. It can be identified with the column qualifier which is having the column family name along with the column name. The columns included in each column family consist of related data. For example if we consider the table of employee details, having personal and professional information as separate column family. Inside the personal column family the name of the employee, phone number, address, email id could be the separate columns while in professional information the project title, salary can be the columns (Table 2).

The table gives the description of the various columns (name, address, phone, project title and salary) that comes under the column family (personal and professional information).

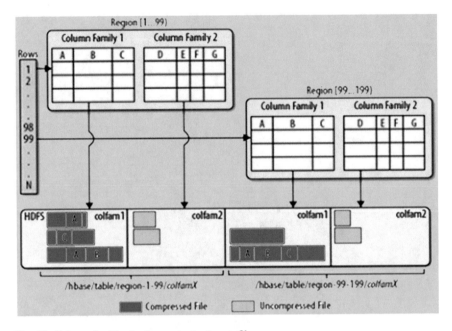

Fig. 18 Column families having separate storage files

Table 2 .

Row id	Personal information	Professional information			
	Name	Address	Ph. no	Project title	Salary
101	Robin	Street4 gandhi nagar (MP)	121–4860	Drug development by analyzing structural data of proteins	56,000
102	Akansha	Street1 apartment no. 41flora apartments.	121–5430	Protein structure prediction	60,000

Cell

Cell is the combination of row, column family and qualifier with their values. Value of cell is bytes of array. The data inside the HBase table is stored within the cells. It's a unique combination of row key, column family and column qualifier. Data in the cells is represented as values. A (row, column, version) defines the cell in the HBase. Rows and columns intersect to form the cell that comprises of the value which is the data stored in the HBase.

References

1. Gopalani, S., & Arora, R. (2015). Comparing apache spark and map reduce with performance analysis using K-means. *International Journal of Computer Applications, 113*(1).
2. Wiewiórka, M. S., et al. (2014). SparkSeq: Fast, scalable, cloud-ready tool for the interactive genomic data analysis with nucleotide precision. *Bioinformatics,* btu343.
3. Shoro, A. G., & Soomro, T. R. (2015). Big data analysis: Apache spark perspective. *Global Journal of Computer Science and Technology,15*(1).
4. Gu, L., & Li, H. (2013). Memory or time: Performance evaluation for iterative operation on hadoop and spark. In *2013 IEEE 10th International Conference on High Performance Computing and Communications & 2013 IEEE International Conference on Embedded and Ubiquitous Computing (HPCC_EUC).* IEEE.
5. Chen, H., et al. (2012). Hog: Distributed hadoop mapreduce on the grid. *High Performance Computing, Networking, Storage and Analysis (SCC), 2012 SC Companion.* IEEE.

Scalable Framework for Cyber Threat Situational Awareness Based on Domain Name Systems Data Analysis

R. Vinayakumar, Prabaharan Poornachandran and K. P. Soman

Abstract There are myriad of security solutions that have been developed to tackle the Cyber Security attacks and malicious activities in digital world. They are fire-walls, intrusion detection and prevention systems, anti-virus systems, honeypots etc. Despite employing these detection measures and protection mechanisms, the number of successful attacks and the level of sophistication of these attacks keep increasing day-by-day. Also, with the advent of Internet-of-Things, the number of devices connected to Internet has risen dramatically. The inability to detect attacks on these devices are due to (1) the lack of computational power for detecting attacks, (2) the lack of interfaces that could potentially indicate a compromise on this devices and (3) the lack of the ability to interact with the system to execute diagnostic tools. This warrants newer approaches such as Tier-1 Internet Service Provider level view of attack patterns to provide situational awareness of Cyber Security threats. We investigate and explore the event data generated by the Internet protocol Domain Name Systems (DNS) for the purpose of Cyber threat situational awareness. Traditional methods such as Static and Binary analysis of Malware are sometimes inadequate to address the proliferation of Malware due to the time taken to obtain and process the individual binaries in order to generate signatures. By the time the Anti-Malware signature is available, there is a chance that a significant

R. Vinayakumar (✉) · K. P. Soman
Amrita School of Engineering, Coimbatore, Centre for Computational Engineering and Networking (CEN), Amrita Vishwa Vidyapeetham, Amrita University, Coimbatore, India
e-mail: r_vinayakumar@cb.amrita.edu; vinayakumarr77@gmail.com

K. P. Soman
e-mail: kp_soman@amrita.edu

P. Poornachandran
Amrita School of Engineering, Centre for Cyber Security Systems and Networks, Amrita Vishwa Vidyapeetham, Amrita University, Amritapuri, Coimbatore, India
e-mail: prabasuja@gmail.com

© Springer Nature Singapore Pte Ltd. 2018
S. S. Roy et al. (eds.), *Big Data in Engineering Applications*,
Studies in Big Data 44, https://doi.org/10.1007/978-981-10-8476-8_6

113

amount of damage might have happened. The traditional Anti-Malware systems may not identify malicious activities. However, it may be detected faster through DNS protocol by analyzing the generated event data in a timely manner. As DNS was not designed with security in mind (or suffers from vulnerabilities), we explore how the vast amount of event data generated by these systems can be leveraged to create Cyber threat situational awareness. The main contributions of the book chapter are two-fold: (1). A scalable framework that can perform web scale analysis in near real-time that provide situational awareness. (2). Detect early warning signals before large scale attacks or malware propagation occurs. We employ deep learning approach to classify and correlate malicious events that are perceived from the protocol usage. To our knowledge this is the first time, a framework that can analyze and correlate the DNS usage information at continent scale or multiple Tier-1 Internet Service Provider scale has been studied and analyzed in real-time to provide situational awareness. Merely using a commodity hardware server, the developed framework is capable of analyzing more than 2 Million events per second and it could detect the malicious activities within them in near real-time. The developed framework can be scaled out to analyze even larger volumes of network event data by adding additional computing resources. The scalability and real-time detection of malicious activities from early warning signals makes the developed framework stand out from any system of similar kind.

Keywords DNS log analysis · Big data analytics · Machine learning Deep learning

1 Introduction

Nowadays, Internet has become the largest critical global communication medium and infrastructure. It connects several billions of nodes enabling them to communicate with each other. At its heart, Internet uses one important protocol—Domain Name System (DNS). The DNS protocol translates, difficult to remember Internet addresses to human readable names and vice versa. With the increasing dependency and usage of the Internet by users and systems, all the malicious activities that used to occur in the physical world moved to the connected digital world of Internet. The vectors of malicious activities mainly include Virus, Worms, Trojans, Cyberattacks, Phishing, Spam and Cyber-intrusions. With the exponential growth of hosts connected to the Internet, the usage of the DNS protocol by the systems and networks is progressively increasing to a very high level. This in turn produces very large volume of event data amounting to trillions of events per minute at the Tier-1 Internet Service Provider (ISP).

2 Related Work

This section discusses the selected research works previously done on cyber threat analysis and detection. In order to analyze the cyber threat or to provide cyber threat situational awareness, we investigated the event data and status information generated by Internet protocols—Domain Name Systems (DNS). The Internet community is facing serious threats from everywhere. Nowadays the behavior and operation of such threats are undergoing a crucial transformation and evolution. One of the widely used attacking strategies is to use botnets. Botnets are generally used for large-scale Cyber Security attacks that include Distributed Denial of Service (DDoS) attacks, large-scale spam campaign, identity theft etc. Botnets try to increase the number of affected hosts by incorporating executable programs that are present in several hundred compromised hosts, which receives its instructions or commands from Command and Control (C&C) server that is operated by the malicious actors known as bot-master [1]. One of the security measures to protect against this threat is to blacklist the C&C server with whom the botnets tries to contact for getting instructions. However, to evade being blacklisted by organizations, the attackers use DNS agility or fluxing approach by changing their domain names or IP addresses frequently in rapid succession [2]. This technique is known as Fast Flux and several families of Botnets use fast flux to hide their criminal activities and evading detection by changing the compromised hosts that acts as proxies for malware distribution. Fast Flux networks can be of two types, they are IP-Flux and Domain-Flux respectively [3]. IP-Flux can be further classified as Single-Flux, and Double-Flux. Constant registering and de-registering of IP addresses for a given malware domain are the main characteristics of Single Flux. In Double-Flux, the SRecord or the name server record for a DNS zone of a malicious site is also changed in addition to 'A' records. These techniques provide additional redundancy for botnet. Unlike IP-flux, Domain-Flux assigns several fully qualified domain names to a single IP address or C&C infrastructure. Usually hundreds of thousands of domain names are employed in a Domain-Flux. Bot-masters achieved this by using Domain Generated Algorithms (DGA) that generates domain names randomly on a large scale for registration. Examples for such botnets are Conflicker [4], Torpig [5], Kraken [6], Murofet [7]. 'Kwyjibo tool' is a special malicious program which uses a sophisticated domain generation algorithm for generating domain names that are similar to English Dictionary words [8]. These generated words cannot be used in botnets because there could be an ambiguity with existing domain names. However, botnets try to contact their C&C server with these randomly generated domain name using hit and trial method. For example, out of several thousand DGA generated domain names, the bot master will procure only a handful of domain names. As a result, lots of Non-Existent (NX) response queries get generated. This process makes the security strategy very expensive for the defender and very economical for the attacker. In this scenario, the attacker has to buy just a handful of domain names, but at the same time, the defender will have to procure/block/sinkhole millions of domains that makes the effort very expensive. A Non-Existent (NX) response or Name Error response are those domain queries for which no IP addresses or any record exists [9]. Pleiades [10] was the first system that was able to detect DGA based domains without reverse engineering the bot malware. The core of Pleiades consists of two modules —DGA discovery module and DGA classification and C&C detection module. DGA discovery module detects and clusters the botnet queries, given all NX domain features. For clustering botnet queries, extracted features from the observed domain names such as n-gram, entropy based features, structural domain features are used with X-means clustering algorithm. The latter module used Alternating Decision Tree learning algorithm for the identification and comparison of NX-Domain clusters, whereas C&C detection module used Hidden Markov Model. It is reported that detection rate of Pleiades' DGA classifier is 99.7% while C&C detection system was said to have detection rate greater than 91% for five out of six tested botnets and 22.67% for the remaining bot. Another work on the detection of botnets is done by Schiavoni et al. [11] and they did an elaborate survey on existing domain flux related C&C server detection systems. They propose an unsupervised algorithm which does not require any prior knowledge of the DGAs and no reverse engineering of the malware samples. In [12], J. Raghuram et al. proposed another unsupervised way for detecting malicious domain names. They used linguistic features and Expectation-Maximization for solving the problem. One of the main reasons for generating the non-pronounceable domain name is to avoid collusion that might occur due to the presence of an existing domain name. While IP-Flux networks make it time consuming and expensive to disrupt, Domain-Flux makes the process of purchasing the malware domains expensive for the purpose of sinkhole, as it will require the purchase or sinkhole several Millions of domain names. In [13], Thomas et al. analyzes DNS traffic from several authoritative name servers to identify strongly connected DGA based domains. Fast fluxing takes the advantage of DNS based load balancing by masking its own activities as it is done in the case of Content Delivery Networks (CDN). Fast Fluxing uses several IP addresses that are hidden behind a single domain name just as large CDNs and Antivirus providers architect their systems. The IP addresses of domains involved in Fast Fluxing changes with extreme frequency in round-robin fashion with very short Time-To-Live (TTL) for each DNS Resource Record (RR).

A dictionary based approach with Smith-Waterman algorithm was proposed for predicting the malicious domain detection [14]. Moreover, the experiments were done with the data captures from real-time systems. Zdrnja et al. [15] proposed a passive DNS anomaly detection project based on data captured at the University of Auckland Internet gateway with a view to detect and correlate domains used for botnet controls using a passive DNS monitor. A passive data capturing sensor was deployed at the network edge and extracted query name, resource record type, resource record data, TTL and first seen time stamp from the DNS data. This information can further be used for analyzing the historical behavior of certain DNS records and how they are linked with each other. Ramachandran and Feamster [16] studied the network-level behavior of spammers such as range of IP addresses which send spams, common bot modes, persistent time of each spam host, spamming bot characteristics etc. From the analysis, they could identify the region of IP addresses from which maximum bot attacks being sent. Anderson et al. [17] proposed a scam-hosting infrastructure in which patterns in emails are identified to track the spam servers. Image shingling method is then applied to find the scams whose webpages are graphically similar to a cluster of spam servers. All these approaches propose botnet detection based on DNS dictionary lookup. These approaches will fail in botnet detection when the number of incoming queries is very high in real-time. In other words, these approaches are not scalable and storage requirements and time requirement for decision making are also very high.

Major drawbacks of this approach are (1) computational complexity is very high with large volume of data. (2) Over fitting may occur with huge data. An old concept of artificial intelligence called as neural network (in recent times typically termed as deep learning) has achieved a significant result in various multitudinous fields namely natural language processing, image processing, speech recognition and many others [18]. Deep learning mechanisms itself facilitated to extract features by taking massive amount of raw data set as input. This provides significant contribution towards big data analytics. Deep learning algorithms are mainly categorized into two types (1) Convolution neural network (CNN) (2) recurrent neural network (RNN); CNN are largely used in the field of image recognition mainly due to the fact that, CNN applies a set of filters on rectangular region to extract complex features by travelling through layer by layer. The complex features are composed from a set of lower level features that forms a hierarchical feature representation. As a result CNN captures the features in various level of abstraction. CNN primarily use a pair of convolution, pooling operations and a non-linear activation functions. [19] adopted CNN in character level towards large scale text classification including several languages. The effectiveness of CNN model is compared with the other traditional mechanisms such as bag of words, n-grams and tf-idf. RNNs are initially proposed for time-series data-modeling [20] and later this has been applied in sequence data modeling tasks in the field of speech processing, natural language processing. As the research goes, [21] found the vanishing and exploding gradient issue when dealing with large sequences. On alleviating the vanishing and exploding gradient issue, researchers worked in the 3 directions; (1) enhancement of optimization methods for example Hessian-free optimization [22], (2) proposing a new architecture; LSTM [23] and its variants GRU [24], (3) appropriate weight initializations for example I-RNN [25, 26]. Moreover, researchers have also used hybrid network, in which the first layer is used as CNN and the CNN outputs are given as input to other recurrent network layers [27].

The effectiveness of various classical and deep learning approaches is studied for various Cyber Security tasks like Android malware detection [28, 29], DGA analysis [30, 31], traffic analysis [32, 33], malicious URL detection [34], intrusion detection [35, 36], anomaly detection [37], ransomware detection [38], encrypted text categorization [39], network traffic prediction [40].

3 Background

3.1 Domain Name System (DNS)

Domain Name System (DNS) is considered as the core Internet protocol. The hierarchical level of DNS is shown in Fig. 1. The present work is based on the analysis and detection of attacks that can be detected by observing the behavior of the DNS protocol and studying the events that are produced when the DNS protocol is used by the systems and networks. The section below gives an overview about the DNS protocol.

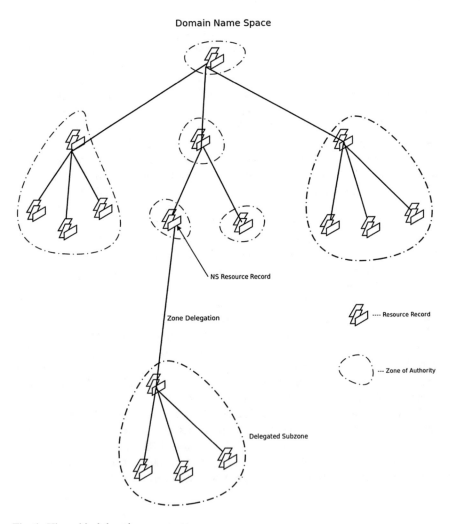

Fig. 1 Hierarchical domain name system

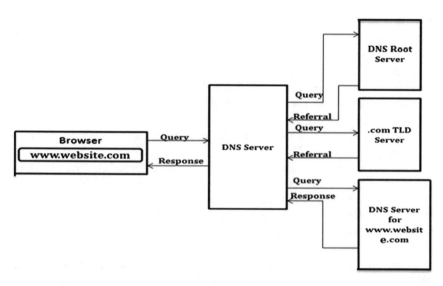

Fig. 2 Recursive DNS query

End users usually access Internet by using a browser that renders the web pages and portals. This is done by typing a domain name in the address bar of the browser. When this is done in a right fashion, Internet helps the users in information exchange and transactions. DNS servers can be classified into two types: Non-recursive/Iterative servers and Recursive servers. Non-recursive DNS servers act as the Start of Authority (SOA). They answer queries inside the governed domains without querying other DNS servers even if they cannot provide the requested answer. Whereas, Recursive DNS servers (an example is shown in Fig. 2) respond to queries of all types of domains by querying other servers and passes the response back to the client. Distributed Denial of Service (DDoS) attacks, DNS cache poisoning, unauthorized use of resources, root name server performance degradation, etc. are some of the major attacks suffered by Recursive DNS servers. DDoS is an attack which makes a network resource unavailable to the legitimate intended users by flooding it with fraudulent or forged requests continuously to the targeted DNS server that overwhelms the system. Cache poisoning/DNS spoofing attacks are the fraudulent activity of inserting a fake address record for an Internet domain into the DNS. The cache is considered as poisoned, if the server accepts this fake record and after which, subsequent requests for the domain will be answered by attacker's server. This cached entry will remain there till it's Time-To-Live (TTL) expires and during this period the subscriber's browser will go to the address provided by the compromised DNS server. This is an attack that diverts Internet traffic away from legitimate servers to fake servers by the malicious actors. Since it spreads across DNS servers, it is considered as a dangerous attack. For example, on 11th October, 2013 Google's Malaysian domains google.com.my and google.my were hijacked and redirected users to a web page that announced that the attack was perpetrated by a Pakistani group called Madleets

[41]. This attack was a DNS cache poisoning attack that lasted a few hours and while it was active, it redirected users to a website hosted in Canada. Hijacking of a server is simple in recursive DNS query enabled servers as attackers can poison the response while it is coming back from the root servers [2]. DNS servers that are not configured correctly are more vulnerable to these types of attacks. In several cases, many organizations do not even know or does not have situational awareness system that can quantify the number of compromised systems in their networks. These organizations could be Internet Service Providers, or small, medium and large enterprise networks [27].

In the case of enterprise networks, the visibility and situational awareness is limited to the anti-virus systems that are installed. And in the case of Internet Service Providers, over 63% of surveyed respondents do not know the proportion of the devices in their network which are compromised and involved in botnet or other malicious activities [27]. One of the sophisticated and popular types of malware that wreaks havoc on the Internet is botnets. When a bot infects a system or host, it can be commandeered to do various automated malicious activities that include sending malwares, stealing private and sensitive information, key stroke logging, and participation in C&C based DDoS attacks. In addition to the above-mentioned attacks, it might affect the local system with various malwares like Clickfraud, Adwares, Spywares, etc. Clickfraud is an automated fraud computer program which will click an advertisement without the knowledge of the actual user with intent to give false clicks to an advertisement in order for the advertiser to make more financial gains illegitimately. A pictorial diagram of the various activities of a Botnet is shown in Fig. 3. Fast flux is termed as one of the malicious activities that are being done using botnets. Fast flux service networks use many IP addresses that are mapped to a single domain name.

Fig. 3 Botnet activities

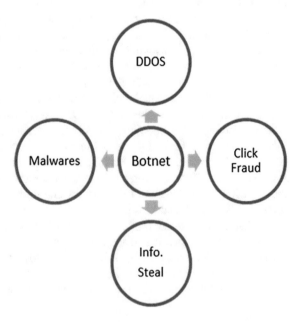

3.2 Scalable Algorithms

In real-time, it is required to process extremely large volume of data that result from the system and network events on the use of DNS system. The algorithms that are computationally efficient and that can be distributed across multiple systems have been studied in the current research work. We use deep learning algorithms for this research work to detect the attacks. The deep learning algorithm contains billions of parameters. In order to train them in the context of detecting the dga generated domain, we use distributed TensorFlow framework. This facilitates parallelism of deep learning models in two aspects; one is within a machine through multi-threading and second one is across machines through message passing. Moreover, the framework also supports the data parallelism where multiple replicas of deep learning models are used to achieve a single task.

4 System Architecture

4.1 Scalable Architecture

Since, DNS and BGP together produce several Billions of data events per minute, a highly scalable framework has been developed that can collect and process the data in real-time. The framework consists of DNS and BGP sensors that collect the data in a distributed manner. These sensors receive data directly from the DNS servers and BGP enabled routers. To ensure that, the introduction of this system does not impact the functionality of the DNS and BGP systems, they are designed to collect the passive information in the form of out-of-band mode. The collected data is parsed and aggregated and then sent to real-time and non-real-time analysis engine that runs highly scalable distributed deep-learning algorithms. The query-router (standalone) services that controls the communication between different modules. The query-router also provides an interface with the message broker. A subsystem known as Front-End Message Router controls the communication between the Interactive Visualization and analysis engine on the processed data. The Figs. 4 and 5 depict the standalone and scalable architecture of the framework respectively.

The present research work proposes to deploy the standalone architecture in individual Tier-1 ISPs. Each standalone system is capable of handling few millions of DNS and BGP data per second without any stability issues. Hence, Terabytes (TB) of data were able to collect within a day. However, monitoring a single ISP might not be enough to get an overall situational awareness of a malware propagating through a zone or country, thus resulting in monitoring and correlating the network activity of several Tier-1 ISPs. The proposed scalable architecture in this research work employs distributed and parallel algorithms with various optimization techniques that make it capable of handling huge volume of data. The scalable

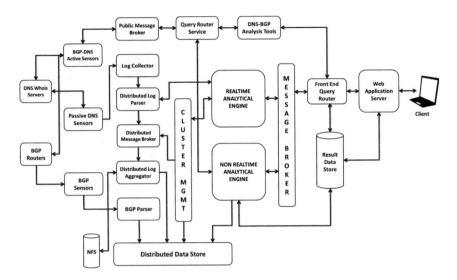

Fig. 4 Architecture of data collection framework

architecture also leverages the processing capability of the General Purpose Graphical Processing Unit (GPGPU) cores for faster and parallel analysis of DNS data. The framework architecture contains two types of analytic engines—real-time and non-real-time analytic engines. The purpose of analytic engine is to detect malicious activities thereby generate an alert in case of threat.

4.2 Supporting Services

The important supporting services needed for this project are explained below.

1. Passive Sensor: The Passive Sensor collects DNS Query/Response from the DNS Servers (Any DNS Server). The passive sensor captures Network Traffic from DNS Servers and passes it to an application. The parser inspects the DNS Response Packet, converts it into human readable format and forwards it to DNS Log Collector. The Passive Sensor could be installed inside the DNS Server itself or any mirrored traffic could be sent to the sensors. Each Sensor could process the data from multiple DNS Servers if needed. It collects data passively from DNS Servers without affecting the DNS Server.

2. Active Sensor: This sub-system performs an active DNS Query related to the given sources, to collect data and perform various analyses. Its main sources are DNS server, WHOIS server, application programming interfaces such as Google Safe browsing Application programming interface (API), and other DNS databases.

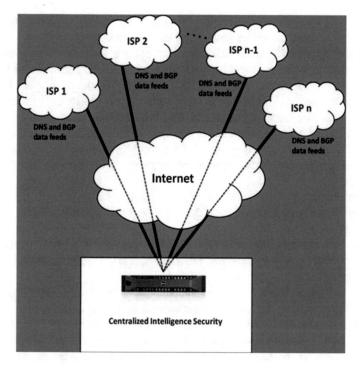

Fig. 5 High-level architecture correlating data from multiple ISPs

3. WHOIS Server: This will provide information associated with a Domain or IP Address. Queries will be forwarded to corresponding WHOIS server for that particular Domain/IP address. The information related to domain and IP are explained below: For a domain, query is sent to retrieve the following information:

- Registrant name, address, email, phone.
- Administrator name, address, email, phone.
- Domain registered date, expiration date, authoritative name server etc.

For an IP address the fields are:

- Owner
- Prefix (Network)
- ASN
- Location
- Expiration Date

4. Online API's: Some of the analysis results could be correlated with the online external systems for validation and cross-correlation.
5. Log Collector and Parser: This subsystem collects the parsed DNS Responses from distributed sensors and forwards it to the collector. The collector looks up the Geo-Location and details of ASN of each IP address (Client IP, DNS Server IP, and A Records in Resource records) in the DNS Responses. The parser uses the Geo IP database to find Geo-location (city, country, latitude, longitude) of an IP Address. ASN database is used for finding details of ASN (AS Number, AS Name) of an IP Address. This data will be appended with original data coming from sensors and publish to queue for real-time and batch analysis.
6. Query Router Service: It is a standalone service, which controls the communication between different modules. It also interfaces with public message broker.
7. Front-End Message Router: As a standalone service, this subsystem controls all the communication to-and-from the front-end UI. It also interfaces with the respective back end subsystems.
8. Internal Router: Internal router software is peered with the BGP Router to get BGP Updates in a configurable time interval such as 5 min etc.
9. BGP Monitor: The BGP monitor subsystem is responsible for collecting BGP Update-messages in real-time. It gets updates from the TCP port as a stream and stores them in an XML file format. The parsers will produce BGP update messages in one single format, after extracting the data from two different sources. A Distributed Log Aggregator collects this parsed data to store it in a distributed database for further analysis.

4.3 Data Collection

DNS data are collected in a passive manner by reading from the mirrored traffic using promiscuous mode on DNS communication information between the DNS server and the DNS clients. The data consists of DNS queries and the corresponding DNS answer made by the DNS client and DNS server respectively. The extracted data is analyzed for malicious events. The BGP data is collected by adding a read-only peer to a BGP speaking router. The read-only peer collects the data that occurs in the form of BGP updates, announcements, neighbor information etc. The BGP data consists of the route and prefix information. To identify the malicious announcements, malware propagation and activities, the prefix announcements, route announcements and updates information can be used.

5 A Sub System for Detecting DNS Anomalies Based on Deep Learning and GPGPU

5.1 Introduction

Recursive DNS servers, responds to queries of all types of domains, by querying other servers and passes the response back to the client as shown in Fig. 6.

DeepBot primarily focuses on identifying botnets using the Domain Flux service. In Domain Flux, the Botmaster changes the domain name that has to be mapped with the IP address of the C&C server frequently. For this, they use Domain Generated Algorithms (DGA) to generate domain names randomly on a large scale for registration. Botnets use different DGA's for domain name generation. For example: Conflicker [4], Torpig [5], Kraken [6], Murofet [7] etc. Kwyjibo [8] uses a sophisticated domain generation algorithm for generating domain names that is similar to English Dictionary words. In Domain Flux, the botnets try to contact their C&C server with the randomly generated domain name using hit and trial method. As a result, in most of the cases, a lot of Non-Existent (NX) response queries get generated. A Non-Existent (NX) response or Name Error response are those domain queries for which no IP addresses or any record exists [9].

The DeepBot framework is based on an assumption that the botnets sitting in an infected system generate large sets of NX queries before actually getting resolved (if successful). Also, since botnets using the same DGA's generate similar domain names, comparing the patterns in NX-Domains and the resolved domain names of a

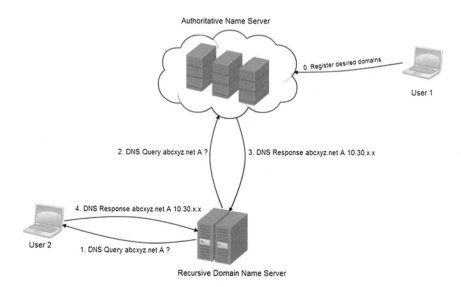

Fig. 6 Working flow of a legitimate DNS query

particular host, one could identify the malicious C&C server. NX-Domains could also get generated due to human errors like spelling mistakes while querying a domain name. Hence, to distinguish between a human error and DGA generated NX-Domains, the framework employ deep learning algorithm. This implicitly obtains optimal feature representations to distinguish the domain name as benign or DGA generated. Once the framework classifies a particular NX-Domain as a DGA domain, the framework assumes that the host is infected and analyses its resolved domain list for finding the corresponding C&C server. Thus by analyzing the NX-Domain queries, the Deep Bot framework is not only able to find out the compromised hosts infected by botnets but also the malicious C&C server.

5.2   System Architecture

This section describes the overall architecture and working of the analysis, along with the methods used for data collection. Figure 7 shows the high level architecture diagram of Domain Flux analysis. The analysis consists of three main modules: Identifying DGA Infected Hosts module, C&C Detection module and finally Time Analysis. The system consists of mainly 2 modes: Training mode and Testing mode.

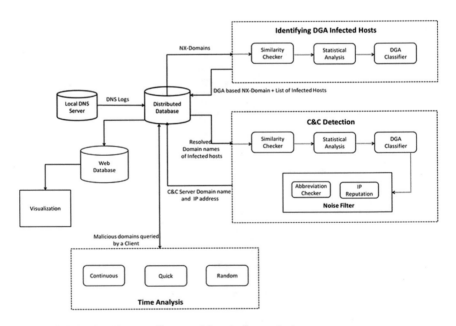

Fig. 7 High level architecture diagram of domain flux analysis

1. Training Mode: The system is trained using the white list and malicious data-sets. In the white-list dataset, there are top 1 million domains provided by Alexa [42] whereas approximately 50 million malicious domains were collected from April, 2012 to 2016. The system is trained using these features in an on-line mode. In due course, thresholds were set for various analyses performed, which are being used for testing.

2. Testing Mode: In this mode, the system analyses DNS log streams which were acquired from the local DNS server. The data coming from the sensors are first stored in a distributed database. The data is fetched from the distributed database periodically with a time interval. The implementation of this module is given bellow.

As mentioned earlier the system has mainly three main modules. Throughout this chapter, for convenience, some terms were used. Suppose a domain d, where d = "abc.example.com". Here, the term "com" is termed as the TLD i.e. the Top Level Domain. Likely, "example.com" is called as 2LD (Second Level Domain) and "abc.example.com" is called as 3LD (Third Level Domain).

5.3 Details of Implementation

This section explains in detail the steps employed to identify the hosts that are already infected by a DGA based bot. The analysis is based on an assumption that an infected host on the process of communicating to a C&C server generates many NX queries. Thus, by analyzing the malicious NX-Domains it could easily be understood that whether the host is infected or not.

However, not all of the NX queries made could be termed as malicious. For example, human typing mistakes also lead to NX queries. Hence, the challenge lies in classifying the human mistake and the malicious queries into different sets. From the distributed database, a set of data were taken which is in the form of a tuple $T(t_s, h_i, d_i, s)$. The tuple t has four values namely, t_s-Timestamp, h_i-host IP address, d_i-domain queried, s-query status. Here, since the focus is only for the NX queries, the status is kept as NX. So NX queries made by all the hosts were recursively analyzed within a time period of 10 min. Once all the 10-min logs were collected, send it to this module in the tuple format mentioned above. Thereafter the domain name is split and only its 2LD label is extracted. The 2LD name thus extracted is sent for Similarity Checker Analysis.

1. Similarity Checker: The objective of this analysis is to find whether any legit-imate domains are present which are quite similar. The 2LD domain label is compared with all of the domains in the 1 million Alexa set with the help of Approximate String matching algorithm [43]. Using the Damerau-Levenshtein [44] Edit Distance algorithm the distance between two strings is calculated. If the distance is more than a threshold value then it is considered that the

particular 2LD domain name is not a typing mistake and it is sent for Statistical Analysis. The advantage is that domain queries like "ggoole" which is caused by typing mistake could be excluded easily.

2. Statistical analysis: The system was developed and deployed with an aim to detect DGA domains automatically in real-time. After detecting the DGA domain, the system monitors its activities in an interval of 5 seconds on a regular basis. Figure 10 presents the architectural diagram of the implemented system which consists of three modules (1) Data Collection (2) Deep learning for detecting DGA (3) Dynamic reputation.

The data for the system is collected by deploying passive sensors across the four geographically distributed University campuses comprising of more than 30,000 unique users. The Passive Sensor collects DNS Query/Response from the deployed DNS Servers (Any DNS Server). Hence in this work data were collected from four DNS servers. The sensor captures the network traffic from DNS servers and passes it to an application, which takes only the traffic that is originated from DNS Server (DNS Response Traffic). It will then dissect the DNS Response Packet, converts it into human readable format and forward it to DNS Log Collector. The Passive Sensor could be installed inside the DNS Server itself (for Linux/Unix Platform) or mirror the traffic to a server which is dedicated for DNS Passive Sensors. The sensors get the data by port mirroring the traffic from the DNS Servers present in the deployed network. The logs received from the 4 campus sensors are received by a distributed log parser. The parser finds Geo Location and ASN Details of each IP address (Client IP, DNS Server IP, and A Records in Resource records) in the DNS Responses. The parser uses the Maxmind Geo IP database to find Geo location (city, country, latitude, longitude) of an IP Address. Maxmind ASN database is used for finding ASN details (AS Number, As Name) of an IP Address. A Query Router Service is then used to control the communication between different modules. The Front End Message Router controls all the communication to and from the front end UI. It also interfaces with the respective back end modules. Dynamic reputation subsystem analyses the DNS traffic in a network and detect and alert the presence of suspicious domains receiving anomalous hits. The architectural details of the developed framework are provided in Fig. 8.

1. Domain names encoding in character level: In recent days, deep learning approaches have achieved a significant performance in various tasks such as language modeling, text classification and many others in the area of natural language processing (NLP) [18]. They have an ability to learn appropriate feature representations by considering the input as in the form of raw data. A primary task in NLP is how to represent the text into numeric vectors. Primarily 2 views of representation are used by researchers (1) Texts are represented as a stream of characters (2) Texts are represented as a sequence of words. Domain name representation is called as domain name encoding. Domain name encoding consists of 2 steps. In first step raw domain names are preprocessed and tokenized to characters. Preprocessing involved in removal of

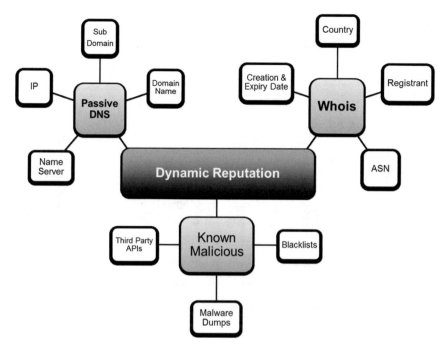

Fig. 8 Ensemble based dynamic reputation system for DNS

top-level domain followed by transforming the characters to lower case, otherwise, results in a regularization issue [19]. In second step, a vocabulary is formed by assigning a unique id to each character. The unknown characters are assigned to default id 0. These unique ids of vectors are passed batch-size of 64 to embedding layer. An embedding layer facilitates to learn the semantics and contextual similarity structures of domain names by coordinating with the other layers in the deep network during optimizing in the backpropogation process. The high-dimensional vectors of embedding layer are passed to t-SNE [45] for visualizing the character clustering. The Fig. 9 showed that the similar characters are clustered together. Most importantly, the special characters and numbers are appeared in a separate cluster. Thus, the embedding layer has learnt the semantic and contextual similarity of domain names. Finally, the embedding layer output is passed to the other layers such as (1) RNN (2) LSTM (3) GRU (4) I-RNN (5) CNN (6) CNN-LSTM. Based on the parameter tuning, the number of units/memory blocks is set to 128 for recurrent hidden layers, 64 filters with filter length 3 for CNN. These layer obtains the optimal feature representation for classifying the domain name as either benign or DGA generated. Finally, the various RNN layers output is passed to the dropout layer. This facilitates to avoid the state of over fitting by randomly removing the neurons with its connections to other neurons.

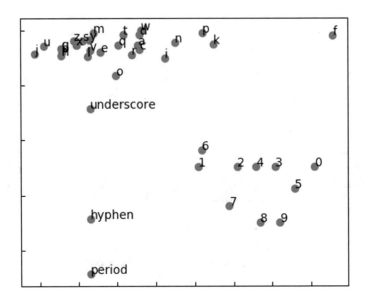

Fig. 9 Embedded character vectors learned by LSTM model is represented using 2-dimensional linear projection (PCA) with t-SNE

2. Classification: The dropout layer output is passed to the dense layer. A dense layer is a fully-connected layer and it composed of two layers. One is dense with unit 1 and followed by an activation layer i.e. *sigmoid* with loss function as binary cross-entropy, as shown below.

$$loss(pr, ex) = -\frac{1}{N}\sum_{j=1}^{N}\left[ex_j \log pr_j + (1 - ex_i)\log(1 - pr_j)\right]$$

Here *ex* is a vector of expected class label, *pr* is a vector of predicted probability for all domain names in testing data set.

3. Evaluation results: All the deep learning architectures are trained using the most recent software framework Google's open source data flow engine, TensorFlow [46]. TensorFlow allows programmers to build numerical systems as unified data flow graphs. The data flow graph has nodes and edges that represent mathematical operations and the tensors respectively. In addition, programmers can also deploy computations on heterogeneous platforms: one or more CPUs, GPU, or mobile devices. To accelerate the gradient descent computations all experiments are run on GPU enabled TensorFlow in single NVidia GK110BGL Tesla k40. The performance of the trained model was evaluated on the testing data set. LSTM and CNN-LSTM have followed improvement in accuracy till

Table 1 Test results for detecting DGA

Algorithm	Accuracy	Precision	Recall	F-score	Loss
LSTM	0.999	0.999	0.998	0.999	0.00
GRU	0.998	0.991	0.996	0.993	0.01
CNN-LSTM	0.997	0.985	0.996	0.990	0.01
RNN	0.968	0.795	0.943	0.863	0.09
I-RNN	0.979	0.866	0.966	0.913	0.06
CNN	0.965	0.777	0.936	0.849	0.10
Bigram-LR	0.937	0.577	0.892	0.701	0.16
Hand-crafted features					
RF	0.926	0.483	0.858	0.618	
DT	0.908	0.564	0.966	0.712	
MT	0.892	0.483	0.951	0.641	
AB	0.924	0.664	0.936	0.777	
NB	0.909	0.564	0.967	0.712	

epochs 700. After, accuracy has seen a sudden decrease due to over fitting. IRNN has performed well till epochs 400. The performance of both CNN and GRU is good till epochs 250 epochs. RNN has started to over fitting once it reaches epochs 800. This infers that each deep layer has required different number of epochs to attain the best performance in classifying the domain name as benign or DGA generated. To compare the performance of deep learning models with the traditional machine learning classifiers [47, 48], we followed the feature engineering approach. The detailed performance of traditional machine learning and deep learning algorithms performance is reported in Table 1. Deep learning models performed well in comparison to the traditional machine learning classifiers. Moreover, LSTM has performed well in comparison to the other deep layer (Fig. 10).

To verify the accuracy of the built model, the binary cross-entropy is calculated. The binary cross-entropy is a standard measure used to identify the inefficiency of the classifier to predict the data based on the premise truth. In other words, binary cross-entropy is a cost function used to measure the inaccurateness of the built model. To minimize binary cross-entropy, the model employs gradient descent algorithm with a learning rate of 0.01. Gradient descent is a first order optimization algorithm, where the model shifts each variable a little bit in the direction that reduces the cost. The classifier is trained with both malicious and non-malicious datasets. If the classifier suggests the domain d_i to be randomly generated, a flag is assigned to that domain d_i as a DGA domain name and mark the host h_i as infected. The domain d_i and the host IP along with the timestamp t_i is then stored in the distributed database for further analysis. Figure 11 illustrates the detailed implementation of the above process.

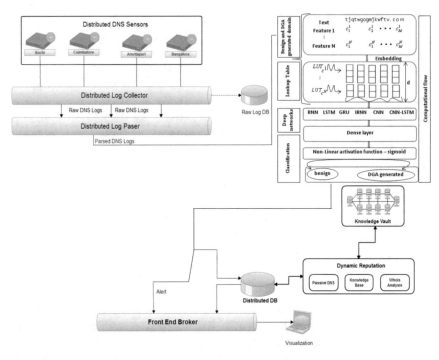

Fig. 10 An intuitive overview of employed deep learning architectures

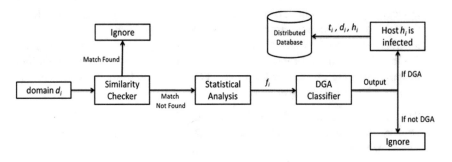

Fig. 11 Detailed diagram of identifying DGA infected hosts module

5.4   C&C Detection

The Deep Bot framework goes a step further by analyzing the response queries of the infected hosts h_i. The idea behind is to find domain names with similar patterns as DGA domains and getting the IP address it actually resolves to. The entire domains that got resolved when queried by an infected host are analyzed. For this analysis, the same procedure of passing it through similarity checker then statistical analyzer followed by the DGA classifier is followed. However, the output of this analysis also includes some false positive domains that could be also termed as noise. The presence of these noisy domains reduces the efficiency of the entire system. To avoid this, a noise filter is employed. In noise filtering two analyzes are performed namely: Abbreviation Checker and 3-Way Scoring, which is explained in detail bellow:

(a) Abbreviation Checker: Here the presences of any abbreviations are checked inside a domain name. Generally, an abbreviation is a short form of a committee or organization. In this analysis, not only check the presence of an abbreviation term but also perform three statistical analyses mentioned bellow:

- Length ratio of the abbreviation to the domain name.
- Position of the abbreviation.
- Presence of any date before or after the abbreviation.

(b) 3-Level Reputation Scoring: Given a domain name or an IP address, a reputation score was annotated to it accordingly. The reputation score is based on a 3-level check which involves spanning through the Passive DNS Intelligence, Malware Knowledge Base and WHOIS Data Base.

1. Malware Knowledge Base: A knowledge base is created by continuous crawling of public block lists, blacklists, online malware dumps and information related to known botnet IPs and domains from open Internet.
2. Passive DNS Intelligence: Passive DNS or pDNS is the methodology of constructing the duplicate copies of zone data from the captured name server responses. For the collection of pDNS data, the preliminary condition is to capture the inter-server DNS message by multiple sensors and then to forward all the data to collection point for analysis. Once the analysis on the captured data is over, every individual DNS record is stored in a database from where data lookup could be performed for any analysis.
3. WHOIS Database: WHOIS provides information associated with a Domain or IP Address. Queries will be forwarded to corresponding WHOIS server for that particular Domain/IP address. The information related to domain and IP are explained below: For a domain, query is sent to retrieve the following information:

- Registrant name, address, email, phone.
- Registrar name, address, email, phone.
- Administrator name, address, email, phone.
- Domain registered date, expiration date, authoritative name server etc.

For an IP address the fields are:

- Owner
- Prefix (Network)
- ASN
- Location
- Expiration Date etc.

Higher the reputation score of a particular domain, lesser are its chances of being malicious. Based on these measure it was understood that the history of this IP and have a say whether it could be malicious or not. Using the noise filter a lot of false positives could be reduced significantly. The final output is the malicious domain name of the C&C server along with the IP address it resolved to. Both these details along the timestamp containing the query time is then stored back to the distributed database.

5.5 Time Based Analysis

Deep Bot additionally performs time based to analyze the behavioral patterns of the C&C server. For this Deep Bot analyzes the timestamps of the entire DGA requests made by an infected host. Based on the query patterns with which a host queries, Deep Bot classify it into three groups namely,

1. Continuous: When the host is generating queries at an equal interval of time.
2. Quick: When a lot of queries are generated in a small period of time.
3. Random: Host queries DGA requests at random time periods.

By classifying the host into these three categories, the nature of the C&C server can be understood. Also one could know the number of hosts belonging in a particular category. Now analyses the domains (both NX and resolved ones) queried by different host residing in the same category. If any similar patterns are found in the domains queried by different hosts, it can be concluded that both the hosts are infected by the same DGA and thus controlled by the same bot-master. All such hosts and the domains are clustered accordingly.

5.6 Case Studies

In this section, three case studies of DGA based domains and its resolved server details as per the DNS data collected on 16th May, 2016 are discussed.

1. Case Study 1: IP—xxx.xxx.213.30 The network activity of all domains contacting a sinkhole server was monitored for a day. Sinkhole servers are generally operated by multiple security organizations whose tasks are to monitor and

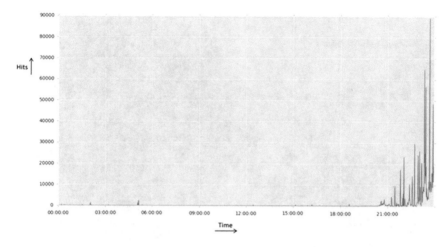

Fig. 12 Hits gained by IP xxx.xxx.213.30 on 16th May, 2016

mitigate the malware infections propagating in a network [49]. This is achieved by redirecting the malicious traffic to trusted and reputed hosts and analyzing them scrupulously. On 16th May, 2016 this sinkhole IP received approximately 1.1 million hits from more than 9 k unique clients. The timeline of the hits received is illustrated in Fig. 12.

There was a sudden surge in the hits received after 9 PM. Figure 13 shows DGA network queries made by a client to this sinkhole IP address. The client also queried a multiple DGA domains, which resolved to a Portugal IP xxx.xxx.26.248. On analysis the DeepBot framework was able to identify the similar patterns between queries made by this client. A vast majority of the domains started with a prefix HLD 'five' followed by a one or two digit number and ended with either '.ru' or '.com' TLD. Some of the DGA domain patterns observed is illustrated in Tables 2 and 3 respectively.

2. Case Study 2: Infected Host IP—xxx.xxx.153.192 The DGA like domains queried by host xxx.xxx.153.192 were monitored. Figure 14 illustrates the DGA domains and its resolved IP addresses identified by DeepBot framework.

The IP address xxx.8.69.25 is a conflicker sinkhole IP hosted in China. Some conflicker domains resolving to this IP address on 16th May, 2016 is provided in Table 4.

3. Case Study 3: Domain Name—hzmksreiuojy.biz: DeepBot framework also found out another conflicker set of domains with the prefix 'hzmksreiuojy' on May 16, 2016. Details of these domains are provided in Fig. 15 and are tabulated in Table 5. DeepBot went also found out many suspicious domains connected to xxx.xxx.249.128. Some of them are shown in Table 6

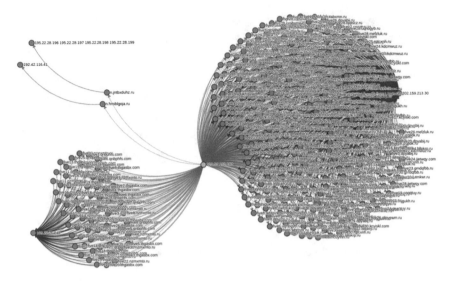

Fig. 13 DGA queries made by a client to the IP xxx.xxx.213.30

Table 2 DGA domains starting with prefix 'five' resolving to 202.159.213.30	five30.lnjgukh.ru five30.jzgjldk.ru five30.mcuyfnh.ru five30. usildbq.ru
	five25.cxabxmn.ru five25.bjqlscz.ru five25.jzgjldk.ru five25. qlpyewm.ru
	five3.klcgduk.ru five3.whtjpzk.ru five3.jjetwqy.com five3.coqqtuy.ru
	five12.usildbq.ru five12.kcyiskl.com five12.gmdqfbb.ru five12. cxabxmn.ru
	five6.jjetwqy.com five6.kcyiskl.com five6.uqhbgyb.ru five6. bjqlscz.ru
	five28.dpyabij.ru five28.whtjpzk.ru five28.kdcmwuz.ru five28. mefzluk.ru

Table 3 DGA domains starting with prefix 'green' resolving to 202.159.213.30	green32.qfmtsvxp.ru green28.qfmtsvxp.ru green5.qfmtsvxp.ru green13.qfmtsvxp.ru
	green32.ztcgyuyh.ru green28.ztcgyuyh.ru green5.ztcgyuyh.ru green13.ztcgyuyh.ru
	green32.entggmuq.ru green28.entggmuq.ru green5.entggmuq. ru green13.entggmuq.ru
	green32.rsfdhhez.ru green28.rsfdhhez.ru green5.rsfdhhez.ru green13.rsfdhhez.ru
	green32.czectdfl.ru green28.czectdfl.ru green5.czectdfl.ru green13.czectdfl.ru

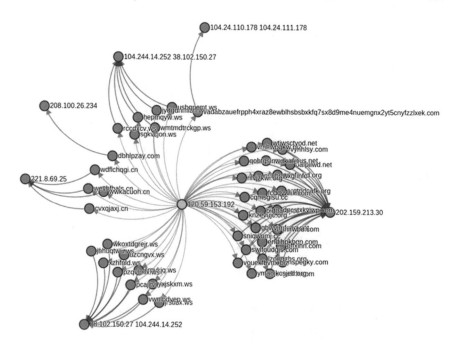

Fig. 14 DGA domains queried by infected host xxx.xxx.153.192

Table 4 Conflicker domains observed on 16th, May 2016

Domain name	Resolved IP address	Name server
abhumk.cn	xxx.xxx.69.25	ns.conflicker-sinkhole.cn
aidmj.cn		
akkbzhqnfv.cn		
aapvv.cn		
adphpkx.cn		
afxddfgcmqq.cn		
bmiusjs.cn		
achrszijz.cn		
agwjrlfycyf.cn		
aidjgjsr.cn		
bdvuzzi.cn		
atdja.cn		
bdgnsjdqcdo.cn		

138   R. Vinayakumar et al.

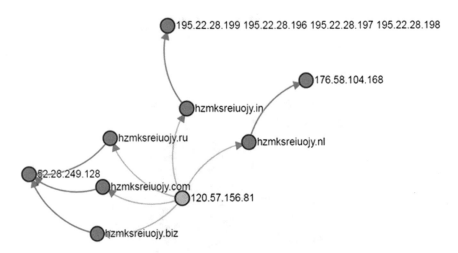

Fig. 15 Conflicker domains with prefix 'hzmksreiuojy'

Table 5 DNS Information of domains with prefix 'hzmksreiuojy' as on 16th May, 2016

Host IP	Domain name	Resolved IP	Name server	Blacklisted
120.57.156.81	hzmksreiuojy.biz	52.28.249.128	ns.conficker-sinkhole.com ns.conficker-sinkhole.net ns.conficker-sinkhole.org	Yes
	hzmksreiuojy.ru		ns1.101domain.com ns2.101domain.com ns5.101domain.com	No
	hzmksreiuojy.com		ns1.dynadot.com ns1.dynadot.com	Yes
	hzmksreiuojy.in	195.22.28.196 195.22.28.197 195.22.28.198 195.22.28.199	a0.in.afilias-nst.in a1.in.afilias-nst.info a2.in.afilias-nst.info b0.in.afilias-nst.org b1.in.afilias-nst.in b2.in.afilias-nst.org c0.in.afilias-nst.info ns1.csof.net ns4.csof.net	Yes
	hzmksreiuojy.nl	176.58.104.168	sinkhole.sidnlabs.nl proteus.sidnlabs.nl	Yes

Table 6 Domains resolving to IP 52.28.249.128 observed on 16th May, 2016

0-0-1-4-5-5-2-5-7-8-0-7-4-2-2-0-2-7-8-0-2-8-4-4-3-4-0-3-8-2-2- .0-0-0-0-0-0-0-0-0-0-0-0-0-49-0-0-0-0-0-0-0-0-0-0-0-0-0-0.info
0-0-2-4-3-5-4-7-3-4-3-7-6-1-5-4-4-3-7-0-8-1-4-2-8-8-7-8-0-5-7- .0-0-0-0-0-0-0-0-0-0-0-0-0-49-0-0-0-0-0-0-0-0-0-0-0-0-0-0.info
0-0-5-6-4-3-0-5-0-6-8-6-2-8-2-7-7-0-2-2-1-6-1-6-5-7-8-2-5-1-4- .0-0-0-0-0-0-0-0-0-0-0-0-0-49-0-0-0-0-0-0-0-0-0-0-0-0-0-0.info
0-0-7-2-1-0-5-7-5-6-2-3-1-5-7-2-2-8-8-7-3-5-5-3-5-7-0-6-7-1-5- .0-0-0-0-0-0-0-0-0-0-0-0-0-53-0-0-0-0-0-0-0-0-0-0-0-0-0-0.info
0-0-8-7-5-7-0-7-8-6-8-3-8-4-4-6-0-0-6-4-0-1-1-3-6-7-7-7-7-3-8- .0-0-0-0-0-0-0-0-0-0-0-0-0-49-0-0-0-0-0-0-0-0-0-0-0-0-0-0.info
0-1-1-6-4-6-0-0-0-1-1-8-3-2-8-2-4-5-0-7-5-8-4-1-8-5-3-1-8-8-6- .0-0-0-0-0-0-0-0-0-0-0-0-0-53-0-0-0-0-0-0-0-0-0-0-0-0-0-0.info
0-1-5-8-7-7-7-4-1-7-1-0-2-7-4-7-6-3-4-5-8-2-8-0-1-5-6-3-8-5-5- .0-0-0-0-0-0-0-0-0-0-0-0-0-49-0-0-0-0-0-0-0-0-0-0-0-0-0-0.info
0-1-6-7-6-4-0-0-4-4-4-6-2-5-8-2-5-6-5-4-6-2-3-2-5-7-5-4-5-3-2- .0-0-0-0-0-0-0-0-0-0-0-0-0-49-0-0-0-0-0-0-0-0-0-0-0-0-0-0.info
0-1-7-1-6-2-4-7-7-1-7-7-6-6-6-4-2-4-7-3-5-1-7-7-7-7-2-3-0-3-6- .0-0-0-0-0-0-0-0-0-0-0-0-0-53-0-0-0-0-0-0-0-0-0-0-0-0-0-0.info
0-1-7-3-7-4-5-6-6-6-7-4-8-1-7-7-2-4-1-4-5-1-1-1-8-5-3-7-3-4-2- .0-0-0-0-0-0-0-0-0-0-0-0-0-49-0-0-0-0-0-0-0-0-0-0-0-0-0-0.info
0-2-8-1-4-1-3-5-8-3-2-4-5-8-0-5-5-8-6-3-1-2-1-7-3-7-7-8-0-6-4- .0-0-0-0-0-0-0-0-0-0-0-0-0-53-0-0-0-0-0-0-0-0-0-0-0-0-0-0.info
0-2-8-1-8-4-6-0-6-1-5-2-4-6-8-2-3-5-4-7-4-8-1-0-5-2-2-5-4-7-2- .0-0-0-0-0-0-0-0-0-0-0-0-0-53-0-0-0-0-0-0-0-0-0-0-0-0-0-0.info
0-3-0-1-7-4-3-0-8-1-2-5-6-7-5-2-2-5-8-8-6-7-7-4-3-4-3-2-3-4-1- .0-0-0-0-0-0-0-0-0-0-0-0-0-49-0-0-0-0-0-0-0-0-0-0-0-0-0-0.info
0-3-0-7-4-1-7-6-5-7-3-0-8-8-6-3-2-6-3-5-8-2-8-1-6-8-7-6-4-8-8- .0-0-0-0-0-0-0-0-0-0-0-0-0-49-0-0-0-0-0-0-0-0-0-0-0-0-0-0.info
0-3-4-6-1-8-8-3-0-7-2-7-5-4-6-1-3-3-3-3-3-0-6-8-4-0-8-8-5-1-5- .0-0-0-0-0-0-0-0-0-0-0-0-0-53-0-0-0-0-0-0-0-0-0-0-0-0-0-0.info
0-4-1-4-4-1-1-5-3-8-3-2-6-1-8-5-5-0-5-3-4-0-5-3-2-1-2-8-7-8-0- .0-0-0-0-0-0-0-0-0-0-0-0-0-49-0-0-0-0-0-0-0-0-0-0-0-0-0-0.info

6 Conclusion, Future Work and Limitations

In the current research work, situational awareness on Cyber Security threats have been thoroughly studied and created a highly scalable framework for the same. This framework analyses DNS event data as its input without making any significant architectural changes to the networking infrastructure. This distributed framework is highly scalable to combine and correlate the attack information from several Tier-1 service provider networks. This is the first of its kind framework to employ deep learning techniques for handling very large scale data to be processed in near real-time. This framework has been compared with other commercially available solutions and found that this framework outperforms (in terms of features and performance) all the existing solutions. Due to the confidential nature of the research, details of the comparison cannot be disclosed. To the best of our knowledge, this is the only framework that works across several Internet Service Providers as a single unified system providing situational awareness across the country.

The current research does not include malware binary analysis that provides detailed information on the structure and behavior of the malware. Since the framework does not have access to the end hosts, the damage done to the end systems cannot be measured. Also, the framework will not be able to detect the malicious communications that avoids DNS by using direct IP addresses for their communication. This limitation can be mitigated by the inclusion Net Flow. As a future direction, the framework will be enhanced by adding multi-lingual Internationalized Domain Names (IDN) domain name support as it does not perform the analysis of IDN based domain names. Though we developed a distributed platform for collecting and correlating the BGP events, the analysis is not done.

Acknowledgements This research was supported in part by Paramount Computer Systems and Ministry of Electronics and Information Technology (MeitY), Government of India. We are also grateful to NVIDIA India, for the GPU hardware support to research grant. We are grateful to Computational Engineering and Networking (CEN) department for encouraging the research.

References

1. Abu Rajab, M., Zarfoss, J., Monrose, F., & Terzis, A. (2006). A multifaceted approach to understanding the botnet phenomenon. In *Proceedings of the 6th ACM SIGCOMM Conference on Internet Measurement* (pp. 41–52). ACM.
2. Antonakakis, M., Perdisci, R., Dagon, D., Lee, W., & Feamster, N. (2010). Building a dynamic reputation system for DNS. In *USENIX Security Symposium* (pp. 273–290).
3. Ollmann, G. (2009). Botnet communication topologies. Retrieved September 30, 2009.
4. Foster, K. (2010). *The conicker worm and variants*.
5. Torpig. (2016). Retrieved January 11, 2016 from http://en.wikipedia.org/wiki/Torpig.
6. Royal, P. (2008). Analysis of the kraken botnet. *Damballa*, Apr 9.
7. Looking back at murofet, a zeusbot variant's active history. (2015). Wikipedia: The Free Encyclopedia. Wikimedia Foundation, Inc. Retrieved August 1, 2014 from https://blog. dambella.com/archives/1008.
8. Crawford, H., & Aycock, J. (2008). Kwyjibo: Automatic domain name generation. *Software: Practice and Experience*, 38(14), 1561–1567.
9. Antonakakis, M., Perdisci, R., Nadji, Y., Vasiloglou, N., Abu-Nimeh, S., Lee, W., & Dagon, D. (2012). From throw-away traffic to bots: Detecting the rise of dga-based malware. In *Presented as part of the 21st USENIX Security Symposium (USENIX Security 12)* (pp. 491–506).
10. Will, C. (2014) Botnet detection with dns monitoring. *Network*, 25.
11. Schiavoni, S., Maggi, F., Cavallaro, L., & Zanero, S. (2014). Phoenix: Dga-based botnet tracking and intelligence. In *International Conference on Detection of Intrusions and Malware, and Vulnerability Assessment* (pp. 192–211). Springer.

12. Raghuram, J., Miller, D. J., & Kesidis, G. (2014). Unsupervised, low latency anomaly detection of algorithmically generated domain names by generative probabilistic modeling. *Journal of Advanced Research, 5*(4), 423433.
13. Thomas, M., & Mohaisen, A. (2014). Kindred domains: detecting and clustering botnet domains using DNS traffic. In *Proceedings of the 23rd International Conference on World Wide Web* (pp. 707–712). ACM.
14. Ashwini, B., Menon, V. K., & Soman, K. P. (2016). Prediction of malicious domains using smith waterman algorithm. In *International Symposium on Security in Computing and Communication* (pp. 369–376). Singapore: Springer.
15. Zdrnja, B., Brownlee, N., & Wessels, D. (2007). Passive monitoring of dns anomalies. In *International Conference on Detection of Intrusions and Malware, and Vulnerability Assessment* (pp. 129–139). Springer.
16. Ramachandran, A., & Feamster, N. (2006). Understanding the network-level behavior of spammers. In *ACM SIGCOMM Computer Communication Review* (vol. 36, no. 4, pp. 291–302). ACM.
17. Anderson, D. S., Fleizach, C., Savage, S., & Voelker, G. M. (2007). Spamscatter: Characterizing internet scam hosting infrastructure. In *Usenix Security* (pp. 1–14).
18. LeCun, Y., Bengio, Y., & Hinton, G. (2015). Deep learning. *Nature, 521*(7553), 436–444.
19. Zhang, X., Zhao, J., & LeCun, Y. (2015). Character-level convolutional networks for text classification. *Advances in Neural Information Processing Systems*.
20. Elman, J. L. (1990). Finding structure in time. *Cognitive Science, 14*(2), 179211.
21. Bengio, Y., Simard, P., & Frasconi, P. (1994). Learning long-term dependencies with gradient descent is difficult. *IEEE Transactions on Neural Networks, 5*(2), 157166.
22. Martens, J. (2010). Deep learning via hessian-free optimization. In *Proceedings of 27th International Conference on Machine Learning*.
23. Hochreiter, S., & Schmidhuber, J. (1997). Long short-term memory. *Neural Computation, 9*(8), 1735–1780.
24. Cho, K., van Merrienboer, B., Gulcehre, C., Bougares, F., Schwenk, H., & Bengio, Y. (2014). *Learning phrase representations using rnn encoderdecoder for statistical machine translation*. arXiv:1406.1078, http://arxiv.org/abs/1406.1078.
25. Le, Q. V., Jaitly, N., & Hinton, G. E. (2015). *A simple way to initialize recurrent networks of rectified linear units*. arXiv:1504.00941 (2015).
26. Talathi, S. S., & Vartak, A. (2015). *Improving performance of recurrent neural network with relu nonlinearity*. arXiv:1511.03771.
27. Anstee Darren, C. F. C. P. B., & Sockrider, G. (2015). *Worldwide infrastructure security report*.
28. Vinayakumar, R., Soman, K. P., Poornachandran, P., & Sachin Kumar, S. Detecting android malware using long short-term memory-LSTM. *Journal of Intelligent and Fuzzy Systems*, IOS Press [In press].
29. Vinayakumar, R., Soman, K. P., & Poornachandran, P. (2017). Deep android malware detection and classification. In *International Conference on Advances in Computing, Communications and Informatics (ICACCI), 2017* (pp. 1677–1683). IEEE.
30. Vinayakumar, R., Soman, K. P., Poornachandran, P., & Sachin Kumar, S. Evaluating deep learning approaches to characterize and classify the DGAs at scale. *Journal of Intelligent and Fuzzy Systems*, IOS Press [In press].
31. Vinayakumar, R., Soman, K. P., & Poornachandran, P. Detecting malicious domain names using deep learning approaches at scale. *Journal of Intelligent and Fuzzy Systems*, IOS Press [In press].
32. Vinayakumar, R., Soman, K. P., & Poornachandran, P. (2017). Evaluating shallow and deep networks for secure shell (ssh) traffic analysis. In *International Conference on Advances in Computing, Communications and Informatics (ICACCI), 2017* (pp. 266–274). IEEE.
33. Vinayakumar, R., Soman, K. P., & Poornachandran, P. (2017). Secure shell (ssh) traffic analysis with flow based features using shallow and deep networks. In *International*

Conference on Advances in Computing, Communications and Informatics (ICACCI), 2017 (pp. 2026–2032). IEEE.

34. Vinayakumar, R., Soman, K. P., & Poornachandran, P. Evaluating deep learning approaches to characterize, signalize and classify malicious URLs. *Journal of Intelligent and Fuzzy Systems*, IOS Press [In press].
35. Vinayakumar, R., Soman, K. P., & Poornachandran, P. (2017). Applying convolutional neural network for network intrusion detection. In *International Conference on Advances in Computing, Communications and Informatics (ICACCI), 2017* (pp. 1222–1228). IEEE.
36. Vinayakumar, R., Soman, K. P., & Poornachandran, P. (2017). Evaluating effectiveness of shallow and deep networks to intrusion detection system. In *International Conference on Advances in Computing, Communications and Informatics (ICACCI), 2017* (pp. 1282–1289). IEEE.
37. Vinayakumar, R., Soman, K. P., & Poornachandran, P. (2017). Long short-term memory based operation log anomaly detection. In *International Conference on Advances in Computing, Communications and Informatics (ICACCI), 2017* (pp. 236–242). IEEE.
38. Vinayakumar, R., Soman, K. P., Velan, K. S., & Ganorkar, S. (2017). Evaluating shallow and deep networks for ransomware detection and classification. In *International Conference on Advances in Computing, Communications and Informatics (ICACCI), 2017* (pp. 259–265). IEEE.
39. Vinayakumar, R., Soman, K. P., & Poornachandran, P. (2017). Deep encrypted text categorization. In *International Conference on Advances in Computing, Communications and Informatics (ICACCI), 2017* (pp. 364–370). IEEE.
40. Vinayakumar, R., Soman, K. P., & Poornachandran, P. (2017). Applying deep learning approaches for network traffic prediction. In *International Conference on Advances in Computing, Communications and Informatics (ICACCI), 2017* (pp. 2353–2358). IEEE.
41. Tripwire, google's malaysian domains hit with DNS cache poisoning attack. (2013). Retrieved October, 2013 from http://www.tripwire.com/state-of-security/top-security-stories/googlesmalaysian-domainshit-dns-cache-poisoning-attack/.
42. Alexa-the top 500 sites on the web. (2014). Retrieved October 10, 2014 from http://www.alexa.com/topsites.
43. Hall, P. A., & Dowling, G. R. (1980). Approximate string matching. *ACM Computing Surveys (CSUR), 12*(4), 381–402.
44. Dameraulevenshtein distance. (2014). Retrieved December 12, 2014 from http://en.wikipedia.org/wiki/DamerauLevenshtein.
45. Van der Maaten, L., & Hinton, G. (2008). Visualizing data using T-Sne. *Journal of Machine Learning Research, 9*(2579–2605), 85.
46. Abadi, M., et al. (2016). TensorFlow: A system for large-scale machine learning. In *OSDI* (Vol. 16).
47. Soman, K. P., Loganathan, R., & Ajay, V. (2009). *Machine learning with SVM and other kernel methods*. Ltd: PHI Learning Pvt.
48. Soman, K. P., Diwakar, S., & Ajay, V. (2006). *Data mining: Theory and practice [WITH CD]*. Ltd: PHI Learning Pvt.
49. Kuhrer, M., Rossow, C., & Holz, T. (2014). Paint it black: Evaluating the effectiveness of malware blacklists. In *International Workshop on Recent Advances in Intrusion Detection* (pp. 1–21). Springer.

Big Data in HealthCare

Margarita Ramírez Ramírez, Hilda Beatriz Ramírez Moreno
and Esperanza Manrique Rojas

Abstract This chapter presents an analysis of the infrastructure of big data, the elements that make it up, the types of data that define it, and the characteristics that distinguish it as a child: Volume, speed, variety, veracity and volatility. In a concrete way, different applications based on this architecture are analyzed, from which it is possible to find health, internet of things, among other applications. A description of the data used in health is performed, which is possible to manage effectively with a model based on big data. Finally, the proposal of a health model for Mexico is presented, based on an infrastructure that allows the integration and sharing of information, the administration of medical histories, public health and research data in the health area, all of them as a basis to carry out data analysis, to support decision-making and to serve as a basis for the creation of Institutional health programs. It concludes with evidence of the significant contribution that a big data model can give to the health sector in Mexico.

1 Introduction

The accelerated growth of the amount of information that is generated every moment through the connected devices, the use of social networks as well as the increase in data consumption, evidences the need to create, design and manipulate a Big Data. In particular a Big Data in health can be very useful to give an adequate follow-up to healthcare issues in a society, as well as store information that once analyzed serves as a basis for decision making and the generation of knowledge

M. R. Ramírez (✉) · H. B. R. Moreno · E. M. Rojas
Facultad de Contaduría y Administración, Universidad Autónoma de Baja California,
Tijuana, BC, México
e-mail: maguiram@uabc.edu.mx

H. B. R. Moreno
e-mail: ramirezmb@uabc.edu.mx

E. M. Rojas
e-mail: emanrique@uabc.edu.mx

© Springer Nature Singapore Pte Ltd. 2018   143
S. S. Roy et al. (eds.), *Big Data in Engineering Applications*,
Studies in Big Data 44, https://doi.org/10.1007/978-981-10-8476-8_7

through research. This chapter contains the description and characteristics of a Big Data, the infrastructure and the elements that make it up, it describes in general the type of applications existing in Big Data and some of the most recognized mobile applications. The topic of information technologies in health is analyzed. A description of the data management in health organizations is presented and finally a model of Clinical Record is presented with the use of Big Data that can be adapted to public and private healthcare in Mexico.

2 Big Data

The term big data was first employed in 1997 by NASA researchers Michael Cox and David Ellsworth, and they defined it as: The management and analysis of massive volume of data that cannot be treated in a conventional way, since they exceed the limits and capacities of software that are commonly used in the gathering, management and processing of data [1].

Today the quantity of data that is generated by everyday practices such as: social networking, email, instant messaging, online profile updates, GPS, among other activities and devices is increasingly larger, however it is possible to store, process and transmit this information through *"Big Data"*. According to one study *"A decade of digital universe of growth" From* 2005 to 2020, the digital universe will grow by a factor of 300, from 130 to 40,000 EB, or 40 trillion gigabytes (more than 5,200 GB for every man, woman, and child in 2020). From now until 2020, the digital universe will double about every two years [2].

This study shows projections of the volume of data generated between 2005 and what is considered will be generated by global users by 2020.

The term Big Data is referred to as the set of massive volume of data which can be composed by heterogeneous formats: Structured, semi-structured and unstructured data:

- Structured data is the type of data which has defined format, type and length, and can be stored in tabular format, such as a relational database or spreadsheet.
- Semi-structured data does not have a set type of data, they contain flags or markers which differentiate its elements, they might contain metadata that describe the objects and the co-relation between them, these can files such as HTML, XML, among others [3].
- Unstructured data is stored just as it is gathered, with no specific format; it cannot be stored in tabular format such as PDF files, emails, multimedia files, images among others.

In Big Data the unit of measure for information is: terabytes, petabytes, exabytes or zettabytes. One terabyte (TB) contains 10^{12} characters, while one petabyte (PB) is integrated by 10^{15} characters, and zettabytes by 10^{24}, and in order to be able to process this volume of data, special units and advanced technological resources are required.

Analysis of large volumes of data also provide large benefits for organizations, big data is a resource that companies were awaiting for according to various authors and while it is true that "Information is power" as stated by a popular saying, it is important to highlight that the one who has the information is not powerful, but the one who uses it properly. The optimal use of data is synonym of innovation, growth and competitiveness [4]. The adoption of big data allows for a wide variety of advantages in areas such as: e-commerce, e-government, healthcare, among others.

Most data scientists and experts define Big Data by the following three main characteristics (called the 3 Vs) [5]. On the other hand there are new concepts that consider big data as a large gathering of information that includes among its characteristics the rule of the 4 Vs or the four dimensions of big data [6]. And there are contributions that consider as much as 5 Vs. We will briefly describe each one of these characteristics.

2.1 Volume

According to Gartner, by 2020 one thousand million devices will be connected to the internet, which will increase the volume of data, it is even possible that in a few years it is multiplied by 10.

A massive amount of data is generated from millions of devices and applications, for instance, in 2013, the total volume of created, replicated and consumed data was of 4.4 zettabytes (ZB) according to the International Data Corporation, and this volume is considered to double every two years. By 2015 data volume grew to 8 ZB [7], and this report also states that the volume of data will reach as much as 40 ZB by 2020 with a 400 times increase from then on [8].

2.2 Velocity

Data is generated every day in ever faster processes, the velocity at which is produced is measurable according to the rhythm in which old data travels from different sources, such as: businesses, social networks and mobile devices, among others. Data integrated in real time can support decision making for organizations, just as research teams can generate new knowledge. We can identify examples such as Wal-Mart's amount of generated data, with as much as 2.5 PB each hour in customer's transactions only, or for instance YouTube, which operates at a very elevated speed of transactions [8].

2.3 Variety

Big Data gathers information provided by different sources and kinds of data, whether they are structured or unstructured. While before, information stored as big data was provided by databases and files generated from tools such as information systems and even spreadsheets, today information stored as big data comes from social networks, emails, videos, sound files, instant messaging and more, this diversification in data creates a huge complexity when storing, managing and analyzing information, due to this problem, special tools have been developed for analyzing data regardless of how diverse and varied it is.

2.4 Veracity

This is about the trust we can have, the bias, the noise and the alteration of data. Ensuring the veracity of the stored information is of extreme importance for the big data responsible, just as the guarantee that data is clean, valid, correct, and precise.

2.5 Volatility

Data is valid by the time it's gathered, but depending on how long it's stored, it can satisfy the needs of the user, for instance, some organizations only store data and transaction from certain periods such as the latest year, due to certain restrictions like limited storage.

2.6 Big Data Infrastructure

Infrastructure used for big data involves technologies that offer services for a specific solution to the processing of various kinds of data, gathered form numerous sources such as: files, networks, sensors, microphones, cameras, scanners, images, videos and others.

Analyzed data becomes information in real time and it is possible to generate new information that allows solving problems that were not possible before. In some cases big data is employed to know information about clients through different ways. We can identify big data infrastructure in the information and its real time processing, the users interacting, the various kinds of data integrated and the tools that enable the visualizing, storing, analysis and processing blocks of data with technologies such as computers for processing and storing, telecommunication networks, and cloud storage. In Fig. 1 we can identify the components of big data.

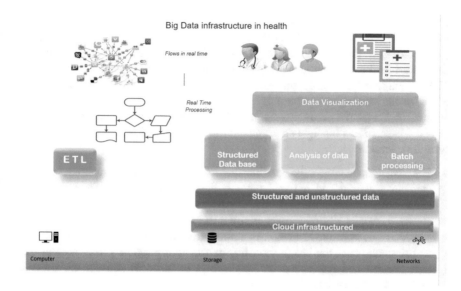

Fig. 1 Big data infrastructure

3 Big Data Applications

The application field for big data is very varied, among them we can consider the smart grid case, E-health, Internet of things, transportation, logistics, E-government services and many others.

Smart Grid Case The connection of millions of devices requires an intelligent network that enables to identify the behavior of connected devices and establish preventive strategic plans to reduce the cost of co reactive actions and the calculation of energy required to plan the use of resources and maximize support [9].

Internet of Things (IoT) Stands as one of the major applications of big data and it involves a wide variety of subjects, this area evolves every day, this application support logistics in companies, with the use of big data it is possible to determine the routes and location of the vehicles through sensors, adapters and GPS, these applications are used to follow up with their employees in the delivery routes, times of travel, and related information [10].

E-Heath Connected health platforms are used to personalize healthcare [11]. Big data is generated from different heterogeneous sources such as laboratory files, medical history, patient's symptoms, and pharmaceutical information among others.

Advanced analysis of joint medical data are of great benefit for healthcare, for instance doctors can monitor symptoms of patients and adjust prescriptions in real time, in the same way public healthcare institutions can create plans and programs about illnesses, advances in treatments and optimization of operation in hospitals, just as a support in creating clinical files.

According to the estimations of the McKinsey Global Institute, the use of big data could increase efficiency and quality of healthcare services by a value worth 220,000 million euros only for American healthcare and some 250,000 million euros for the whole public sector in Europe [12].

In 2016, Langkafel presents a diagram in which he shows different scenarios that can be presented in a big data environment (See Fig. 2), which is developed starting from three coordinates':

- Horizontal, in which it is possible to visualize information generated by the patients, (internal and external patient), information gathered from the treatments history and attention that includes ambulance, hospitalization and recovery, just as any other meaningful part of the process.
- Vertical, this coordinate integrates information generated by the medical unit and research from the health industry, information generated from administrative, clinical and research data. This allows to maintain strong, reliable data updates for a permanent support for experts.
- Diagonal, allows integrating information proceeding from different moments and making a full analysis of data, information obtained from the past that can be used to make better decisions, just as information generated in real time to come up with a more accurate forecast.

Health Information Technologies (HIT)

Information technologies are required in all areas, but in health it becomes essential due to the progress that offers along with the benefits of creating a better healthcare environment, efficiency and financial [13]. When speaking of medicine we refer to the science that cures and prevents human illness where health is a detonating element.

Health information technologies are used in surgery, clinical laboratories, imaging files, such as magnetic resonance imaging, management software, medical research data, medical logs, etc.; enabling access to data for an immediate response and better decision making [14].

The clinical data that one person can generate from birth to death are of critical importance to prevent and to take care of health, some of them can be stored in a *medical history* where others may not. Medical History is composed by various documents that are generated in medical appointments, integrating medical notes, logs, treatments, reports, and follow up process by the doctor [15]. It is important to highlight that one person can have various medical history records if they've visited different healthcare professionals. All information that is generated can amount to large volumes of data, so complex that it cannot be successfully processed using conventional management software.

This problem gives way for Big Data in health, as mentioned by Bonnie Feldman and other colleagues in October 2012 [16] "Big data is becoming an increasing force of change in the health industry scenario, and it's potential resides in the possibility of combining structured, unstructured or semi-structured data to an individual or population level". In the same year Forbes magazine published an article "Big data the future of health" [17] represents one opportunity for innovators and all those interested in healthcare, rising substantially the possibility of obtaining more effective information of all data processed and lower mortality rates.

To process all data contained in the **Medical history** or **medical file** big data is required and technologic tools that can support its management, since recent studies remark it is the key for healthcare [18], and to prevent some illnesses. Other scientific studies have demonstrated that the majority of diseases that have afflicted humanity can be avoided by modifying lifestyle and environment [19].

Analyzing the information that we have on health and the information technologies available make it possible for global statistics, since they provide a comparison on traditional data and national, regional, and even global trends. The institution overseeing this information is the Global Health Observatory (GHO) [20].

3.1 Mobile Healthcare

With new technologies we can have access to all kinds of information anytime, anywhere, due to internet connection through a smartphone. These devices have become the basis of the day to day communication, to be informed, for social networks, to transmit personal or work information.

In recent years, the use of smartphones has increased more and more and with it the use of apps. An app is a mobile Smartphone program with a specific end, apps can be downloaded online and the number of apps available in the market has increased enormously. According to Statista, there are more than 5 million apps available between the largest providers such as Google Play, Apple Store, Windows, Amazon and Blackberry [21]. App Annie comments that in less than 5 years the amount of downloads will be over 352 billion compared to the 197 billion estimated for this 2017. The impact of revenue estimated figures around 139 billion dollars as opposed to the 82 billion dollars generated this year. See Fig. 3.

Health applications are the third most downloaded, only after games and utilities and an increase of 23% is expected in the following years according to Deloitte, in its study mHealth in an mWorld: How mobile technology is transforming health care [22]. Another important aspect is provided by IMS Health Institute, one of the largest evaluation institutions on health, where it is mentioned that the number of mobile applications dedicated to health surpasses 165,000 [23]. 70% is for general public through wellness and exercise and the rest are for a more specific target: health professionals and their patients. The main use of these applications is related to prevention or lifestyles (nutrition, physical activities, rest, relaxation, addiction control, etc.).

It is very hard to classify health applications but we will describe 5 Apps of a study made by the *THE APP intelligence, inform of the* 50 *best health APPS in Spanish* by level of quality and content, design and utility to offer solutions to concrete issues (See Fig. 4).

Doctoralia: both for iOS and Android, allows to search healthcare facilities and medical professionals, it even allows to filter according to healthcare provider or by their insurance, management of appointments, and users can rate their specialist, it was awarded by the App circurs 2012.

Mobile MIM: for iOS only, it is for professionals and allows visualization, exchange and register of images, such as: SPECT, PET, CT, MRI, X-rays, and ultrasounds. It can be used to review images, contours, insulated glass and isodose curves of the radiation treatments. It was awarded with the Apple Design Award to the best health App in 2008.

Ablah: for both iOS and Android. It's main goal is to contribute to the better communication between people with language barriers such as autism, down syndrome, or adults with ictus or strokes, among others. It enables the interaction of these people with their relatives, physicians and social environment. It has been awarded as the best app in the wellness category by the Smart Accessibility Awards 2012 of the Vodafone foundation, also as the best Spaniard app 2010 by The App Date.

Social Diabetes: available on iOS and Android alike, this is an aid in controlling type 1 and 2 diabetes, it can calculate the dose of hydrates and the dosage of insulin, just as the physicians can control remotely its patients parameters. It has the award @appsaludable and was among the finalist for Ideas Saintes 2013 awards.

Endomondo: available on iOS, Android, WP, BB, Symbian. It promotes a healthy lifestyle through physical activity, acting as a personal trainer; it registers tracks, calorie intake, heart rate, among other parameters, motivating users to stay active. It also promotes social workout, planning challenges with friends and sharing through social networks.

In the universe of mobile apps we must be informed about the function and utility of them because not all of them meet the expected quality, besides the lack of knowledge on how to use them or the poor support by healthcare professionals, it is important to highlight that the apps developed so far do not correspond to areas of major expense, such as the ones related to chronic disease, opening a scenario of opportunity for software developers or future professionals.

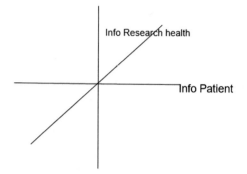

Fig. 2 Environment of big data in health

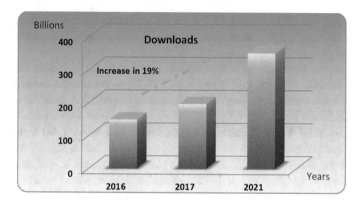

Fig. 3 Application downloads forecast

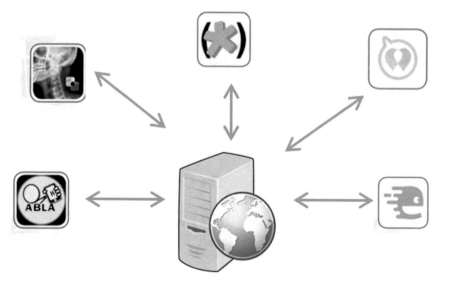

Fig. 4 Healthcare applications

4 Data Management in Healthcare Organizations

Within the health industry there are numerous sources of heterogeneous data that issue a significant amount of co-related information about patients, disease and healthcare centers. When this information is analyzed, it is of great utility to healthcare professionals (doctors, nurses and pharmaceuticals) in decision making, which allows a more adequate healthcare service [24].

The incorporation of an electronic universal medical history in Mexico will improve healthcare services in the country. Having an electronic medical history file not only replaces paper when it comes to registering patient's updates, but it takes it further, allowing an electronic database with patient's information readily available to strengthen activities. With the support of technology and Big Data, a full analysis is possible due to patterns found in data, creating a more solid service and becoming the starting point for healthcare institutions in having an integration that connects not only the areas of a hospital, but also public and private services once for all [25].

According to Siemens, in Mexico, there are about 23,260 healthcare units, of which 86.8% belong to the public sector and 17.2% to the private sector and while every healthcare unit has various devices to collect information about its users, 80% of that data is not structured [26].

Through the NORMA Official Mexicana, NOM-168-SSA1-1998 legislation the Mexican health department defines medical history or clinical history as the technical and legal document which is fundamental for healthcare on patients and also the organization of its users. It is an indispensable instrument to systematically register the healthcare process and the clinical evolution that reflects the different forms and moments in which healthcare specialists intervene. These file must be kept for at least 5 years, starting from the date of the last medical appointment. Meeting the scientific and ethical standards that govern healthcare institute or the physician responsible for safeguarding this information [27].

On November 30th of 2012, the NORMA Official Mexicana NOM-024-SSA3-2012 legislation establishes that Electronic information registry systems for health, exchange of information have the objective of regulating these electronic systems, as well as establishing the mechanism for healthcare providers to register, exchange and consolidate information in the public, private and social sectors of the national healthcare system in Mexico [28].

5 Big Data Model on Health. The Case of Mexico

As mentioned before, a Big Data model is of great utility to the health industry. In Mexico, extraordinary efforts are made to improve day after day healthcare services; however, it has not been possible to integrate collected information by the numerous healthcare providers.

We propose a model that allows to integrate the needed information for the treatment of a patient during the various stages of life, as well as the gathering of useful information in creating programs that support health conditions in Mexico.

Big Data model proposal integrates user's medical history information in one single clinical file with an identifier for each user. The file would be created at the date of birth, when the birth certificate is generated. This file will integrate basic general information such as: registration number, name, nationality, date of birth, religion, address, telephone number, weight, date of file annexation, physiological data, personal background, family background and later, information collected through anamnesis, or heteroanamnesis, information like vaccine registry, medical history, diagnose of illnesses, treatments, integration of clinical results, X-rays, tomography and magnetic resonance, etc.

Below, the areas of the Big Data model in health, the case of Mexico, are described. In Fig. 5, it is possible to observe the distribution and interrelation of each of these areas.

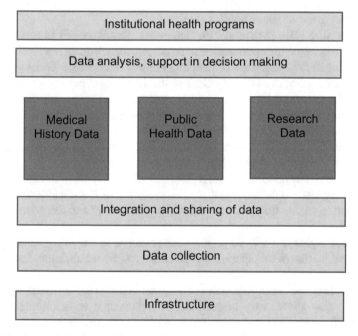

Fig. 5 Big data model on health. Mexico case

6 Institutional Health Programs

This phase of the model will allow the integration of institutional health programs, which will receive information and data from different programs generated for specific purposes through the health area. This system in Mexico is comprised of two sectors: public and private.

In the public sector there are social security institutions such as: Mexican Institute of Social Security (IMSS), Institute of Security and Social Services of State Workers (ISSSTE), Mexican Petroleums (PEMEX), Secretariat of National Defense (SEDENA), Naval Secretariat (SEMAR) and institutions that provide care to the population without social security: Ministry of Health (SSA), Popular Health Insurance, among others, in the private sector there are insurance companies, clinics and private hospitals.

In Mexico it is common that users take advantage of the services of several institutions and can be attended by different specialists or medical areas. The information generated by the patient in each clinic or office is stored exclusively in its archives; however, it is important and it may be necessary to share this information among the different health agencies.

A standardized Big Data model that integrates clinical records in health services in Mexico will allow the optimization of resources, the adequate monitoring of each case, as well as the integration of data that are raw material for decision making.

A case of application of a Big Data in the health system may be the national cancer registry, created by the Ministry of Health in Mexico, which aims to collect information on cancer cases that occur in different regions of the country Through this program it is possible to follow up the cases for at least five years, to know behaviors of this disease, number of patients, real situation of the disease and thus be able to undertake actions such as prevention and timely care.

This registry has national coverage, integrates data related to the identity of the patients, diagnostic data and specific characteristics of the condition, as well as information regarding the treatment and follow-up received and the current status of the patient.

With this registry, it will be possible for scientists to analyze medical histories and medical databases to achieve better research and, above all, better treatment for patients.

Another proposed institutional health program is the e-card program, which is an application that allows easy accessibility to vaccination card information in all public health entities in the country, which will serve as a support to the population and to the staff of these institutions so that it is feasible to consult from any device with an internet connection the child's vaccination status.

6.1 Data Analysis, Support for Decision Making

In this phase, the analysis of the data coming from the different sources of data is carried out, such as: clinical files, public health and research data. Having all these data for processing, storage and optimal management, provides an additional value beyond what they have on their own to support decision making. The sources of data are described below.

6.2 Medical History Data

This area of Big Data integrates with structured data (general information, physiological data, family background, vaccine registry, clinical results, etc.).

Semi-structured Data: information gathered through interview (anamnesis, diagnose, treatments).

Unstructured Data: Imaging, topographies, X-rays, diagnose files or treatment description, pdf files, etc.

6.3 Public Health Data

This area will integrate information generated by public healthcare programs in the country, information from ongoing programs such as national development plan and sectorial health plan 2013–2018.

Information or indicators that can serve as the base for creating public regulations that contribute in the achievement of health goals defined by national programs. Structured, unstructured and semi-structured are included, data resulting by statistics generated by healthcare services, prevalence of chronic diseases, main causes of death, mortality rate in children, causes for maternal death, information that supports decision making, as well as the creation of plans and new health programs.

7 Research Data

Once the information from different areas has been integrated in a medical history and public health data, it is possible then to make an analysis and integration of information that will be the material for research and generation of knowledge in the health industry.

Each component of the model will allow to integrate a solid base of information that supports decision making and the creation of institutional programs in benefit of health.

7.1 Integration and Sharing of Data

These data are analyzed to obtain statistical information from which new knowledge is derived in the health area. With this new information, decisions can be made in the areas of public health, patient care or new research. Figure 6 indicates the way in which (structured and unstructured) data are generated from a population.

Fig. 6 Representation of information generated by the population in the healthcare area

7.2   Data Collection

Data can be obtained through smartphones, wearables and cutting-edge technologies. Other data can be obtained from X-ray studies, clinical and case reports, and clinical records.

7.3   Infrastructure

An important aspect that stands out in the infrastructure is the technological platform with organizations or institutions must have, such as: hardware, software, mobile technology, communication technology, and networks, to name a few, since they form the basis for working with large-scale data and the analysis of huge amounts of information that cannot be treated in a conventional manner.

8   Conclusions

The information generated by medical history or clinical file, correspond to the different types of data mentioned above, which is why Big Data has become a crucial element for the processing of these large amounts of information and to obtain a more detailed and comprehensive analysis for the prevention, diagnosis and monitoring of healthcare.

With the implementation of the Big Data Model in Healthcare, support is generated in the decision making process as well as in the development of predictive and probabilistic models in the health area. It is feasible to have the possibility of saving millions of pesos in the public healthcare sector, through the different types of data that could be generated for analysis in a region of the country [8]. All the information that can be obtained for your manipulation of a Big Data can be: by individual, by city, by state, by country, by gender, etc. so it would be a powerful technological tool in the prevention and care of human health.

In this context, basic features of Big Data health applications go beyond volume, variety and speed, as they incorporate important aspects such as veracity, allowing a reuse by adding new information to the data history. These new ways of storing, processing and analyzing information as well as the use of technological tools can improve health care and provide opportunities for professionals in both health and technology areas.

© All rights reserved, the marks, logos, content, design and photographs shown are only illustrative, are registered and are property of each manufacturer.

References

1. Salazar, A. (2016). Infraestructura para big data. Revista Digital Universitaria. Retrieved September, 2017, from http://www.revista.unam.mx/vol.17/num11/art77/.
2. Gantz, J. The digital universe in 2020: Big data, bigger digital shadows, and biggest growth in the far east. Retrieved August, 2017, from https://www.emc.com/collateral/analyst-reports/idc-the-digital-universe-in-2020.pdf.
3. López-Messa, J. Tipos de big data. Revista Electrónica de Medicina Intensiva. Retrieved August, 2017, from http://www.medicina-intensiva.com/2017/07/A232.html.
4. Arvizu, L. Big Data Lo que las empresas esperabas, México Forbes. Retrieved January, 2017, from https://www.forbes.com.mx/big-data-lo-que-las-empresas-esperaban/.
5. Furht, B., & Villanustre, F. (2016). Introduction to big data. Big data technologies and applications (pp. 3–11). Cham: Springer.
6. Gagnon, C., John, E., & Theunissen, R. (2017). Organizational health: A fast track to performance improvement.
7. Rajaraman, V. (2016). Big data analytics. Resonance, 21, 695–715.
8. Oussous, A., et al. (2017). Big Data technologies: A survey. Journal of King Saud University-Computer and Information Sciences. https://doi.org/10.1016/j.jksuci.2017.06.001.
9. Stimmel, C. L. (2014). Big data analytics strategies for the smart grid. CRC Press.
10. Chen, C. P., & Zhang, C.-Y. (2014). Data-intensive applications, challenges, techniques and technologies: A survey on big data. Information Sciences, 275, 314–347.
11. Namblar, Bhardwaj, Sethi, & Vargheese. A look at challenges and opportunities of big data analytics in healthcare. In 2013 IEEE International Conference of Big Data, IEEE (pp. 17–22).
12. Langkafel, P. (2016). Big data in medical science and healthcare management: Diagnosis, therapy, side effects. De Gruyter.
13. Koppers, B., & Thorogood, A. (2017). Ethics and big data in health. Current Opinion in Systems Biology, 4, 53–57.
14. Ramirez, M., Ramirez, H., Osuna, N., Salgado, M., & Alanis, A. (2017). In Big Data and Health "Clinical Records", Innovation in Medicine and Healthcare 2017, Smart Innovation, Systems and Technologies. Springer.
15. Martínez, H. J. (2006). Historia Clínica, Sistema de Información Científica Redalyc Red de Revistas Científicas de América Latina y el Caribe, España y Portugal, vol. XVII, núm. 1, pp. 57–68.
16. Bonnie, F., Ellen, M., & Tobi, S. (2012). Big data in healthcare hype and hope. Dr. Bonnie 360° business development for digital health. Retrieved June, 2017, from http://es.scribd.com/doc/107279699/Big-Data-in-Healthcare-Hype-and-Hope.
17. Riskin, D. The next revolution in healthcare. Retrieved June, 2017, from en Forbes. www.forbes.com/sites/singularity/2012/10/01/the-next-revolution-in-healthcare.
18. Castillo, Y. L. (2017). Big Data, clave en el cuidado de la salud, Saludiario el Medico para Médicos. Retrieved June, 2017, from http://saludiario.com/big-data-clave-en-el-cuidado-de-la-salud-estudio/.
19. Secretaria de Salud de Baja California, Tu Salud Promoción de la Salud. Retrieved January, 2017, from http://www.saludbc.gob.mx/tu-salud.
20. Organización Mundial de la Salud. Datos del Observatorio mundial de la salud. Retrieved January, 2017, from http://www.who.int/gho/es/pdf.
21. e-growing online and mobile marketing, Evolución del mercado de las apps hasta 2021. Retrieved August, 2017, from http://e-growing.com/evolucion-mercado-de-las-apps-2021/.
22. The APP inteligence, Informe 50 mejores APPS de salud en Español. Retrieved August, 2017, from http://boletines.prisadigital.com/Informe-TAD-50-Mejores-Apps-de-Salud.pdf.
23. QuintilesIMS Formerly Known as IMS Health. Retrieved August, 2017, http://www.imshealth.com/en/thought-leadership/quintilesims-institute#.

24. Instituto de Ingeniería del conocimiento, Big Data en Salud. Retrieved September, 2017, from http://www.iic.uam.es/soluciones/salud/.
25. Medina Alejandro, A México le urge el Expediente Clínico Electrónico Universal, Forbes México.
26. Ramírez, A., Abraham, Retos del Big Data para el sector salud. Retrieved September, 2017, from https://www.the-emag.com/theitmag/blog/2017/04/14/retos-del-big-data-sector-salud.
27. Secretaria de Salud en México, NORMA Oficial Mexicana, NOM-168-SSA1-1998, Del Expediente Clinico. Retrieved September, 2017, from http://www.salud.gob.mx/unidades/cdi/nom/168ssa18.html.
28. Secretaria de Salud en México, NORMA Oficial Mexicana NOM-024-SSA3-2012. Retrieved September, 2017, from http://www.dgis.salud.gob.mx/descargas/pdf/NOM-024-SSA3-2012.pdf.

Facing Up to Nomophobia: A Systematic Review of Mobile Phone Apps that Reduce Smartphone Usage

David Bychkov and Sean D. Young

Abstract Excessive smartphone use has been linked to adverse health outcomes including distracted driving, sleep disorders, and depression. Responding to this growing trend, apps have been developed to support users in overcoming their dependency on smartphones. In that vein, our investigation explored the "big data" available on these types of apps to gain insights about them. We narrowed our search of apps, then reviewed content and functionality of 125 Android and iOS apps that purport to reduce device usage in the United States and elsewhere. This sample was curated based on popularity through the market research tool, *App Annie* (which indicates revenue and downloads per category of app and by country). The apps fell into 13 broad categories, each of which contained several different features related to filters, usage controls, and monitoring programs. Findings suggest that social media technologies, including smartphone apps, are being attempted for use for health behavior change. We discuss methods of sorting through "big data" generated by apps that purport to curb smartphone addiction. Finally, we propose data-driven features, such as social facilitation and gamification, that developers might use to enhance the effectiveness of these apps.

Keywords Smartphone · Phantom vibration syndrome · Social cognitive theory · Cognitive behavioral therapy

D. Bychkov
Whiting School of Engineering, Institute for Nano-Bio Technology in Health Measurement Corps, Johns Hopkins University, Baltimore, MD, USA

S. D. Young (✉)
Department of Family Medicine, University of California, Los Angeles, 10880 Wilshire Blvd., Ste. 1800, Los Angeles, CA 90024, USA
e-mail: sdyoung@mednet.ucla.edu

© Springer Nature Singapore Pte Ltd. 2018
S. S. Roy et al. (eds.), *Big Data in Engineering Applications*,
Studies in Big Data 44, https://doi.org/10.1007/978-981-10-8476-8_8

1   Introduction

The proliferation of smartphones and access to wireless and data networks has enabled people to learn, connect, and navigate worldwide. Despite the clear advantages of these "smart" devices, their near-constant use has given rise to negative social and health consequences such as smartphone addiction, lowered sleep quality [15], and decreased road safety [18].

This is not surprising, given a 2013 Harris poll which revealed that 72% of United States (US) adults claimed they were so dependent on their smartphones that they kept them within 5 feet of their bodies the majority of the time. In fact, some respondents noted they kept their smartphone nearby while showering, during sexual intercourse, and/or attending religious services [19]. Findings like these have given rise to a new term, "nomophobia." The fear of being without one's phone is so pervasive that it not only has a name, it was recently proposed for inclusion into the *DSM*-V [4, 12]. Symptoms of excessive smartphone use and nomophobia include compulsive text messaging and phantom vibration syndrome, wherein a user feels the sensation of a device's vibration even when it is not in use [17, 20].

Fortunately, these symptoms may be resolved through behavioral interventions, as the technology itself may be a useful tool for that very behavior therapy. For example, public health announcements have been conveyed by television and radio advertisements, text messages, and mobile apps [24].

More precisely, behavior change techniques include gaining end-user acceptance and commitment ("commitment") [5], goal-setting and promotion of standards ("setting standards") [21], offering tools for self-awareness and monitoring ("self-tracking") [22], customization of feedback based on the audience and situation ("tailoring messages") [11], encouraging the end user to strengthen relationships with family, friends, and peers ("social facilitation") [2, 7, 23, 26, 27] 蚱 and implementation of rewards, competition, or other game elements ("gamification") [8, 16]. Therefore, although smartphones may be associated with addiction, they may also serve as a platform that enables positive behavior change [9].

We provide a review of the "big data" from these types of apps based on data from the data aggregator, *App Annie*. This review aimed at: (i) illustrating the content and functionality of the new generation of apps that serve to mitigate the problems associated with smartphone addiction and excessive device usage; and (ii) exploring (indirectly) ideas on how to enhance the effectiveness of these "detox apps".

## 2	Methods

### 2.1	Keyword Search Terms

Prior to initiating the review process, we conducted research via *Google Trends* to identify relevant keyword terms related to mobile phone "over-usage." First, we entered the expression "smartphone addiction" into *Google Trends* on September 1, 2016. Three related search terms were returned, all of which were relevant, namely: "smartphone addiction test," "smartphone addiction scale," and "phone addiction." Entering the term "phone addiction" yielded a total of 11 results, of which 3 were relevant, i.e. "cell phone addiction," "internet addiction," and "social media addiction." Entering "smartphone addiction scale" and "smartphone addiction test" did not generate additional results.

### 2.2	App Selection Process

Google Play and *iTunes App Store* were used to conduct a systematic review of apps on September 1, 2016. Our inclusion criteria were composed of the noted search terms. Exclusion criteria were: (i) the app description did not feature content related to excessive device usage, (ii) the app was a duplicate, or slight adaptation, of another app already under investigation, (iii) the app had no English-language user interfaces, (iv) the app claimed to increase device usage, or (v) the app could harm (even judge) the end user.

### 2.3	Ranking

Popularity was assessed using *App Annie* [1]. AppAnnie.com is a market research website that collects aggregate data from its clients (comprised of approximately 94% of the top smartphone app publishers), and then extrapolates rankings for all apps [6]. Given that neither *Google* nor *Apple* provide precise data about the total revenue or downloads achieved by an app, *App Annie*'s estimates are a valuable resource for contextualizing and interpreting the data provided on these platforms. Furthermore, *App Annie*'s data have high credibility, in that it has been successfully applied in other research studies for evaluating the market visibility of health apps [3, 25].

2.4   Statistical Analysis

Data were expressed as percentages for categorical variables and as median (interquartile range [IQR]) for ordinal/continuous variables. Comparisons between groups were performed by Chi-squared or Fisher's exact tests for categorical variables, and by the Mann-Whitney test or Kruskal-Wallis with Dunn's post hoc test for continuous variables.

A logistic regression model was built so that we could explore the relationship between ranking and the characteristics of the selected apps (e.g. platform and features). Because the ranking did not have a normal distribution, the variable was included in the model as a dichotomous variable (i.e. ranking less than or greater than 500). The independent variables were chosen based on an exploratory analysis and a hierarchical backward approach was then used to determine which variables to keep or drop from the model [13]. A generalized variance inflation-factor (VIF) greater than 5 served to identify multicollinearity. Following this, the Durbin Watson test was applied to evaluate the model's validity. From this foundation, statistical analysis was performed using STATA-10. A p value <0.05 was considered significant.

3   Results

The *Google Play* and *iTunes* filtering search resulted in 125 apps that were available on Android (n = 67), iOS (n = 37) or both platforms (n = 21). While the search term "smartphone addiction" returned the greatest number of results across both operating systems, the *iTunes* search engine provided, on average, 56.0% more relevant results to our queries than *Google Play*. The flow chart of the selection process is shown in Fig. 1.

3.1   Content and Functionality of Selected Apps in the US and the Rest of the World

The selected apps fell into the following categories: productivity (52.0%), tools (12.0%), lifestyle (9.6%), and health and fitness (8.8%). The remaining 17.6% is represented by business, communication, education, medical, parenting, personalization, photography, social networking, and utilities apps. Each app had at least one of the following features: blocking, parental control, tracking, rewards, reminders, coaching, and/or social media. As demonstrated in Table 1, reminders and tracking were the most represented (90.4% and 82.4%, respectively). Half of the categories included at least 3 features and there were no significant differences among app categories in terms of average number of features (KW = 7.75, p = 0.101).

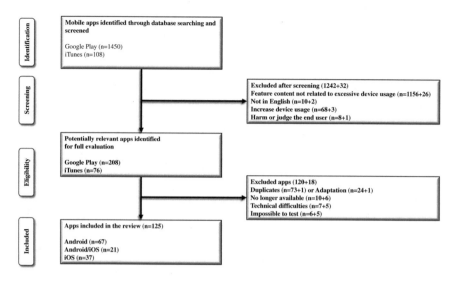

Fig. 1 Flow chart of the app selection process

Since roughly half of the data (56.7%) originated in the US, we focused our analysis on the possible differences in app preferences and usage between countries. Specifically, Android and iOS platforms showed a different distribution between the US and the rest of the world. In the US, the prevalence of iOS users (40.3%) was comparable to the prevalence of Android users (41.7%), while in the rest of the world the choice of the platform was polarized ($\chi^2 = 11.2$, p = 0.004) toward Android (69.8%) rather than iOS (15.1%). Tracking and reminder features were the most represented, at 80.0% (Table 1).

Because sample size was not always representative when stratified by category, the least numerically relevant categories were grouped in "other" (e.g. business, communication, education, medical, parenting, personalization, photography, social networking, and utility). Android and iOS showed a statistically significant difference in apps related to productivity ($\chi^2 = 7.94$, p = 0.019) and tools (p < 0.001). These categories were the least and most represented on the Android platform, respectively (Table 2).

3.2 Popularity of Apps in the US and the Rest of the World

Ranking was available for 91 out of 125 of the apps. On average, *Annie's* rank was significantly higher (KW = 22.50, adjusted p < 0.001) on the iOS platform (917; 577.3–1310.0) than on Android (477.5; 332.3–526.5). The lowest ranked 25% of iOS apps were still better ranked than Android's top 25% (Fig. 2). Within app

Table 1 Prevalence distribution of the 125 selected apps into the seven different features

Feature	Apps among platforms %(N)			Apps between country blocks %(N)		Overall apps %(N)
	Android (n = 67)	Android/iOS (n = 21)	iOS (n = 37)	US (n = 72)	Rest of the world (n = 53)	
Blocking	55.2 (37)	33.3 (7)	18.9 (7)	37.5 (27)	45.3 (24)	40.8 (50)
Parental control	13.4 (9)	14.3 (3)	0.0 (0)	9.7 (7)	9.4 (5)	9.6 (12)
Tracking	80.6 (54)	90.5 (19)	81.1 (30)	80.6 (58)	84.9 (45)	82.4 (103)
Rewards	3.0 (2)	14.3 (3)	8.1 (3)	5.6 (4)	7.5 (4)	6.4 (8)
Reminders	86.6 (58)	95.2 (20)	94.6 (35)	87.5 (63)	94.3 (50)	90.4 (113)
Coaching	13.4 (9)	9.5 (2)	24.3 (9)	13.9 (10)	18.9 (10)	16.0 (20)
Social media	7.5 (5)	4.8 (1)	13.5 (5)	8.3 (6)	9.4 (5)	8.8 (11)

Note The data refer to the total number of apps and to the apps stratified among platform and country blocks; the prevalence is expressed as percentage and absolute frequency (N) in brackets

Table 2 Prevalence distribution of the 125 selected apps into the 13 different categories

Category	Apps among platforms %(N)			Apps between country blocks %(N)		Overall apps %(N)
	Android (n = 67)	Android/iOS (n = 21)	iOS (n = 37)	US (n = 72)	Rest of the World (n = 53)	
Productivity	40.3 (27)	66.7 (14)	64.9 (24)	62.5 (45)	37.7 (20)	52.0 (65)
Tools	22.4 (15)	0.0 (0)	0.0 (0)	12.5 (9)	11.3 (6)	12.0 (15)
Lifestyle	14.9 (10)	4.8 (1)	2.7 (1)	5.5 (4)	15.1 (8)	9.6 (12)
Health and fitness	7.5 (5)	4.8 (1)	13.5 (5)	8.3 (6)	9.4 (5)	8.8 (11)
Business	0.0 (0)	4.8 (1)	0.0 (0)	0.0 (0)	1.9 (1)	0.8 (1)
Communication	4.5 (3)	0.0 (0)	0.0 (0)	2.8 (2)	1.9 (1)	2.4 (3)
Education	1.5 (1)	0.0 (0)	5.4 (2)	0.0 (0)	5.7 (3)	2.4 (3)
Medical	0.0 (0)	0.0 (0)	5.4 (2)	1.4 (1)	1.9 (1)	1.6 (2)
Parenting	1.5 (1)	0.0 (0)	0.0 (0)	1.4 (1)	0.0 (0)	0.8 (1)
Personalization	4.5 (3)	0.0 (0)	0.0 (0)	1.4 (1)	3.8 (2)	2.4 (3)
Photography	0.0 (0)	0.0 (0)	2.7 (1)	1.4 (1)	0.0 (0)	0.8 (1)
Social networking	2.9 (2)	4.8 (1)	2.7 (1)	1.4 (1)	5.7 (3)	3.2 (4)
Utilities	0.0 (0)	14.3 (3)	2.7 (1)	1.4 (1)	5.7 (3)	3.2 (4)

Note The data refer to the total number of apps and to the apps stratified among platform and country blocks; the prevalence is expressed as percentage and absolute frequency (N) in brackets

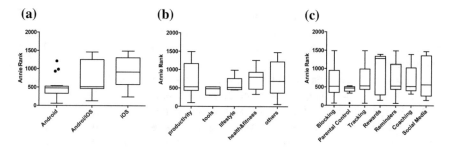

Fig. 2 Boxplot showing *App Annie* rank of apps developed in the three groups of platforms, Android, iOS and the shared one's (**a**), stratified by category (**b**) or by feature (**c**). Whiskers were drawn based on Tukey method

categories and features, ranking did not show any statistical difference (KW = 3.09, p = 0.543 and KW = 5.54, p = 0.476, respectively).

Based on the results attained in the explorative phase, we built a logistic regression model of *Annie's* rank on platforms as a significant, independent variable (β = 1.46, standard error = 0.535; p = 0.006). Note, "parental features" was included as a covariate, since it was differently distributed within the three platform groups. Our model suggested that of the apps available for iOS, the opportunity to have a medium/high rank (>500) was approximately four times greater than apps available for Android.

3.3   Review of Three Randomly Selected Smartphone Apps

We randomly selected three apps with a broad range of features. The apps *BreakFree* and *Unplug* were downloaded to an iPhone 5S, and *QualityTime* to a Samsung Galaxy S5. *Breakfree* included 3 features and presented an average rank (Annie rank = 491), *Unplug* included the greatest number of features (n = 5) and presented a high rank (Annie rank = 1394) and, finally, *QualityTime* included 2 features and presented a low rank (Annie rank = 316). Then, we explored whether the apps implemented any of the 6 behavior change methods that were noted in the introduction, that is: commitment, setting standards, self-tracking, tailoring, social-facilitation, and/or gamification.

Based on our assessments, we highlight features and functions in what follows, with the expectation that it may address outstanding issues for end users seeking to reduce their device usage.

QualityTime, BreakFree, and *Unplug* include key features consistent with at least 4 behavior change principles. Two apps—*BreakFree* and *Unplug*—contain key features consistent with gamification (i.e. "achievement titles" and "last high score"). The only app to provide key features consistent with social facilitation (i.e. "parental notification" and "family time") is *BreakFree*. This is also the only app of the group to require the use of GPS to deliver a key feature, while *Unplug* is the only app to integrate the "Airplane Mode" feature of iOS as one of its key functions.

QualityTime provides several features that require users to commit to behavior change, such as auto-blocking inbound calls and sending an auto-reply to inbound text messages. It also features self-tracking, in the form of summaries that report usage and standard settings in the form of a configurable screen and app auto-locks. Still, *QualityTime* does support tailoring, insofar as it enables the user to tune auto-locks, auto-replies, and restriction periods (with exceptions).

Unplug immediately, repeatedly, and colorfully prompts the user to set their phone to "Airplane Mode" and to put the phone down until the user complies. It is important to note that *Unplug* also blocks graphics and alerts. It is perhaps best described as a nomophobia version of the "cold turkey" method.

BreakFree embraces gamification by using a points-and-title system. Unlike *Unplug* and *QualityTime*, *BreakFree* features an animated character, "Sato" that notifies the user to "slow down" after an hour of phone usage. Moreover, users can schedule "family time" hours in advance, during which the Internet and sound are disabled on the device. This app also comes with parental control settings that enable users to monitor their children's usage. The latter feature differentiates *BreakFree* from the other apps investigated, as supporting "social facilitation."

4   Discussion

To our knowledge from this analysis, there are no apps for nomophobia that rely on evidence-based research. In fact, the most promising evidence-informed apps, such as *PTSD Coach*—an app used to mitigate acute distress in veterans who experience post-traumatic stress disorder—has no long-term data regarding its effectiveness [14]. Nevertheless, the Veterans Administration has successfully used PTSD Coach to generate "big data" related to patient satisfaction and VA mental health outreach services on the order of 130,000 downloads (as of 2013). On the clinical side, several smartphone apps have been effective in studies for managing depression and anxiety, though their examination is outside the scope of this paper [10]. Generally speaking, these apps help users commit to new behaviors; learn about and set quantifiable goals and standards; self-monitor and track trends; receive tailored messages from professional counselors; enjoy elements of games; and they bring users closer to friends and family.

Given that we were unable to directly explore associations between the use of behavior change techniques and changes in health behavior, or other health-related outcomes, we posit some key observations in app features and usage patterns between the US and other countries. Foremost, there were no significant differences in app categories as a function of their average number of features, or the most used types of apps in the US as compared to other countries. Platforms, however, are used differently across countries. For instance, apps showed, in some cases, a different distribution among platforms in terms of feature and category. Of note, the platform seems to play a crucial role in defining the popularity of an app, insofar as iOS ranks significantly better than Android.

4.1 Recommendations for Developers to Use Big Data from Apps

Big data from apps, such as aggregate data from *App Annie*, can be used to inform developers about new features they can use to improve apps, such as apps to address nomophobia. Smartphone app developers interested in addressing the challenges of nomophobia might focus on features that address the key principles of behavior change that are underrepresented in the apps featured in this review, namely: commitment, setting standards, self-tracking, tailoring, gamification, and social facilitation. "Big Data" from nomophobia apps can also inform app publishers whether their users find these apps acceptable. They can evaluate this, for example, by analyzing data on device usage provided by Google and Apple, daily information from *App Annie*, potential data from in-app advertisers, as well as data that end users may choose to volunteer on their experiences with features.

4.2 Limitations

One of the feasibility and time-challenges of this review was that individual apps in *Google Play* and *iTunes* had to be opened to determine whether they were intended to reduce smartphone usage. To minimize the harm to end users who are using devices while searching for apps that address nomophobia, developers can create logos that clarify the app's key features and/or behavior change techniques.

A further limitation is that *App Annie* could not be used for the purposes of an academic research article to unlock raw or "premium" content, such as estimated downloads and lifetime advertising dollars earned by each app. To date, there is no more timely way to conduct a systematic review of nomophobia apps without using the leading provider of such data.

Finally, it was not possible to directly determine the effectiveness of "detox apps" against a range of processes and health-related outcomes. Additional well-designed studies are needed to explore associations between the use of these behavior change techniques and smartphone addiction recovery.

5   Conclusion

Big data from apps can be used to study the usability and acceptability of apps. This information can incorporated into analyses to determine how apps are being used, and for example, in the case of nomophobia apps, whether they are based on scientific research on how to reduce nomophobia. *QualityTime, BreakFree,* and *Unplug* are nomophobia-focused apps that are marketed to smartphone users seeking to reduce their usage. Each of these apps has achieved notable success among non-nomophobia users within the "productivity" app store category, as validated by *App Annie.* Of the 3 apps, *BreakFree* includes the most features consistent with the principles of behavior change. However, this review using data found no evidence-based smartphone apps to thoroughly address nomophobia, and so recommends further research efforts devoted to this area. Future research using smartphone data can help to guide the integration of scientific research findings into nomophobia apps to help address nomophobia.

References

1. App Annie (2017) About. https://www.appannie.com/en/about/.
2. Bandura, A. (1998). Health promotion from the perspective of social cognitive theory. *Psychology and Health, 13*(4), 623–649. https://doi.org/10.1080/08870449808407422.
3. BinDhim, N. F., Freeman, B., & Trevena, L. (2015). Pro-smoking apps: Where, how and who are most at risk. *Tobacco Control, 24*(2), 159–161. https://doi.org/10.1136/tobaccocontrol-2013-051189.
4. Bragazzi, N. L., & Del Puente, G. (2014). A proposal for including nomophobia in the new DSM-V. *Psychology Research and Behavior Management, 7,* 155–160. https://doi.org/10.2147/prbm.s41386.
5. Bricker, J. B., Mull, K., Kientz, J. A., Vilardaga, R. M., Mercer, L. D., Akioka, K., et al. (2014). Randomized, controlled pilot trial of a smartphone app for smoking cessation using acceptance and commitment therapy. *Drug and Alcohol Dependence, 143,* 87–94. https://doi.org/10.1016/j.drugalcdep.2014.07.006.
6. Clancy H (2016) Mobile insights firm App Annie adds new investor, director. Forbes. http://fortune.com/2016/01/14/mobile-insights-app-annie-director-financing/.
7. Cohen, S. (2004). Social relationships and health. *American Psychologist, 59*(8), 676–684. https://doi.org/10.1037/0003-066x.59.8.676.
8. Cugelman, B. (2013). Gamification: What it is and why it matters to digital health behavior change developers. *JMIR Serious Games, 1*(1), e3. https://doi.org/10.2196/games.3139.
9. Dennison, L., Morrison, L., Conway, G., & Yardley, L. (2013). Opportunities and challenges for smartphone applications in supporting health behavior change: Qualitative study. *Journal of Medical Internet Research, 15*(4), e86. https://doi.org/10.2196/jmir.2583.
10. Donker, T., Petrie, K., Proudfoot, J., Clarke, J., Birch, M. R., & Christensen, H. (2013). Smartphones for smarter delivery of mental health programs: A systematic review. *Journal of Medical Internet Research, 15*(11), e247. https://doi.org/10.2196/jmir.2791.
11. Fjeldsoe, B. S., Marshall, A. L., & Miller, Y. D. (2009). Behavior change interventions delivered by mobile telephone short-message service. *American Journal of Preventive Medicine, 36*(2), 165–173. https://doi.org/10.1016/j.amepre.2008.09.040.

12. King, A. L. S., Valença, A. M., Silva, A. C. O., Baczynski, T., Carvalho, M. R., & Nardi, A. E. (2013). Nomophobia: Dependency on virtual environments or social phobia? *Computers in Human Behavior, 29*(1), 140–144. https://doi.org/10.1016/j.chb.2012.07.025.
13. Kleinbaum, D.G., & Klein, M. (2010) Logistic Regression, Statistics for Biology and Health (3rd ed.). Springer.
14. Kuhn, E., Greene, C., Hoffman, J., Nguyen, T., Wald, L., Schmidt, J., et al. (2014). Preliminary evaluation of PTSD Coach, a smartphone app for post-traumatic stress symptoms. *Military Medicine, 179*(1), 12–18. https://doi.org/10.7205/milmed-d-13-00271.
15. Lanaj, K., Johnson, R. E., & Barnes, C. M. (2014). Beginning the workday yet already depleted? Consequences of late-night smartphone use and sleep. *Organizational Behavior and Human Decision Processes, 124,* 11–23. https://doi.org/10.1016/j.obhdp.2014.01.001.
16. Lewis, Z. H., Swartz, M. C., & Lyons, E. J. (2016). What's the point? A review of reward systems implemented in gamification interventions. *Games for Health Journal, 5*(2), 93–99. https://doi.org/10.1089/g4h.2015.0078.
17. Lin, Y. H., Chang, L. R., Lee, Y. H., Tseng, H. W., Kuo, T. B., & Chen, S. H. (2014). Development and validation of the smartphone addiction inventory (SPAI). *PLoS One, 9*(6), e98312. https://doi.org/10.1371/journal.pone.0098312.
18. National Highway Transportation and Safety Administration (2015). Distracted driving. https://www.nhtsa.gov/risky-driving/distracted-driving.
19. Roberts, D. J. A. (2015). *Too much of a good thing: Are you addicted to your smartphone?* Austin: Sentia Publishing Company.
20. Rothberg, M. B., Arora, A., Hermann, J., Kleppel, R., St Marie, P., & Visintainer, P. (2010). Phantom vibration syndrome among medical staff: A cross sectional survey. *BMJ (Clinical Research Ed.), 314,* c6914. https://doi.org/10.1136/bmj.c6914.
21. Tate, D. F., Wing, R. R., & Winett, R. A. (2001). Using Internet technology to deliver a behavioral weight loss program. *JAMA, 285*(9), 1172–1177. https://doi.org/10.1001/jama.285.9.1172.
22. Toscos, T., Faber, A., An, S., & Gandhi, M. P. (2006). Chick clique: Persuasive technology to motivate teenage girls to exercise. In: *CHI '06 Extended Abstracts on Human Factors in Computing Systems* (pp. 1873–1878).
23. Umberson, D., Crosnoe, R., & Reczek, C. (2010). Social relationships and health behavior across life course. *Annual Review of Sociology, 36,* 139–157. https://doi.org/10.1146/annurev-soc-070308-120011.
24. Webb, T.L., Joseph, J., Yardley, L. & Michie, S. (2010). Using the Internet to promote health behavior change: A systematic review and meta-analysis of the impact of theoretical basis, use of behavior change techniques, and mode of delivery on efficacy. *Journal of Medical Internet Research* (1), e4. https://doi.org/10.2196/jmir.1376.
25. Winestock, C., & Jeong, Y. K. (2014). An analysis of the smartphone dictionary app market. *Lexicography, 1*(1), 109–119.
26. Young, S. D., Cumberland, W. G., Lee, S. J., Jaganath, D., Szekeres, G., & Coates, T. (2013). social networking technologies as an emerging tool for HIV prevention: A cluster randomized trial. *Annals of Internal Medicine, 159*(5), 318–324. https://doi.org/10.7326/0003-4819-159-5-201309030-00005.
27. Young, S. D., Cumberland, W. G., Nianogo, R., Menacho, L. A., Galea, J. T., & Coates, T. (2015). The HOPE social media intervention for global HIV prevention in Peru: A cluster randomised controlled trial. *Lancet HIV, 2*(1), e27–e32. https://doi.org/10.1016/s2352-3018(14)00006-x.

A Fast DBSCAN Algorithm with Spark Implementation

Dianwei Han, Ankit Agrawal, Wei-keng Liao and Alok Choudhary

Abstract DBSCAN is a well-known clustering algorithm which is based on density and is able to identify arbitrary shaped clusters and eliminate noise data. Parallelization of DBSCAN is a challenging work because there is an inherent sequential data access order and based on MPI or OpenMP environments, there exist the issues of lack of fault-tolerance and there is no guarantee that workload is balanced. Moreover, programming with MPI requires data scientists to handle communication between nodes which is a big challenge. We present a new parallel DBSCAN algorithm using Spark. kd-tree technique is applied in our algorithm to reduce search time. More specifically, a novel merge approach is used so that no communication between executors is required while partial clusters are generated. Appropriate and efficient data structures are carefully used in our study: Using Queue to contain neighbors of the data point, and using Hashtable when checking the status of and processing the data points. Also other advanced data structures from Spark are applied to make our implementation more effective. We implement the algorithm in Java and evaluate its scalability by using different number of

D. Han (✉) · A. Agrawal (✉) · W. Liao (✉) · A. Choudhary (✉)
EECS Department, Northwestern University, Evanston, IL 60208, USA
e-mail: dianweih@eecs.northwestern.edu

A. Agrawal
e-mail: ankitag@eecs.northwestern.edu

W. Liao
e-mail: wkliao@eecs.northwestern.edu

A. Choudhary
e-mail: choudhar@eecs.northwestern.edu

© Springer Nature Singapore Pte Ltd. 2018
S. S. Roy et al. (eds.), *Big Data in Engineering Applications*,
Studies in Big Data 44, https://doi.org/10.1007/978-981-10-8476-8_9

173

processing cores. Our experiments demonstrate that the algorithm we propose scales up very well. Using data sets containing up to 1 million high-dimensional points, we show that our proposed algorithm achieves speedups up to 6 using 8 cores (10 k), 10 using 32 cores (100 k), and 137 using 512 cores (1 m). Another experiment using 10 k data points is conducted and the result shows that the algorithm with MapReduce achieves speedups to 1.3 using 2 cores, 2.0 using 4 cores, and 3.2 using 8 cores.

Keywords DBSCAN · Scalable data mining · Big data · Spark framework

1 Introduction

Clustering is a data mining approach that divides data into different categories that are meaningful, useful, or both [20]. Cluster analysis has been successfully applied to many fields: bioinformatics, machine learning, information retrieval, and statistics [20]. Well-known algorithms include K-means [13], BIRCH [24], WaveCluster [19], and DBSCAN [6]. Current clustering algorithms haven been categorized into four types: partitioning based, hierarchy-based, grid-based, and density-based [6]. Density Based Spatial Clustering of Applications with Noise (DBSCAN) is a density based clustering algorithm [6].

Parallel DBSCAN has been implemented with MPI and OpenMP [4, 7, 15, 25]. Generally, an MPI implementation can obtain better performance but it requires the programmers to take care of implementation in detail, such as how to partition the data, how to deal with communication, synchronization, file location, and workload balancing. Besides parallelization with MPI, MapReduce-based approach is presented as well [7, 9, 14].

We propose a new distributed parallel algorithm with Spark that implements DBSCAN. A master-slave based approach is as follows. The algorithm first reads data from the Hadoop Distributed File System (HDFS) and forms Resilient Distributed Datasets (RDDs), transforming them into data points. Certainly, this process is done in Spark driver. It then sends the RDDs into multiple executors. Within each executor, partial clusters are generated and sent to driver at the end of *foreach* statement. Each executor just performs its computation without communicating with others. This way we avoid shuffle operations that are very expensive. So we place some additional points (SEEDs: the new term we introduce in our paper) in each partial cluster. After all the partial clusters are collected through shared variable accumulator, the algorithm identifies the clusters that are supposed to be merged by SEEDs. Merging is done in driver code too. In our new design and implementation, we use the power of shared variables of Spark framework:

broadcast and accumulator. Also, in order to shorten the search time for points' neighbors, we implement Java-based kd-tree [3] to reduce complexity from $O(n^2)$ to $O(nlogn)$. The experiments performed on a distributed-memory machine show that the proposed algorithm can obtain scalable performance.

The organization of the paper is as follows: In Sect. 2, we briefly give an overview of two frameworks based on big data: Map Reduce and Spark, and the basic idea of DBSCAN algorithm. Our proposed DBSCAN algorithm is introduced in Sect. 3. In Sect. 4, we present the parallel implementation with Spark. The experiments and the results are presented in Sect. 5, followed by some concluding remarks in Sect. 6.

2 Background

In this section, we first briefly review the basic idea of DBSCAN algorithm. And then we introduce two distributed computation frameworks that are very powerful and widely used in big data applications.

2.1 DBSCAN Algorithm

DBSCAN is a clustering algorithm proposed by Ester et al. [6]. And it has become one of the most common clustering algorithms because it is capable of discovering arbitrary shaped clusters and eliminating noise data [6]. The basic idea of this algorithm is finding all the *core points* and forming the clusters by clustering core points with all points (core or non-core) that are *reachable* from them. Essentially, DBSCAN algorithm is based on three basic definitions: core points, directly density-reachable, and density-reachable [25]. Given a data set D, of points.

eps-neighborhood of a point p is the neighborhood of $p \in D$ within a radius *eps*.

Definition 1 A point p is a *core point* if it has neighbors within a given radius (*eps*), and the number of neighbors is at least *minpts* (which is a threshold). In this case, the number of neighbors is called *density*.

Definition 2 A point y is *directly density-reachable* from x if y is within *eps-neighborhood* of x and x is a *core point*.

Definition 3 A point y is *density-reachable* from x if there is a chain of points p1, p2,..., pn, with p1 = x, pn = y and pi + 1 is directly density-reachable from pi for all $1 <= i < n$, pi $\in$ D.

Algorithm 1 The DBSCAN algorithm

Input (*eps, minpts, D*)
Output (a set of clusters)
1.   initialize all points as unvisited
2.   **for** each unvisted point p ∈ D **do**
3.   mark p as visited
4.   Let *N* be *eps*-neighborhood of p
5.   **if** the size of *N* < *minpts* points then
6.   mark p as noise
7.   **else**
8.   create a new cluster C, and add p to C
9.   **for** each point p′ ∈ N
10.   **if** p′ is unvisited then
11.   mark p′ as visited
12.   let N′ be the *eps*-neighborhood of p′
13.   **if** the size of N′ is >= *minpts* then
14.   add those points to N
15.   **endif**
16.   **endif**
17.   **if** p′ is not yet a member of any cluster
18.   **add p to C**
19.   **endif**
20.   **endfor**
21.   **endif**
22. **endfor**

The pseudocode of the DBSCAN algorithm is given in Algorithm 1 [8]. The algorithm starts with an arbitrary point p ∈ D and checks its *eps*-neighborhood (Line 4). If the *eps*-neighborhood size is bigger than pre-defined number minpts, the code generates a new cluster C. The algorithm then retrieves all density reachable points from p in D, and add them to the cluster C (Line 8–20). Otherwise, if the *eps*-neighborhood contains less than minpts points, then p is marked as noise (Line 6). The computational complexity of Algorithm is $O(n^2)$ where n is the number of data points. If we use spatial indexing, the complexity reduces to $O(nlogn)$ [3].

2.2   Two Powerful Frameworks Based on Big Data: MapReduce and Spark

In Hadoop version 1, MapReduce is the only data processing framework that is available for distributed computation. But in Hadoop version 2, based on Yarn (resource manager), MapReduce, Spark, and other data processing frameworks are

Fig. 1 An overview of data flow in MapReduce

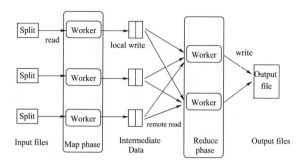

available. MapReduce and Spark may share the same HDFS, but it should be pointed out that Spark jobs can be run with or without Yarn (Standalone mode).

(1) *Map Reduce:* In big data domain, MapReduce is a simple but powerful framework which makes programmer easily implement parallel processing. It is based on Hadoop Distributed File System (HDFS), which allows programmers to focus mainly on the problem itself instead of the low level implementation details. Figure 1 tells us about how this programming model works. MAP workers read data from HDFS and process the data based on the business logic and then write intermediate data to local disk for sorting and shuffling process. It is also in the form of key-value pair. After a reduce worker is notified by master, it uses remote procedure call to read data from local disk of MAP workers, and then sorts data so that all occurrences of the same key are grouped together. The output of reduce function will be appended to final output files (generally HDFS).

Compared with the other distributed computation frame-work, MapReduce has the following advantages:

- Extremely Scalable. It does not require the support from centralized RAID-based SAN or NAS storage systems. Every node has its own local hard-drives. The nodes are loosely coupled and connected with standard network devices. So adding and removing nodes to a cluster becomes very easy and convenient, and has no impact to running MapReduce jobs [17].
- Highly Parallel and Abstracted. Based on the frame-work's principle, programmers do not have to take care of low level implementation details such as message transferring between master and workers, file location, and workload balancing. They only need focus on the problem itself. One of the major contributions of MapReduce is that it supports parallelization automatically. The programmers only need to implement map() method of Mapper class and reduce () method of Reducer class and the framework will do the rest. However, for

complicated job, the programmers still need to figure out the number of Mappers and how to split the input data.

- Highly Reliable and Fault-tolerant. From the data source perspective, HDFS uses the replication strategy to handle data source reliability. A single process failure in MPI will cause the whole job to fail. In MapReduce framework, another task will be automatically launched if one task fails and the job will continue running. This feature is especially useful and important for long-running jobs.

(2) *Spark:* At a high level, a running Spark application has one driver process talking to many executor processes, sending them work to do and collecting the results of that work. The first thing a Spark program must do is to create a SparkContext object in driver code, which tells Spark how to access a cluster. Then it reads one file or multiple files in HDFS and processes them as Distributed Datasets (RDD), which is a collection of elements partitioned across the nodes and can be operated on in parallel. RDD is the main abstraction Spark provides, and RDDs can be created from a file in the Hadoop file system or by transforming other RDDs. We want to point out that Spark can use not only its own APIs to read data but also Hadoop API to read data (newAPIHadoopFile method in this case). TaskScheduler launches tasks to executors via Resource manager, which in this case, is YARN. After executors complete their tasks, they will send the results back to the driver (see Fig. 2) (if it is the final RDD of an action such as count()) [12], or write output to external storage. Spark framework captures all the important features that MapReduce have. In addition, it has the following new features.

- In-memory computations. In Spark, Resilient Distributed Datasets (RDDs) are the first abstraction that allows programmers to perform in-memory computations on large clusters. RDDs are motivated by two types of applications that MapReduce handle inefficiently: iterative algorithms and interactive data mining [22]. Figure 1 depicts that MapReduce frameworks does not fit iterative algorithms. In order to use MapReduce model to tackle iterative algorithms, many rounds of map-reduce executions will be performed which is not very efficient because map's intermediate results should be written to local disks and then they are remotely read to reduce workers, and disk I/O operations are very expensive in this case. In Spark, the benefit of keeping everything in memory is the ability to perform iterative computations at blazing fast speeds.

- Supporting Streaming data, complex analytics, and real time analysis. MapReduce offers a very simple but powerful programming model that are efficient for data-intensive algorithms [11]. But we can not use MapReduce to perform real time analysis and implementing complex graph based algorithms in an efficient manner.
- Fast fault recovery. In MapReduce old version, if the JobTracker does not receive any heartbeat from a TaskTracker for a specified period of time, the JobTracker understands that the worker associated to that TaskTracker has failed. When this situation happens, the JobTracker needs to reschedule all pending and in-progress tasks to another TaskTracker, because the inter-mediate data belonging to the failed TaskTracker may not be available anymore [21]. After hadoop-0.21, checkpointing was added where Job-Tracker records its progress in a file. When a JobTracker starts, it can restart work from where it left off. MapReduce uses replication strategy to handle fault recovery. On the other hand, Spark reconstructs RDDs via lineage to handle this issue. Compared to the replication method, which consumes more memory, reconstruction of RDDs takes shorter time [23].

Even though spark is very efficient, offers parallelization automatically, we still need to put much effort to avoid shuffle operation. So in our implementation of DBSCAN we avoid all-to-all communication.

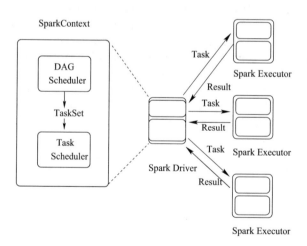

Fig. 2 An overview of data flow in spark

3   Novel DBSCAN with Spark Implementation

To our best knowledge, there are many DBSCAN implementations with Hadoop's MapReduce [9, 14, 17]. But very few people implement DBSCAN with Spark because the programmers need to design a new algorithm to avoid shuffle operations to make their parallelization more efficient. For example, after one data point's state is updated in one executor we need to spread this updating across the cluster. So this will introduce shuffle operations which are very expensive in Spark. Let us take a look at the pseudocode of our new DBSCAN's algorithm.

3.1   DBSCAN Algorithm with Spark

The pseudocode of the DBSCAN algorithm with Spark implementation is given in Algorithm 2. The algorithm starts with the code in Spark driver, which reads data, generates RDDs and transforms them into appropriate RDDs (Line 1, Line 2, and Line 3). The code in Spark executor is in Lines 4 through 32. After comparing with Algorithm 1, we can see that two places are new: Line 15 and Lines 29 through 31. We assume each executor only deals with the points that belong to it. Otherwise, there would be a lot of overlap of computation between different executors. Placing SEEDs is in Line 15. The detailed description regarding it will be given in next Section. The partial clusters are sent back to driver right before the executor finishes its task by accumulator, which also will be explained in detail in the next Section. This implementation is meant for ensuring that merging process will not be started until all the executors finish their tasks. Lines 33 through 34 perform merging partial clusters and produce the final global clusters (see Algorithm 4). So the code [1–3] is run in Driver mode, code [4–32] is run in Executor mode, and code [33–34] is run in Driver mode.

Algorithm 2 DBSCAN algorithm with Spark

Input(*eps, minpts, D*)

Output (a set of clusters)

1.   read an input file from HDFS and generate RDDs from the read data
2.   transform the existing RDDs into appropriate RDDs with Point type
3.   distribute those RDDs into executors
4.   **foreach** (closure'start)
5.   **if** point p is not in hashtable **then**
6.   get the neighbors of p using eps and kdtree
7.   push all appropriate neighbors into Queue N
8.   **if** the size of N < minpts points **then**
9.   mark p as noise
10.   **else**
11.   create a new cluster C, and add p to C
12.   **while** N is not empty
13.   let p' be the removed point from N
14.   put the index of p' into the hashtable
15.   place SEEDs processing
16.   **if** p' is unvisited **then**
17.   mark p' as visited
18.   let N' be the *eps*-neighborhood of p'
19.   **if** the size of N' is >= *minpts* **then**
20.   add appropriate points to N
21.   **endif**
22.   **endif**
23.   **if** p' is not yet a member of any cluster
24.   add p' to C
25.   **endif**
26.   **endwhile**
27.   **endif**
28.   **endif**
29.   **if** current point is the last one in closure **then**
30.   send partial clusters to driver through accumulator
31.   **endif**
32. **endforeach**
33. analyze partial clusters based on the placed SEEDs
34. search for all partial clusters and merge them if necessary

---

3.2 Two Important Data Structures Affecting Performance

Using Java as the programming language in our implementation, we need to consider using the appropriate data structures for efficiency. Here, two data structures Hashtable and Queue are discussed.

If we take a look at Line 14, this operation should be *put(key, value)*, which is usually $O(1 + n/K)$ where K is the hash table size. If K is large enough, the result is effectively $O(1)$. Method *containsKey(key)* is performed in Line 5, Line 7, and Line 20. Again, under normal circumstances, it is $O(1)$. The add operations on Queue are performed in Line 7 and Line 20, and remove operation on Queue is performed in Line 13. The number of add operations should be the same as the number of remove operations according to the condition in Line 12 (while loop will not terminate until it is empty). Among LinkedList, ArrayList, and Vector, the best performance on both add and remove operations is obtained using LinkedList. In our code, we thus use LinkedList to implement Queue.

4   Novel Techniques in Parallel DBSCAN with Spark

In this Section, we will present the implementation details of our parallel DBSCAN algorithm with Spark. The pseudocode of algorithms is given in the first part. Then we analyze the time complexity of the whole algorithm.

Algorithm 3 Placing SEEDs in Executors

```
1.    identify the current partition of this executor as par_A
2.    initialize the place_flg for all the partitions
3.    while N is not empty
4.        let p' be the removed point from N
5.        put the index of p' into the hashtable
6.        for j = 0.. all the possible partitions
7.            if p' is in par_A
8.                let continue_flg be true
9.                break
10.           else
11.               if place one seed already
12.                   let continue flg be false
13.                   break
14.               else
15.                   place_flg=1
16.                   let continue flg be true
17.                   break
18.               endif
19.           endif
20.       endfor
21.       if continue flg = false
22.           continue
23.       endif
24.       if place_flg is 1
25.           place a seed
26.       endif
27.   endwhile
```

4.1 New Clustering Algorithm Without Communication Between Executors

We need to update data points' state by map function if we apply the traditional method, and then propagate this update to other executors. However, that implementation will introduce a shuffle operation in order to make this update visible by other executors. Here, we propose a novel clustering algorithm to get around the shuffle operation. After data points have been partitioned to each executor, we just let each executor compute the partial clusters locally for data points that are assigned to this executor. The merging process is deferred until all the partial clusters have been sent back to the driver. This new design, however, introduces new challenges: how to create the partial clusters in executors so that they can be merged in the driver? And how to identify those partial clusters which are supposed to be merged into one cluster? The pseudocode of algorithms and an example are given as follows.

Algorithm 3 gives the basic idea of our design. In order to avoid overlap of computation of partial clusters, we would let individual executors only deal with the points that belong to this partition so that the executors do not have to communicate to spread points' updated states across the clusters. However, we could not merge the partial clusters into the global clusters after all the partial clusters are collected in driver because there are no global states of these partial clusters. Therefore, we come up with a new idea, using SEEDs, which are points that do not belong to the current partition. And these SEEDs serve as something like markers so that we can easily identify the outer master partial clusters by using them and merge them into a bigger cluster. The SEEDs are not related to the locations. If the current point's index is beyond the range of current partition it is taken as a SEED. So the main goal on executor side is to place SEEDs, and on driver side, we find out SEEDs and identify master partial clusters and merge them.

Before moving on to the algorithm of digging out SEEDs from partial clusters in Spark driver, we would like to use an example to display how to identify SEEDs and search for master partial clusters. Figure 3a shows that there are 2 partial clusters from 2 partitions. SEEDs are those points whose indexes are beyond the partition's range. For example, for C[0], its range is from 0 to 2499. So the point whose indexe is greater than 2499 is 3000. Then the algorithm will identify the master partial clusters. Obviously, for 3000, the master partial cluster is C[5] because it contains 3000 and 3000 is a regular element in this cluster. When we merge two partial clusters we need to remove duplicate elements. Figure 3b show the resulting cluster C[0].

Fig. 3 An example showing
the proposed merging cluster
algorithm at different stages.
a There are two partitions and
two partial clusters. Integers
in squares are SEEDs. **b** After
C[0] merges C[5], C[0] status
is updated as "finished" from
"unfinished"

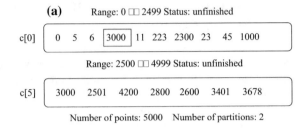

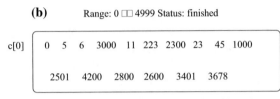

In Spark driver, Algorithm 4 shows how to use SEEDs to merge partial clusters into global clusters. First of all, it identifies the SEEDs by comparing elements with its range. In general, the number of SEEDs should be equal to or greater than the number of partitions. So we obtain an array of seeds (see Line 3). Lines from 4 through 8 form a for loop, which finds the master cluster that contains the seed as a regular element, then merges the two clusters, and finally, updates the status of master cluster. When the for loop terminates the status of current cluster is updated from 'unfinished' to 'finished'.

Algorithm 4 Using SEEDs and merge partial clusters in Driver

```
1.   for i = 0.. all partial clusters
2.       if the status of current partial cluster is 'unfinished'
3.           seed = identify seeds from current partial cluster
4.           for j = 0..seed.size()
5.               rtn_index = find master partial cluster index
6.               merge current with master cluster
7.               update the status of master cluster to 'finished'
8.           endfor
9.           update the status of current cluster to 'finished'
10.      endif
11.  endfor
```

4.2 Time Complexity Analysis

We define some related notations as follows:

n the number of data points;
p the number of partitions;

m	the number of partial clusters;
K	the maximum size of partial clusters;
$t_{straggling}$	the average wait time for framework to allow all stragglers to finish.
T_s	the average time complexity of the sequential algorithm;
T_p	the average time complexity of the parallel algorithm;
$Save$	the average speed-up.

Basically, there are three parts in our algorithm.

In the first part, the driver reads data points from HDFS and transforms the data points into appropriate form that can be processed in executors and constructs the *kd-tree*. The time for this phase includes reading file, transforming RDDs, and building *kd-tree*. We assume we use Δ for the first two items. For *kd-tree* construction, we use $O(nlogn)$ [10]. So summing them up, we use $\Delta + O(n * logn)$.

In the second part, the local partial clusters are generated in executors. Basically, in the best case, searching a point from a balanced *kd-tree* takes $O(logn)$ time. In the worst case, the time could be n. Some researchers have reported that (near neighbor) range search's upper bound is $O(n1 - 1/d + k)$ [10]. So we use V to represent the search time, which is between $logn$ and $n1 - 1/d + k$; If we use parallel processing, we need to add the time for SEEDs placement part. Let us assume an additional $O(m * V)$ time is added. So in parallel processing, we would spend $O((n/p * V) + (m * V)) + t_{straggling}$ time in our case.

In the last part, after all the partial clusters have been sent back from executors to the driver, the driver merges them and produces the global clusters. Based on our Algorithm 4, the search operations takes $O(n)$ time at most if we check each element in the partial clusters. For merging phase, it takes Km times which is less than n. So we use $O(n + Km)$ time.

To sum up: $T_s = O(\Delta + n*logn + n*V + n + Km)$.

$$T_p = O(\Delta + n*logn + (n/p)*V + m*V + t_{straggling} + n + Km)$$

$$Save = T_s/T_p.$$

5 Experiments and Analysis

A series of experimental tests are conducted to evaluate the effectiveness and efficiency of our DBSCAN algorithm with Spark and MapReduce's implementations. We need to note that all parallel executions generate the same result as the serial execution. The dimension of data is relevant to the computational cost of querying the kd-tree. We do not perform tests based on varying number of attributes because we focus on Spark implementation instead of kd-tree implementation in our work. The tests are done on different sizes of data points with multiple dimensions. Our experimental results have been reported in terms of the CPU times.

After comparing with the results from Patwary et al. [15], we find that our results match them so we do not list the accuracy in our paper.

5.1 Experimental Setup

To perform the experiment for our DBSCAN's parallel implementation with Spark, we use Edison (operated by Lawrence Berkeley National Laboratory and the Department of Energy Office of Science), a Cray XC30 distributed memory parallel computer. It has 5576 compute nodes, 133,824 cores in total. Each node has two 12-core Intel "Ivy Bridge" processors at 2.4 GHz and 64 GB DDR3 1866 MHz memory. Each core has its own L1 and L2 caches, with 64 KB (32 KB instruction cache, 32 KB data) and 256 KB, respectively; A 30 MB L3 cache shared between 12 cores on the "Ivy Bridge" processor [5]. The algorithms have been implemented in Java (1.7) using the Spark (1.5) and Hadoop (2.4).

Our testbed consists of 5 datasets, which are divided two groups: (c10 k, c100 k), and (r10 k, r100 k, r1 m). Both groups of datasets (*synthetic-cluster*) have been generated synthetically using the IBM synthetic data generator [1, 16]. Table 1 lists the properties of our test data.

5.2 Comparison of the Time Taken by MapReduce and Spark

As we are not able to get source code from the other research teams [7, 9, 14], we have implemented our own DBSCAN with MapReduce approach. From Fig. 4, it is seen that 9–16 times faster performance is obtained from Spark than MapReduce. Due to the length of time taken by MapReduce, we have not conducted further tests on medium scale and large scale data sets.

Table 1 Properties of test data

Name	Points	d	eps	Minpts
c10 k	10,000	10	25	5
c100 k	102,400	10	25	5
r10 k	10,000	10	25	5
r100 k	102,400	10	25	5
r1 m	1,024,000	10	25	5

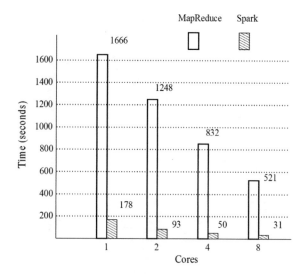

Fig. 4 Time used by MapReduce and spark. Number of points: 10,000, dimension: 10, eps, 25.0, minPnts: 5

5.3 Comparison of the Time Spent in Driver and in Executors

In this part, we discuss the time taken in our program. Figure 5a–d shows the time taken between executors and driver according to our experiments. Based on the Algorithm 2, we expect to see more time will be spent in driver with the number of partial clusters increasing. Let us take a look at Fig. 5a first. When we use more cores (1–8) to run our program, we see the number of partial clusters becomes bigger (10–392), but the time spent in driver does not change very much. That is because the data set is too small. Take a look at Fig. 5c, d, their patterns are exactly the same. When using more cores (4–32), more partial clusters are produced (from 720 to 9279), and the time spent in driver gradually becomes more. This is consistent with our analysis on the time complexity that we conduct in Sect. 4, where when the number of partial clusters m increases, the time $n + Km$ becomes large as well. Figure 5b follows the complexity analysis as well.

5.4 Scalability of Parallel DBSCAN with Spark

Before we discuss the scalability of our algorithm, we need to mention that for large data sets ($>=1$ million data points), we use *kd-tree* with pruning branches to shorten search time.

The speedup obtained by our DBSCAN algorithm with Spark is given in Fig. 6. The left column in Fig. 6 shows the speedup considering only the computation in executors while the right column shows the results considering the computation in

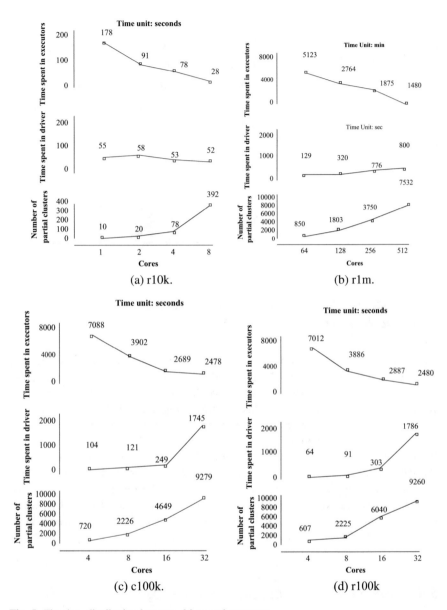

Fig. 5 The time distribution between driver and executors

executors and driver. It is obvious that the local computation in executors scales better than the whole computation since their computations are independent. For 10 k data sets, we obtain speedup up to 1.9, 3.6, and 6.2 respectively using 2, 4, and 8 cores. For 100 k data sets, speedup up to 3.3, 6.0, 8.8, and 10.2 respectively using

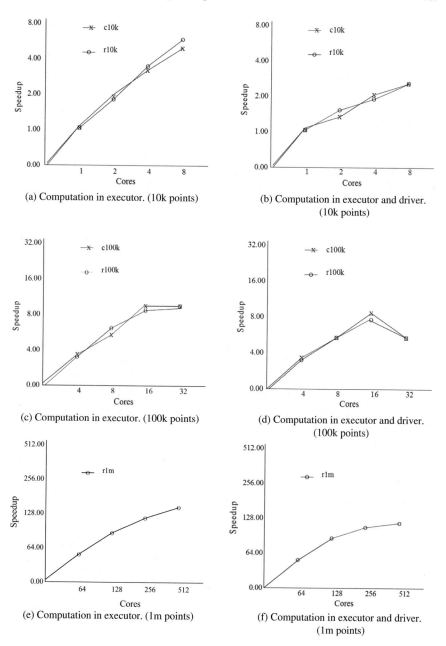

(a) Computation in executor. (10k points)

(b) Computation in executor and driver. (10k points)

(c) Computation in executor. (100k points)

(d) Computation in executor and driver. (100k points)

(e) Computation in executor. (1m points)

(f) Computation in executor and driver. (1m points)

Fig. 6 Speedup of DBSCAN algorithm with spark. Left side: time spent in executor. Right side: time spent in driver and executor

4, 8, 16, and 32 cores. For 1 m data set, speedup up to 58, 83, 110, and 137 respectively using 64, 128, 256, and 512 cores.

Take a look at right column, Fig. 6b, d, f show the speedup when total time is considered. The curves seem more flat compared with the ones in left column. For 10 k data sets, because the total time is less, the merging time is not significant. For 100 k data sets, more partial clusters are collected in driver. When using 4, 8, and 16 cores, the local computation time still dominates the total time, so speedup does not change very much. When using 32 cores, 9279 partial clusters are generated in executors and collected in driver. So the speedup drops to 5.6 from 10.2.

For r1 m, we use pruning branches technique, and thus the neighbor size of each point is decreased. Also we filter out those partial clusters whose size is too small, and their removal does not impact the accuracy significantly. Therefore, the speedup of total time does not change a lot compared with local computation.

6 Conclusions

DBSCAN algorithm has been very powerful and popular because it is able to identify arbitrary shaped clusters as well as handle noisy data. However, parallelization of DBSCAN based on MPI and OpenMP suffers from lack of fault-tolerance. Moreover, in order to implement parallelization with MPI or OpenMP, data scientists need to take care of implementation in detail, such as handling communication, dealing with synchronization, and so forth, which can pose a challenge for many users. In this paper, we proposed a new Parallel DBSCAN algorithm with Spark, which avoids the communication between executors and thus leads to a better scalable performance. The results of these experiments demonstrate that our new DBSCAN algorithm with Spark is scalable and outperforms the implementation based on MapReduce by a factor of more than 10 in terms of efficiency. In the future, we would try to apply partitioning strategy with Spark implementation and try to use larger datasets in our study.

Acknowledgements This work is supported in part by the following grants: NSF awards CCF-1409601, IIS-1343639, and CCF-1029166; DOE awards DESC0007456 and DE-SC0014330; AFOSR award FA9550-12-1-0458; NIST award 70NANB14H012. This research used Edison Cray XC30 computer of the National Energy Research Scientific Computing Center, which is supported by the Office of Science of the U.S. Department of Energy under Contract No. DE-AC02-05CH11231.

References

1. Agrawal, R., & Srikant, R. (1994). Quest synthetic data generator, *IBM Almaden Research Center*.
2. Beckmann, N., et al. (1990). The r*-tree: An efficient and robust access method for points and rectangles. In: *Proceedings of the 1990 ACM SIGMOD International Conference on Management of Data* (Vol. 19, no. 2, pp. 323–331).
3. Bentley, J. (1975). Multidimensional binary search trees used for associative searching. *Communications of the ACM, 18*(9), 509–517.
4. Brecheisen, S., et al. (2006). Parallel density-based clustering of complex objects. *Advances in Knowledge Discovery and Data Mining,* pp. 179–188.
5. DOE Office of Science (2015, September 17). Edison Configuration (Online). https://www.nersc.gov/users/computational-systems/edison/configuration/.
6. Ester, M., et al. (1996). A density-based algorithm for discovering clusters in large spatial databases with noise. In: *Proceedings of the 2nd International Conference on Knowledge Discovery and Data Mining* (Vol. 1996, pp. 226–231). AAAI Press.
7. Fu, Y., et al. (2011). Research on parallel DBSCAN algorithm design based on mapreduce. *Advanced Materials Research 301*, 1133–1138.
8. Han, J., et al. (2011). *Data mining: Concepts and Techniques*. Morgan Kaufmann.
9. He, Y., et al. (2014). MR-DBSCAN: A scalable mapreduce-based DBSCAN algorithm for heavily skewed data. *Frontiers of Computer Science, 8*(1), 83–99.
10. Kakde, H. M. (2005, August 25). *Range Searching using Kd Tree* (Online). http://www.cs.utah.edu/lifeifei/cs6931/kdtree.pdf.
11. Kang, S. J., et al. (2015). Performance comparison of OpenMP, MPI, and MapReduce in practical problems. *Advances in Multimedia 2015*.
12. Karau, H., et al. (2015). *Learning Spark: Lightning-fast Data Analysis*. O'Reilly Media.
13. MacQueen, J., et al. (1967). Some methods for classification and analysis of multivariate observations. In *Proceedings of 5th Berkeley Symposium on Mathematical Statistics and Probability* (Vol. 1, pp. 281–297). USA.
14. Noticewala, M., & Vaghela, D. (2014). MR-IDBSCAN: Efficient parallel incremental DBSCAN algorithm using mapreduce. *International Journal of Computer Applications 93*(4), 13–17.
15. Patwary, M. M. A., et al. (2012). A new scalable parallel DBSCAN algorithm using the disjoint-set data structure. In *Proceedings of the International Conference on High Performance Computing, Networking, Storage and Analysis, SC 2012*, pp. 62:1–62:11. IEEE Computer Society Press.
16. Pisharath, J., et al. (2010). *NU-MineBench 3.0*. Technical Report CUCIS-2005-08-01, Northwestern University (Technical Report).
17. Sakr, S., & Gaber, M. M. (2014). *Large Scale and Big Data: Processing and Management*. CRC Press.
18. Spark, A. (2015). *Spark Programming Guide* (Online). http://spark.apache.org/docs/latest/programming-guide.html.
19. Sheikholeslami, G., et al. (2000). WaveCluster: A wavelet based clustering approach for spatial data in very large databases. *The VLDB Journal, 8*(3), 289–304.
20. Tan, P., et al. (2005). *Introduction to Data Mining*. Pearson.
21. White, T. (2011). *Hadoop: The Definitive Guide*. O'Reilly Media.
22. Zaharia, M., et al. (2012) Resilient distributed datasets: A fault-tolerant abstraction for in-memory cluster computing. In *Proceedings of the 9th USENIX conference on Networked Systems Design and Implementation* (pp. 2–2). USENIX Association.

23. Zaharia, M. (2014). *An Architecture for Fast and General Data Processing on Large Clusters*. Technical Report UCB/EECS-2014-12, University of California, Berkeley (Technical Report).
24. Zhang, T., et al. (1996). BIRCH: An efficient data clustering method for very large databases. In *ACM SIGMOD Record* (Vol. 25, Issue. 2, pp. 103–114). ACM.
25. Zhou, et al. (2000). Approaches for scaling DBSCAN algorithm to large spatial databases. *Journal of Computer Science and Technology, 15*(6), 509–526.

Understanding How Big Data Leads to Social Networking Vulnerability

Romany F. Mansour

Abstract Although the term "Big Data" is often used to refer to large datasets generated by science and engineering or business analytics efforts, increasingly it is used to refer to social networking websites and the enormous quantities of personal information, posts, and networking activities contained therein. The quantity and sensitive nature of this information constitutes both a fascinating means of inferring sociological parameters and a grave risk for security of privacy. The present study aimed to find evidence in the literature that malware has already adapted, to a significant degree, to this specific form of Big Data. Evidence of the potential for abuse of personal information was found: predictive models for personal traits of Facebook users are alarmingly effective with only a minimal depth of information, "Likes". It is likely that more complex forms of information (e.g. posts, photos, connections, statuses) could lead to an unprecedented level of intrusiveness and familiarity with sensitive personal information. Support for the view that this potential for abuse of private information is being exploited was found in research describing the rapid adaptation of malware to social networking sites, for the purposes of social engineering and involuntary surrendering of personal information.

1 Introduction

Exactly how much can be known from a user's online social networking profile or profiles? These days, more and more people are spending significant portions of time every day on social networking. In 2011, the worldwide average for Facebook was 40 min for 800 million users, according to Los Angeles Times (2011). In fact, the sheer quantity of social interaction now occurring over social networking is such that a qualitative shift is taking place in our globalized society. This shift is

R. F. Mansour (✉)
Faculty of Science, Department of Mathematics, New Valley,
Assiut University, Asyut, Egypt
e-mail: romanyf@aun.edu.eg

© Springer Nature Singapore Pte Ltd. 2018
S. S. Roy et al. (eds.), *Big Data in Engineering Applications*,
Studies in Big Data 44, https://doi.org/10.1007/978-981-10-8476-8_10

towards a replacement, in many ways, of in-person social interaction with interaction over social networking [1]. For example, social rituals or rites such as deciding whether a person would be suitable for dating now often occur first over Facebook or other social networking sites. There is a rise in the studies in social networking through its development, its effect to the global economy and the human psychology behind the use of these social networks [2]. Employers are also likely to screen prospective employees through an examination of their social networking profiles, especially Facebook and LinkedIn. The iniquitousness of social networking websites makes them immense repositories of personal information, carrying grave risks for abuse of privacy at the hands of malware. It is important that malware that depends on such Big Data techniques to perform social engineering and other unethical or socially compromising activities be more fully identified, characterized, and ultimately addressed.

Objectives of the study

I. To find out how much personal information can be obtained from the social networking sites.
II. To find out the privacy risks associated with personal information on social networking sites.

Significance of the study

Understanding social networking is an important aspect for users. This study will help the users identify the risks that are associated with the exposure of their personal information and how well they can mitigate these risks. Maintaining privacy of the users in the social networks is a necessary agenda for the users of the social networks.

2 Methods

The literature was examined for two separate lines of evidence related to the risk of dire loss of privacy as a result of Big Data—based mining of social networking website information. First, literature dealing with the theoretical potential for inferring personal details of users of social networking websites was searched for. Searches were performed on Google Scholar and Web of Science, using the terms "social networking", "social engineering", "big data", and "predictive models".

The second line of literature research aimed to discover evidence the malware is already adapting to exploit the potential of social networking websites and degrading privacy of users. Again, Google Scholar and Web of Science were used. However, in this case, the search terms were extended to include "malware", "phishing", and "hacking".

For both lines in literature research and inquiry, only articles from the last 5 years (2010–2015) were considered.

3 Results

3.1 Potential of Big Data Techniques for the Inference of Sensitive Personal Information

The study by [3] used six different features of a sizeable sample of 180,000 Facebook users' profiles to predict personality traits. The personality trait measurement method used was the standard Five Factor Model, which measures the level of the following personality traits: Extraversion, Neuroticism, Agreeableness, Openness, and Conscientiousness. The six features used by [3], summarized in Table 1, are numbers of: Facebook friends, associations with groups, Facebook "likes", photos uploaded by user, status updates by users, and times others "tagged" user in photos. The 180,000 volunteers who provided information from their facebook profiles also completed the Five Factor Model personality test. Therefore, it was possible to compare predictions from the Facebook model to objective results from the Five Factor Model. Using multiple regression, the authors found that predictions from the Facebook model could be generated that were very accurate, assuming that the results from the Five Factor Model did not incorporate any misrepresentations of personality. These findings supported findings from an earlier work that social networking profiles do not present an idealized or skewed version of a user's persona, but rather a realistic and fairly objective summary [4]. The [3] study did find, however, that the traits of "Agreeableness" and "Openness" were significantly ($p < 0.05$) less accurately predicted than were the other three traits. A somewhat later, but similar, study reported the ability to predict personality traits using a natural-language parsing model to automatically analyze individuals' statuses [5]. This model was trained on a corpus of over 700 essays that had been manually curated and assigned labels with the appropriate amounts of the five favors (Openness, Agreeableness, Extraversion, Neuroticism, and Conscientiousness) assigned. This study corroborated the findings of the [3] study that personality traits could be accurately inferred.

Perhaps the most recent transformative research on the subject of inferring personal details from facebook or other social networking information was reported by [6]. This group took a sample of 58,000 volunteers who had made part of their Facebook information available (Facebook 'Likes"). The authors were able to show that a list of a person's likes, which are highly visible as they are generally

Table 1 Features used by [3] to predict Facebook user personality traits (according to Five Factor Model)

Feature	Details
Friends	Number of Facebook friends
Groups	Number of associations with groups
Likes	Number of Facebook "likes"
Photos	Number of photos uploaded by user
Statuses	Number of status updates by user
Tags	Number of times others "tagged" user in photos

publically available, can be used to predict certain demographic and personal pieces of information with great accuracy. The categories of personal information that were predicted were diverse, but among those that could be predicted with high accuracy were sexual orientation, ethnicity, religion, political orientation, personality, IQ, drug use and various other pieces of personal and family information. The most accurately predicted demographic and personal factors were sexual orientation in men (88%), African American versus Caucasian American (95%), and political orientation (Democrat or Republican) (85%). Thus, a large amount of personal information of great relevance to potential employers can be predicted from an individual's collection of "Likes" on Facebook [4, 7], argue that such information on the web can be used in carrying bout advertisements targeting a specific group of people. Social networks can provide useful information about the users that can be useful to the marketers in laying down their marketing strategies.

Jernigan and Mistree [8], carried out a study among MIT students based on the hypothesis that the number of an individual's Facebook friends can be used to determine the sexual orientation group of the user. A thorough analysis was carried out on the students who used the MIT browsers. The study revealed that the number of friends that an individual has can be used to predict the sex orientation of the user. For instance is a user has more homosexual friends then the likeliness that the individual is homosexual is very high. The findings are summarized in the Table 2.

It has also been found that people's social strategies, and therefore possibly even the underlying social motivations, can be inferred from a careful analysis of Facebook and social networking patterns. For example, through an analysis of the evolution of Facebook connections over time, [9] were able to differentiate non-social capital seeking from social capital seeking friends. The researchers developed a predictive model based on the patterns of connectivity over time, and found that these patterns only differed significantly from normal when an individual was making connections with the intentional goal of seeking social capital. For

Table 2 Percentage friends per sex orientation group

Sex orientation group	Percentage friends per group					
	Heterosexual (%)		Bisexual (%)		Homosexual (%)	
Heterosexual						
Female	19.0	22.4	0.7	0.5	0.4	0.8
Male	13.9	28.3	0.5	0.4	0.3	0.7
Bisexual						
Female	15.5	20.7	1.4	1.1	0.3	1.2
Male	12.6	22.3	0.8	0.6	0.3	1.9
Homosexual						
Female	18.0	23.6	0.9	0.7	0.2	0.8
Male	13.1	21.4	1.1	1.1	0.4	4.6

Retrieved from: [8]

example, if an individual has recently been introduced to a new group, he or she is likely to first connect with a central hub in the Facebook environment for the group of people, and then rapidly add connections (which then become mutual connections between the individual and the central hub). This central hub is often someone in a position of power or privilege. Not only can analysis of a person's friend connection patterns reveal social intent, but it can reveal who a person's real friends are more likely to be, and who a person has "friended" merely as acquaintances. In a reversal of the predictive methodology, [10] used measured traits of personality to predict Facebook usage. Specifically, the researchers were able to find that certain personality traits (neuroticism foremost) were strong predictors of wall posting "regret", or the tendency to remove a posting on a user's own, or a friend's wall. Hughes et al. reviewed work done to create predictors with data from Facebook versus from Twitter, finding the two sites to be very similar overall.

Ross et al. [1], carried out a similar study on how personality traits and competency influenced the way in which university students utilized Facebook for social purposes and came up with different results centrally to the ones discussed above. The research utilized 97 students from the Southwestern Ontario as the respondents, in which 85 were women and 25 were men. A 28 item questionnaire was used as the study tool. The authors found out that some personality characteristics influenced the Facebook use but their level on the impact of Facebook use differed greatly. The students who scored highly in the extraversion characteristics belonged to more Facebook groups but had very few friends. The reason behind this is that some of the users prefer instant contact with the friends a feature which is not enabled with the Facebook. Therefore they choose not to use Facebook as their primary source of interaction. Those high in the neuroticism character preferred their walls as compared to those low in the same personality as they preferred photos. Those who preferred wall posting are associated with the ability to think out well before positing. Those who post photos are exposed to privacy intrusion as such photos contain some personal information such as place where the photo was posted from. Openness and experience in utilization of Facebook was associated with the ability to understand on how to use the several elements if Facebook, how to comment and how to use other Facebook feature. However more agreeable individuals contained lesser online contacts, also there was no significant relationship between conscientiousness with the utilization of Facebook.

In their study [11] carried out a comparison between the social culture of interaction between the two social virtual world of China, Uworld and HiPiHi. Unlike Uworld, HiPiHi makes the use of the social networking to promote its business products. This has become another strategy for promoting business activities through the virtual games online. On the other hand Uworld provides entertainment games which are in different forums that are not related to businesses. In these forums. In the Uworld the users can make and chat with friends in the virtual room and also playing games. This makes it possible to create more friends in the Uworld than in the HiPiHi.

In summary, it appears that there is currently a surprising amount of information that can be inferred from a user's social networking profile. As emphasized by [12], people social agendas could be revealed. Indeed, in some cases, it seems possible that models based on Facebook "Likes", for instance, might be able to correctly make predictions that an individual himself would never have known. This is possible thanks to the vast sample size available (nearly a billion users worldwide, just for Facebook), as well as the richness, standardization, and quantity of information that is routinely deposited on Facebook by users. One cannot help but speculate that this diversity and potency of information could be used to intelligently craft tools and traps to manipulate users of social networking websites, or indeed, other websites (after saving information from the users' profiles).

3.2 Social Networking Sites and Malware Risk

Social engineering occurs perhaps most directly on websites where malware and phishing programs are able to induce internet browsers and users into places where security is less available or effective. Quite often, bright-colored ads or links artificially placed high in the results from search engines lead users stray into areas where their ability to detect malware is reduced [13], as a result of a weaker firewall, less visible pop-ups, or the leverage of anti-anti-malware tools.

Tracking the behavior and attack styles of these socially-engineering forms of malware could be a very interesting and compelling, modern and promising way to go about thesis research. A recent article collected information on the intensity and frequency of malware [14]. This article found, for example, that the pervasiveness of malware is generally due to the use of common avenues of attack. The group further found that such malware relies on two primary strategies, technological and psychological manipulation. Technological manipulation includes placing fake versions of functional navigational buttons over the actual buttons on the graphical user interface of social networking websites, or having the link pop up the instant the user clicks. Psychological manipulations involve listing unsponsored pop-ups in the side bar that supposed the user finds appealing enough to want to click on, regardless of prior plans on the website. The advantage, from the perspective of the malware, of hijacking personal information on social networking websites, is that users are often rather less rushed or focused in their browsing habits, and therefore can be more easily led astray [14]. Further elucidating sub-types of these two primary types of social engineering (technological and psychological manipulation) could be a compelling goal for thesis research. The limiting factor in this case might be access to sufficient user profiles, and the resistance one would likely encounter when trying to avail oneself of the user profiles when the users are informed that a virus is to be run on their system or targeted at their user profile.

A number of other areas exist in the internet wherein fraud in its various forms takes place. In general, whenever a great deal of technological competence is required, it becomes easier for malware to defraud an individual by false or alternative navigation around the website(s). In general, any area or circumstance in which the individual is suddenly faced with a request or demand seeming to emanate from a technically-knowledge authority are far more likely than average to lead to incidences of internet fraud [15].

3.3 Social Engineering on Social Networking Sites

Social networking sites are particularly prone to unknowingly or unwillingly giving a platform for the attack of such malware. Social networking sites are some of the biggest and most popular, and although incredible amounts of data exist, the study of social networking is still in its infancy. Because most social-engineering types of viruses are found on social networking sites, it would be fairly direct and intuitive to design a thesis around the habits of users who fall victim to more malware (or to generate and provide evidence for/against other hypotheses [13]. This malware could take a number of forms, as the information on Facebook is sufficient, for nearly all individuals, to infer a great deal of additional very personal and sensitive information.

Not only are social websites ideal for leading user astray, but by virtue of their sheer size and versatile functionality these websites also contain unprecedented amounts of valuable personal information. Even if such personal information is not directly provided by the user on the website himself or herself, it may still be obtainable for malware, by dint of tunneling through privacy restrictions and reading, e.g., information from instant message conversations [16]. Through these conversations unauthentic messages can be sent. Often, these messages are not obviously "robotic" in nature, but rather have greetings from supposed people (users on facebook) as their first line of attack to disarm and socially position the victim for further information attacks. "Bots", for example, may replicate themselves and even generate false pictures and histories, and by first friending a victim and then posting indirectly related material, induce the victim to actually make first contact and assume himself/herself to be in charge of the social situation. In fact, this trust and "belief" in the legitimacy of the communication disarms the user, compelling him or her to surrender valuable personal information or even money.

4 Conclusion and Discussion

Even without soliciting information directly from a user of a social networking site, hackers, malware distributors, or other internet social engineers could quite easily infer a great deal of personal information from users, based simply on the users' profiles and networking behavior. The potential for abuse is clear—[3] show that analysis of profile information about Facebook users at the Big Data level (thousands of users) can lead to profiling of personal characteristics across a broad range of factors. More profoundly, [6] find that just using Facebook "likes" allows for the creation of predictive models that indicate an individual's range, sexual orientation, and other sensitive demographic and personal details with alarming accuracy, up to 98% in the case of race. Undoubtedly, models can only be made stronger with the addition of more complex and rich data, e.g. from the mining of status updates, history, social connections, groups, and even pictures. Facebook is already capable of identifying facial features and other features of environs presented in photos.

Equally importantly, it is clear that bots and malware have already evolved that take advantage of the social milieu and at least some personal details of users to lead users astray, e.g. into less secure sites where further personal information can be stripped away. These bots and malware take advantage of the high level of activity the users engage in, when navigating through social networking sites. Mimicry of more legitimate ads targeted to users makes malware difficult to spot, especially for a distracted and enthusiastic user. It is important that these trends are recognized and reversed, before they can become even more powerful and insidious.

4.1 Recommendations

Enhancing privacy settings is a key strategies in mitigating privacy risks in the social networks. Setting privacy settings and cookies that can detect malwares and block them automatically help in dealing with vulnerabilities in social networks [17]. Authentication mechanisms can also be used to avoid hijackers or non-authorized users from login in into an individual's account [18]. The operators have also provided internal protection mechanisms that protect and detect spams or other such messages which are designed to collect user's personal information secretly [19]. Commercial solutions too can work by purchasing specialized softwares that have ability to defend user against any form of cyber-attacks [20].

4.2 Future Research

The need for personal security in the social networks has become increasingly important. Several solutions have been suggested and implemented but still the problem persists and several people have lost a lot of their resources due to these attacks. There is need to carry out research on how effective the adopted solutions are in helping solve these problems in the ever dynamic field of technology and need for coming up with new strategies of solving the problem.

References

1. Ross, C., Orr, E. S., Sisic, M., Arseneault, J. M., Simmering, M. G., & Orr, R. R. (2009). Personality and motivations associated with Facebook use. *Computers in Human Behavior, 25*(2), 578–586.
2. Zhang, X., Wang, W., de Pablos, P., Tang, J., & Yan, X. (2015). Mapping development of social media research through different disciplines: Collaborative learning in management and computer science. *Computers in Human Behaviour, 51,* 1142–1153.
3. Bachrach, Y., Kosinski, M., Graepel, T., Kohli, P., & Stillwell, D. (2012, June). Personality and patterns of Facebook usage. In *Proceedings of the 3rd Annual ACM Web Science Conference* (pp. 24–32). ACM.
4. Back, M. D., Stopfer, J. M., Vazire, S., Gaddis, S., Schmukle, S. C., Egloff, B., et al. (2010). Facebook profiles reflect actual personality, not self-idealization. *Psychological Science, 21* (3), 372–374.
5. Farnadi, G., Zoghbi, S., Moens, M. F., & De Cock, M. (2013). How well do your Facebook status updates express your personality? In *Proceedings of the 22nd Edition of the Annual Belgian-Dutch Conference on Machine Learning (BENELEARN)*.
6. Kosinski, M., Stillwell, D., & Graepel, T. (2013). Private traits and attributes are predictable from digital records of human behavior. *Proceedings of the National Academy of Sciences, 110*(15), 5802–5805.
7. De Bock, K., & Van Den Poel, D. (2010). Predicting website audience demographics for Web advertising targeting using multi-website clickstream data. *Fundamenta Informaticae, 98*(1), 49–70.
8. Jernigan, C., & Mistree, B. F. (2009). Gaydar: Facebook friendships expose sexual orientation. *First Monday, 14*(10).
9. Ellison, N. B., Steinfield, C., & Lampe, C. (2011). Connection strategies: Social capital implications of Facebook-enabled communication practices. *New Media & Society, 13*(6), 873–892.
10. Moore, K., & McElroy, J. C. (2012). The influence of personality on Facebook usage, wall postings, and regret. *Computers in Human Behavior, 28,* 267–274.
11. Zhang, X., de Pablos, P., Wang, X., Wang, W., & Sun, Y. (2014). Understanding the users' continuous adoption of 3D social virtual World in China: A comparative case study. *Computers in Human Behaviour, 35,* 578–585.
12. Butler, D. (2007). Data sharing threatens privacy. *Nature, 449*(7163), 644–645.
13. Algarni, A., Xu, Y., Chan, T., & Tian, Y.-C. (2013). Social engineering in social networking sites: Affect-based model. In *Proceedings of the 8th IEEE International Conference for Internet Technology and Secured Transactions (ICITST-2013)* (pp. 508–515). London: The Institute of Electrical and Electronics Engineering, Inc.
14. Abraham, S., & Chengalur-Smith. (2010, August). An overview of social engineering malware: Trends, tactics, and implications. *Technology in Society, 32*(3), 183–196.
15. Rusch, J. J. (1999). *The "social engineering" of Internet fraud.* USA: United States Department of Justice.
16. Laszka, A., Felegyhazi, M., & Buttyan, L. (2014). A survey of interdependent information security games. *ACM Computing Surveys (CSUR), 47*(2), 23.
17. Tipton, H. F., & Krause, M. (2012). *Information security management handbook.* CRC Press.
18. Whitman, M., & Mattord, H. (2011). *Principles of information security.* Cengage Learning.
19. Rasool, M. A., & Jamal, A. (2011). *Quality of freeware antivirus software.*
20. Sukwong, O., Kim, H. S., & Hoe, J. C. (2011). Commercial antivirus software effectiveness: An empirical study. *Computer, 44*(3), 0063–70.

Big Data Applications in Health Care and Education

B. K. Tripathy

Abstract Technology plays a major role in all spheres of life and higher education and health care are no exceptions. The use of big data in higher education and health care are relatively new. The dynamics of higher education is passing through a phase of rapid changes. Also, the amount of data available in this field and proper analytics can reap the benefits and highlight on future techniques to be followed in handling the complex situations arisen from pressure exerted by accrediting agencies, governments and other stake holders. Higher education is becoming more and more complex with several institutes entering into the market with more and more diversified approaches. This makes the functionalities of all institutes of higher education to revise their approaches frequently to cope up with this pressure. The educational institutes have to ensure that the quality of learning programmes is at par with that of their counterparts at the national and global level. Analysis of vast data sources generated in this connection being more often not available for analysis is a major concern. The analysis of these volumes of data plays a major role in understanding and ensuring that institutions are aware of the changes occurring everywhere and they are taking care of their social responsibilities. Due to digitization of medical records in an attempt to make them available for research and development over the past ten to fifteen years, there is a huge amount of data, which besides being voluminous are complex, diverse and temporal which is collected by healthcare stockholders. An analysis of these data could collectively help the healthcare industry to find out problems related to variability in healthcare quality and escalating healthcare expenditure. In this chapter we shall make a critical analysis of these aspects of higher education and healthcare with respect to big data analysis and make some recommendations in this direction.

Keywords Big data · Health care · Higher education · Learning management systems · Analytics

B. K. Tripathy (✉)
SCOPE, VIT University, Vellore 632014, Tamil Nadu, India
e-mail: tripathybk@vit.ac.in

© Springer Nature Singapore Pte Ltd. 2018
S. S. Roy et al. (eds.), *Big Data in Engineering Applications*,
Studies in Big Data 44, https://doi.org/10.1007/978-981-10-8476-8_11

1 Introduction

Big data has no universally accepted definition. It can be classified into two sources; physical data, which is obtained through sensors, scientific experiments and observations and human centric data, which is the data acquired from social networks, internet, health, finance, economics and transportation. Big data has now become equally important in industry and academia. To differentiate big data from small data, we can say that small data are mainly sampled whereas big data are automatically harvested using different techniques such as data crawling or application programming interfaces from a large population of users. Big data are defined at an individual level rather than an aggregated level. Technology plays a major role in all spheres of life and higher education and health care are no exceptions [1].

So, instead of trying to define big data, it is worthwhile to state their characteristics [2]. Even here, there are several approaches and the number of V's used as a characteristic of big data is in its ever increasing trend. But mainly, researchers focus on four of the V's; Volume, Variety, Velocity and Veracity. Volume refers to size of the data set under consideration and now has reached zettabytes (10^{21} bytes). Variety deals with the structure of the data under consideration, which can be structured, semi-structured or unstructured. The rate at which data is generated and the rate at which it needs to be attended define velocity. It extends from any-time batch processing to real-time streaming. Veracity deals with quality of data, which is very much essential in getting useful information from it. It also means the relevance of the data with the context, its predictive value and its semantics.

The sources responsible for the huge growth of data are Internet, Internet of Things (IoT) and Cloud Computing. It has occurred in almost every institution, business house and industry. As a result we find a sudden resurgence in Big data research among the academicians, government organisations and private industrial set ups [3].

Modern day higher education is becoming more and more complex with several institutes being coming up in the market with more and more diversified approaches. This makes the functionalities of institutes of higher education to revise their approaches frequently to cope up with this pressure. They have not only to stand up to the competition from their peers but also have to keep up an eye from the government policies. A lot of economic and political changes are coming up at the national and global level and also the social changes are occurring at fast pace. So, the educational institutes have to ensure that the quality of learning programmes is at par with that of their counterparts at the national and global level.

In addition to making decisions to face the rapid changes occurring in their environment, which is complex in character, analysis of vast data sources generated in this connection being more often not available for analysis is a major concern. The analysis of these volumes of data plays a major role in understanding and ensuring that institutions are aware of the changes occurring everywhere and they are taking care of their social responsibilities [4].

Big Data analytics has emerged recently and like other data mining activities it helps in getting patterns in these volumes of data, generating rules and drawing decisions as and when required. Some of the thrust areas where it is currently being explored are business, government and health care. The reason for these fields being thrust area is the collection of large amount of data generated in these environments. Research in Big Data is focused to find out methods for efficiently aggregating and correlating massive volumes of data in order to find behavioral patterns and meaningful trends [4].

Digitization of existing hard copies available in healthcare systems has led to the accumulation and hence increasing the volume of data there in [5]. Already there is a large volume of such data in the form of personal medical records, clinical trial data, genomic sequence, radiology images FDA submission, human genetics and population data etc. There were a lot of healthcare data available in the form of records of patients in hospitals, the prescriptions written by doctors and nurses, records maintained by the medical record persons at the time of admission and at the time of leaving the hospitals, reports generated from various testing devices like scanners in the form of MRI, CT etc. Every now and then volumes of data are added to the healthcare system, which are both structured and unstructured from the devices used to maintain fitness, several social media, genetics and genomics, innovations and sources related to this. From the utility point of view, only a small fraction of this volume of data is being used by computer scientists through simulation or otherwise and are analysed to generate useful information. There is a strong necessity to transform the structured data to unstructured ones, combine the different data sources efficiently and doing it automatically. This will lead to efficient analysis of such data, by the way generating useful information. Also, there is problem in combining data collected and stored at different point of time due to different formats being used. This should be handled carefully such that their utility increases. The combination process can be done at the individual or population level as per the requirement. The various features of large data can be properly addressed once such transformation is done efficiently and seamlessly. We cannot avoid the complexities of big data generated in healthcare systems, but only have to find technologies to handle them efficiently including their storage and manipulation.

The constant flow of data accumulating at unprecedented rates presents new challenges. The velocities at which it is generated and the speed needed to retrieve, analyze, compare and make decisions using the output. According to speed, the healthcare data can be categorized into three categories. The static data covers paper files, X-ray films etc. The medium velocity data include diabetic glucose measurements, blood pressure readings and EKGs. The on-line data comprises of trauma monitoring for blood pressure, operating room monitors for anesthesia and heart monitors etc. Data quality for healthcare data is precious as it leads to life-death decisions and the unstructured data like the handwritten prescriptions are too often incorrect. It is very much required that the diagnoses, treatments, prescriptions, procedures should be correctly captured. Although it requires data cleaning, high velocity and high veracity becomes the hurdles for it. Moreover use of traditional IT issues like data management, warehousing, compliance, audit, fraud prevention, error reporting and regulatory compliance prevents the organisations to be more vigilant on veracity of data.

So, big data plays a vital role in the two important components of our society, higher education and healthcare. It is the purpose of this chapter to analyse and present the origin, chronological development of research on these two topics and propose some directions of research as well. In the present section, we introduced big data, its role in higher education and healthcare. In the following sections we shall make a deeper journey into the fields on hand and critically present the beads which have so far been responsible for the preparation of the necklace termed as big data in higher education and healthcare.

Traditional methods are too complicated for processing big data; these methods consume a lot of time and also not economic. So, several approaches have been proposed to handle the situation. One popular method is to apply parallel computing; using which large problems are divided into smaller components, individually solved and then the solutions are combined to get the final solution. The most frequently used frameworks in big data are MapReduce, Hadoop, Hadoop Distributed File System (HDFS), Apache Hive and NoSQL. In MapReduce a large number of computers are used to process large data with parallelizable problems. The three steps used under this architecture are Map, Reduce and Combine. Hadoop is an open source software project which enables processing of data across clusters of computers. HDFS stores a large number of files using a large amount of servers. Apache Hive is a data warehouse infrastructure built on the top of Hadoop. A NoSQL database is non-relational one which provides a mechanism for storage and retrieval of data that is modelled by using representations other than the tabular relations used in relational databases.

2 Big Data in Higher Education

Big Data as an emerging field, research in it has been broadly focused on fields like business, government and health care as the amount of data collected in these fields is very high. The research efforts for application of big data in education or for that matter on higher education are low, although the amount of data collected has started to increase heavily [6].

All the organisations handling higher education have their student forums for present and past students. Analysing their behaviour will definitely lead to generate useful information for the future. Identification of the views of the alumni and their suggestions for betterment needs to be critically analysed. They are the torch bearers for any organisation and are also the ambassadors. Their views bear a lot of importance for the incoming students. The suitability of different programs offered by an institution, their future prospects in the form of placements for both on and off campus jobs, the comparative analysis of similar kind of programs offered by other competitive institutions also have stronger effects on the intake in the coming years than any other source. Also, the administrators of an institute can get advices from the alumni to rectify their existing liabilities. A lot of factors are responsible for students to transfer and the reasons and trends need to be analysed. The suggestions from industry experts, who are directly, involved with the students' future of the bears a distinct edge over the other sources of information. Enrolment management and time-to-degree continue to motivate higher education institutions to search for better solutions [7].

A major step for application of big data techniques in higher education is the corporate-academic partnership. Its strength depends upon the assurance given by the educational institutes for utilization and development of advanced technologies that are likely to support applied research outputs and potentials for knowledge transfer and commercialization [8].

Several technological changes have occurred in higher education, like ubiquitous computing devices, flexible class room design and massive open online courses [9–11]. Education and research have become dependent upon these new found technologies. Even though such technological changes have occurred, the role of data is highly important and enough care has not been taken to realize this. In fact these data if fully utilized is likely to provide higher education institutes with important solutions to handle changes occurring inside and outside their environment.

One such area which can be used to handle large data sets available is Knowledge Discovery in Databases (KDD), which has the primary goal to generate meaningful and useful patterns from large datasets. The techniques from KDD, which can be used, are clustering, classification and association algorithms, regression models, predictive methods and factor analysis.

Data mining techniques can be used to find and solve the challenges posed by the students or alumni through analysis and presentation of data. As is its characteristics Data mining helps organisations to use their current reporting capabilities to generate hidden patterns and understand them. Data mining models are used to capture these patterns to study and predict behaviour of individuals with high accuracy. This helps the institutions to optimize the allocation of their resources in the form of teaching and non-teaching faculty and infrastructure. It is highly desirable for the institutes to have such a facility so that they can make a budget accordingly and implement it.

The data mining approaches used in educational institutions are different in their characteristics as instead of using the traditional tools these techniques obtained in the form of innovative methods to discover patterns and analyse large data sets instead of simply automating the process as is done in most application areas [4, 12].

Decision making tools are used by some organizations to generate better decisions from data available with them about their strategic and operational directions. Using data to make decisions is not new; business organizations have been storing and analyzing large volumes of data since the advent of data warehouses systems in the early 1990s. However, the nature of data available to most organizations is changing and the changes bring with them complexity in managing the volumes and analysis of these data. It was observed in [12] that most businesses today run on structured data (numerical and categorical). However, this does not reflect the complexity on the nature of available corporate data and their untapped hidden business value. But most of the recent data are unstructured and are in the form of text, audio, diagrams, images and combinations of more than one of these.

Recent developments in database technologies made it possible to collect and maintain complex and large amounts of data in many forms and from multiple sources. In addition, there are analytical tools available that can turn this complex data into meaningful patterns and value, a phenomenon referred to as Big Data. Concept architecture for processing big data was proposed in [13]. The flow chart of the algorithm is as described in Fig. 1.

Data collection, data analysis and data visualization are the three essential stages of Big Data analysis. Data to be collected should have useful information which is valuable after their identification. The collected data are to be filtered to remove some unwanted components and characteristics. This will fine tune the data set and make it ready for the next step. In the analysis phase data linking, connecting data and finding inter relationship among data is carried out. It helps in grasping the information the collected dataset is to reflect. This data should be made available to the users in the form they desire to have it. This process is called data visualization.

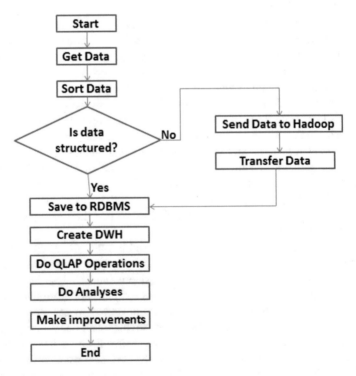

Fig. 1 Algorithm for processing big data [13]

Table 1 Higher education equivalents of private sector questions [16]

Private sector questions	Higher education equivalents
Who are my most profitable customers?	Which students are taking the most credit hours?
Who are my repeat website visitors?	Which students are most likely to return for more classes?
Who are my loyal customers?	Who are the "persistors" at my university/college?
Who is likely to increase his/her purchases?	Which alumni are likely to make larger donations?
Which customers are likely to defect to competitors?	What types of courses will attract more students?

Also, it helps the interpretation of data and integrates them into the existing processes. Finally, it is used as a guide for decision making.

Institutional databases comprise of volumes of data collected over the years and are categorized as large. Online courses are of recent origin and its popularity has become high over the past few years. A good number of data repositories are available now along with digital libraries and other associated tools [14, 15]. These repositories comprise of student related data in various forms and kinds like social media usage data, learning management systems, student library usage, individual computers and administrative systems holding information on Programme completion rates and learning pathways.

While handling data mining applications in higher education, it is mentioned in an executive report of SPSS [16] the following equivalent activities in higher education to that in private sector (Table 1).

There are several areas like administrative and instructional applications, financial planning, donor tracking, recruitment, admission processing and student performance monitoring where big data and analytics can be applied within higher education [17–19].

3 Big Data in Healthcare

Due to digitization of medical records in an attempt to make them available for research and development over the past ten to fifteen years, there is a huge amount of data, which besides being voluminous are complex, diverse and temporal which is collected by healthcare stockholders. An analysis of these data could collectively help the healthcare industry to find out problems related to variability in healthcare quality and escalating healthcare spend. To put it as illustrations, researchers can find the specific medicines or treatments which are most effective for specific conditions; identify patterns related to drug side effects and gain additional information that can help patients and reduce costs. Stakeholders also need to focus on the non-disclosure of sensitive attributes of patients by may be anonymizing these databases before their release. There are a lot of anonymisation techniques available for the so called small databases. The two basic approaches followed in literature are generalisation and suppression. However, there are problems in suppression as much of the information will be lost in it. Even in generalisation, it should be done judiciously as otherwise it may lead to the problems of information loss or the anonymisation being insufficient. Hence other methods are followed which are different from these two basic techniques. However, in spite of their drawbacks, the basic techniques can be followed for any dataset. The other approaches are needed to be enhanced for making them applicable for large datasets.

The amount of patient data is growing exponentially during the past few years. In addition to clinical data there are sources like claims and cost data, Pharmaceutical R and D data that provides therapeutic mechanism of action of drugs and patient behaviour and sentiment data, which describe patient activities and preferences.

Advances in technological systems like the electronic medical records (EMRs) are now more affordable and data exchanges have become easy. Sometimes the technological advantages like the preservation of patient privacy by deleting the identifiers like names or otherwise has made it possible to compile, store and share information in a secured manner. There is a strong necessity to follow this because of Health Insurance Portability and Accountability (HIPAA) patient confidentiality standards.

Several industrial efforts have been made for big data either through their collaboration or commercialization. The organization "Premier" offers membership-based service to providers of all types, which contribute their information. Private payers like OptumInsight, Active Health and HealthCare operate their stand-alone analytics divisions for United Health, Aetna and WellPoint respectively. The group TransCelerate Biopharma which is formed by ten global pharmaceutical companies has the intention to simplify and accelerate drug development [20].

The Human development Index (HDI) which is a composite statistic of life expectancy, education and per capita income indicators, facilitates release of information from Hyperosmolar Hyperglycemic state (HHS) through its Health Data.gov Website. HDI has started a conference in 2010 and it has become an annualaffair which incorporates companies that are investigating innovative strategies for using health data in tools and applications. It also conducts a "code-a-thon" event in which the innovators collaborate besides working on showcasing and demonstrating their products.

Keeping the changes in healthcare systems, a holistic framework was proposed in [21], which considers five key pathways to value. The five new pathways are; Right living, Right care, Right provider, Right value and Right innovation. The right-living pathway focuses on encouraging patients to make lifestyle choices that help them remain healthy through proper diet and exercise and also taking their own care if they fall sick. The right care pathway involves ensuring timely and appropriate treatment available to the patients. It brings a coordinated approach across settings and providers to have same information for all the caregivers towards the same goal so that duplication of efforts is avoided and optimal strategies are adopted. The right provider pathway requires that the patients should receive the services by high-performing professionals. High-performing professionals whose expertise matches with the requirement for the treatment of the patient and other associates like the nurses with proven outcome are to be appointed. The right value requires ensuring cost-efficient care, eliminating fraud or abuse of the system. Invention of new therapies and taking care of all characteristics of the system is the right way in this direction.

It has been observed that evolution of new directions in health care innovation has taken place because of big data. A study on Rock Health and Capital IQ databases has indicated that the insurgence of big data is responsible for the origin of fresh innovators. For example, the usage of inhalers by asthmatic patients is being monitored by trackers supported by GPS-system. Some mobile applications has offered agreement between the patients and their providers that the patients can be tracked and assisted with behavioural health therapies. Patients are supported with chronic care medication such that through an interactive system patients can be provided knowledge and enhanced treatment. Also some other organisations are getting sources for collecting data and their organisation through exchange of information [22].

In health sciences, there are many problems that can be addressed with big data technologies, such as recommendation system in health care, Internet based epidemic surveillance, sensor based health condition and food safety monitoring.

3.1 Big Data Studies in Health Sciences

Health science has been enriched by big data technologies being applied successfully in it. The prevalence of several epidemics can be monitored by gathering information from people of the affected areas. It usually takes a lot of time when one tries to gather this information from the organisations like CDC and WHO and by the time the data is obtained the effect of the epidemic may have either ceased or it has turned out to be uncontrollable.

3.2 Recommendation System in Health Care

Many researchers have applied recommendation systems techniques to health information systems. Duan et al. [23] proposed a nursing care plan recommendation system to provide clinical decision support, nursing education and clinical quality control. Hoens et al. [24] proposed a reliable privacy-preserved medical recommendation system.

There are many problems which can be solved with big data technologies including healthcare. To solve these problems many advanced computing technologies are used. The applications of big data in health sciences are in collection of data from search engines and social networks can help to gather people's reactions and monitor the conditions of epidemic diseases. Several case studies have been performed in this direction.

4 Case Studies

We present some case studies in applications of big data in health sciences in this section. These are

4.1 Recommendation System in Health Care

Some of these are mentioned above. Duan et al. [23] proposed a nursing care plan recommendation system to provide clinical decision support, nursing education and clinical quality control. It serves as a complement to existing practice guidelines. Hoens et al. [24] proposed a reliable privacy preserved medical recommendation system. In this medical system, the patients can contribute their secured ratings of the physicians on different health conditions based on their satisfactions. This system can recommend a list of physicians who best suit to their needs. Wiesner and Daniel [25] proposed a health recommendation system in the context of personal health record system. In their health recommendation system, the items are non-confidential, scientifically proven or at least generally accepted medical information. The goal of HRS is to provide information to the patients which are highly relevant to the patient's personal health record [25].

4.2 Internet Based Epidemic Surveillance

With the assumption that when the number of people having the symptoms of an epidemic is high, the search for the epidemic related topics will be high. One such instance is the study on influenza provided by http://www.google.com/flutrends of Google.

Twitter is widely used social network and news-sharing platform. The tweets reflect opinions of people and their judgements about public event, especially the outbreak of epidemics. Signorini et al. [26] discuss on the use of twitter to track levels of disease activity and public concern in the U.S. Another such paper is that of Paul et al. [27], where they analyze twitter for public health. In fact, they followed the tweets in connection with H1N1 by searching tweets through key words like flu, influenza and H1N1. The tweets involving public concern were filtered by using key words like travel, flight and ship for disease transmission. Keywords like wash, hygiene and mask were used to identify measures carried out for countering the disease.

4.2.1 Classification Model to Analyze the Spread and Emerging Trends of the Zika Virus

Starting in 2015 the Zika virus continues to boost a dreaded disease and being an epidemical virus besides being a global health issue. A study of twitter data shared through this social network was the inspiration behind the work in Tripathy et al. [28] where the authors proposed a classification model which was used to divide the Zika related tweets into similar groups which could provide useful information to people and thus enabling them to extract helpful insights. World Health Organization (WHO) on February 1, 2016 declared Zika virus as a Public Health Emergency of International Concern (PHEIC). This disease has become a global issue as there is no vaccine or any other form of treatment has been developed for it so far. In the study in [28], Twitter Streaming API was used to collect the most recent tweets. The tweets collected by the API are then pre-processed initially to make the later analysis easier. The URLs, hashtags, and user mentions are separated from the text in the original tweet. An analysis had generated an ordering of the countries with respect to the frequency of twits coming from them. An architecture for classification of Zika virus was proposed [28] (Fig. 2).

Since the classification in this study was to process is to classify texts and as has been supported by literature, Support vector machine (SVM) algorithm and Naïve Bayes algorithm were used. Other characteristics in support of using SVM are that the number of classes is small (3), high number of features (around 2804) and the process being non-probabilistic. The characteristic features are the commonly occurring words inside all the tweets in connection with Zika virus. It was observed that SVM classification generates an accuracy of nearly 90%.

According to the analysis, it was inferred that in the tweets gathered 36.50%, 24.94%, and 38.54% of tweets belonged to 'fight and prevention,' 'cure,' and

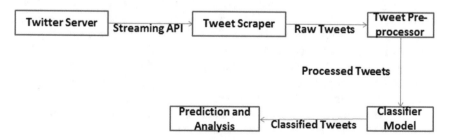

Fig. 2 Zika virus classification model architecture

'infected and death,' respectively. These values also provide a statistical evidence of social community support and awareness available for Zika presently.

4.3 Sensor Based Health Condition and Food Safety Monitoring

Software and hardware combination such as sensors being used, create enough of wonderful applications which take control of food safety and health condition. In the market we find some such products like the Apple Watch from Apple which measures the heart rate, Smart chopsticks which measure PH levels, temperature, calories and freshness of cooking oil. The tread mills produced by several companies which measuring the walking rate, mileage and pulse. Testo's 270 allows you to determine within seconds whether cooking oil needs changing. This will not only make sure you're cooking with clean oil, but is also proven to save you money on cooking oil costs. **HACCP** (Hazard Analysis Critical Control Points) guidelines have applied internationally for many years in terms of monitoring in the food sector. HACCP is a method of risk management which is used to improve monitoring and food safety. One of the most frequent causes of food poisoning is inadequate cooling or heating of foods. Food safety is a scientific discipline describing handling, preparation, and storage of food in ways that prevent foodborne illness.

4.4 Genome Wide Association Studies (GWAS) and Expression Quantitative Trait Loci (EQTLs)

In genetics, a genome-wide association study (GWA study, or GWAS), also known as whole genome association study (WGA study, or WGAS), is an examination of a genome-wide set of genetic variants in different individuals to see if any variant is associated with a trait. GWASs typically focus on associations between single-nucleotide polymorphisms (SNPs) and traits like major human diseases, but can equally be applied to any other organism. GWAS is gaining popularity due to the cost of genotyping coming down over the past few years. The results are available in the GWAS databases like GWAS catalog and GWASdb. However because of the unclear nature of the GWAS identified SNPs, researchers have been thinking to improve the design of GWAS. The paper of Freedman et al. [29] is an useful source for discussion on GWAS. Expression traits differ from most other classical complex traits in one important respect—the measured mRNA or protein trait is almost always the product of a single gene with a specific chromosomal location. eQTLs that map to the approximate location of their gene-of-origin are referred to as local eQTLs. In contrast, those that map far from the location of their gene of origin, often on different chromosomes, are referred to as distant eQTLs. Often, these two types of eQTLs are referred to as cis and trans, respectively, but these terms are best reserved for instances when the regulatory mechanism (cis vs. trans) of the underlying sequence has been established.

4.5   Inferring Air Quality Using Big Data

Air pollution can cause several serious diseases like lung cancer and cardiovascular disease. Monitoring stations have been established world over to know the quality of air and traditional air quality monitoring methods need to establish and maintain the physical monitoring stations. Recently, several approaches are proposed to estimate air pollution from the perspective of big data. The approach relies on other data sources than the monitoring stations. For example, Zheng et al. [30] inferred the air quality information in big cities in China by combining existing monitor station data with information obtained from meteorology, traffic flow, human mobility and road networks. Mei et al. [31] estimated air quality from social media posts. Honicky et al. [32] suggested to attach sensors to GPS-enabled cell phones and used them to collect air pollution information. Chen et al. [33] introduced an indoor air quality monitoring system.

4.6   Metabolomics and Ionomics for Nutritionists

Metabolomics is the scientific study of chemical processes involving metabolites. Specifically, metabolomics is the "systematic study of the unique chemical fingerprints that specific cellular processes leave behind", the study of their small-molecule metabolite profiles [34]. The metabolome represents the collection of all metabolites in a biological cell, tissue, organ or organism, which are the end products of cellular processes [35]. mRNA gene expression data and proteomic analyses reveal the set of gene products being produced in the cell, data that represents one aspect of cellular function. Conversely, metabolic profiling can give an instantaneous snapshot of the physiology of that cell. One of the challenges of systems biology and functional genomics is to integrate proteomic, transcriptomic, and metabolomic information to provide a better understanding of cellular biology. To identify and quantify all metabolites with in a system using Nuclear Magnetic resonance (NMR) and Mass Spectroscopy (MS) is the primary goal of Metabolomics [36]. This has been changed from identification and quantification to associating it to diseases. It has been used to study diabetes [37] and Toxicology [38].

5 Some Research Directions

In this section we present some possible directions of research.

5.1 Many of the open problems in the field of big data management and analysis are associated with healthcare fields. Researchers who require big data solutions in order to manage large medical datasets are now being assisted by hospitals in getting them. That with in clinical environments big data will be able to revolutionize pharmaceutical research and development was proposed by McKinsey & Company. According to it, the target will be the diverse roles played by users, physicians, consumers, insurers and regulators [39]. Big data can be helpful in reducing the cost of research and development for pharmaceutical industry to a large extent [40]. Drug makers, healthcare providers and health analyst companies are collaborating on this topic. Also, they are working on private cloud for pharmaceutical industry sharing securely anonymized data [41]. So, big data research is required to be made more focused and more efficient to handle these issues and come up with solutions.

5.2 It is required to identify and establish policies that specify who is accountable for various portions or aspects of institutional data and information including its accuracy, accessibility, consistency, completeness and maintenance [4].
Also, it is desirable to defining processes concerning how data and information are stored, achieved, backed up and protected. Also, it is required to determine developing standards and procedures that define how the data and information are used by authorized personnel and implement a set of audit and control procedures to ensure ongoing compliance with governmental regulations and industrial standards.

5.3 Educational topics like performance in scientific research, correlations between the knowledge of students and the competencies required, academic failure, to realizing the learning gaps, improve teaching methods and educational management processes require to be researched for better solutions using big data techniques.

5.4 An analysis of the research articles published between 2010 and 2015 was carried out in [14]. The major trends could be put in the three categories of "development of academic analytics and introduction of learning analytics, its concepts, implication and impact to higher education and e-learning", "Use of datasets to improve learning analytics [42–44], especially through communication and collaboration between educational data mining and learning analytics communities" and "the use of learning analytics in social learning and MOOCs". It was observed in [14] that less study has been conducted on evaluating learning outcomes by analyzing natural language text. So, more focus can be on finding out the techniques for prediction of student performance in learning environments where students interact through forums [45].

6 Conclusions

Now day enormous amounts of data are generated in several fields. Two very important fields in this direction are education and health care. Techniques to analyse the data sets related to these two fields will help these sectors in rectifying their traditional approaches, which are inadequate for such analysis. In this chapter we have presented the different issues involved in dealing these data sets and some solutions obtained by different sources and organisations so far. Finally we have proposed some problems for further studies in this direction.

References

1. Daniel, B. K., & Butson, R. (2013). Technology enhanced analytics (TEA) in higher education. In *Proceedings of the International Conference on Educational Technologies, 29 November–1 December, 2013, Kuala Lumpur, Malaysia* (pp. 89–96).
2. Mauro, A. D., Greco, M., & Grimaldi, M. (2015). What is big data? A consensual definition and a review of key research topics. In *AIP Proceedings of the International Conference on Integrated Information (IC-ININFO 2014)* (Vol. 1644, pp. 97–104).
3. Jin, X., Wah, B. W., Cheng, X., & Wang, Y. (2015). Significance and challenges of big data research. *Big Data Research, 2,* 59–64.
4. Daniel, B. (2014). Big data and analytics in higher education: Opportunities and challenges. *British Journal of Education Technology,* 1–17.
5. Huang, T., Lan, L., Fang, X., An, P., Min, J., & Wang, F. (2015). Promises and challenges of big data computing in health sciences. *Big Data Research, 2,* 2–11.
6. Tulasi, B. (2013). Significance of big data and analytics in higher education. *International Journal of Computer Applications, 68*(14), 21–23.
7. Sin, K., & Muthu, L. (2015). Application of big data in education data mining and learning analytics—A literature review. *ICTACT Journal on Soft Computing* (Special Issue on Soft Computing Models for Big Data), 1035–1049.
8. Hilbert, M. (2014). Big data for development: From information to knowledge societies (January 15, 2013). Retrieved October 30, 2014 from http://ssrn.com/abstract=2205145 or https://doi.org/10.2139/ssrn.2205145.
9. Kumar, V., & Chadha, A. (2011). An empirical study of data mining techniques in higher education. *International Journal of Advanced Computer Science and Applications, 2*(3), 80–84.
10. Pandey, U.K., & Pal, S. (2011). A data mining view on class room teaching language. *International Journal of Computer Science and Information Technologies, 2*(2), 686–690.
11. Pal, S. (2012). Mining educational data to reduce dropout rates of engineering students. *International Journal of Information Engineering and Electronic Business, 4*(2), 1.
12. Luan, J. (2012). Data mining and its application in higher education. In A. Serban, & J. Luan (Eds.), *Knowledge management: Building a competitive advantage in higher education* (pp. 17–36).
13. Michalik, P., Stofa, J., & Zolotova, I. (2014). Concept definition for big data architecture in the education system. In *IEEE 12th International Symposium on Applied Machine Intelligence and Informatics* (pp. 3321–334).
14. Kalota, F. (2015). Applications of big data in education, world academy of science, engineering and technology. *International Journal of Social, Behavioral, Educational, Economic, Business and Industrial Engineering, 9*(5), 1607–1611.

15. Romero, C. R., & Ventura, S. (2010). Educational data mining: A review of the state of the art. *IEEE Transactions on Systems, Man, and Cybernetics Part C: Applications and Reviews, 40*(6), 601–618.
16. Bresfelean, V. P. (2008). Data mining applications in higher education and academic intelligence management. In J. E. Meng, & Z. Yi (Eds.), *Theory and novel applications of machine learning* (pp. 209–228).
17. Bhardwaj, B. K., & Pal, S. (2011). Mining educational data to analyze students' performance. *International Journal of Advanced Computer Science and Applications, 2*(6), 63–69.
18. Bhardwaj, B. K., & Pal, S. (2012). Data mining: A prediction for performance improvement using classification. *International Journal of Computer Science and Information Security, 9*(4).
19. Minaei-Bidgoli, B., Kashy, D., Kortmeyer, G., & Punch, W. (2003). Predicting student performance: An application of data mining methods with an educational web-based system. In *Proceedings of 33rd Annual Frontiers in Education Conference FIE 2003* (pp. T2A13–T2A18).
20. Downs, E. N. (2014). UF hires bioinformatics expert. https://m.ufhealth.org/news/2014/uf-hires-bioinformatics-report.
21. Merelli, I., Perez-Sanchez, H., Gesing, S. & D'Agostino, D. (2014). *Managing, analyzing and integrating big data in medical bioinformatics: Open problems and future perspectives* (pp. 1–13). Hindawi Publishing Corporation, Biomed Research International.
22. Feldman, B., Martin, E. M., & Skotnes, T. (2012). Big data in healthcare, hype and hope. In Dr. Bonnie, *Business development for digital health* (Vol. 360, pp. 1–56).
23. Duan, L., Street, W. N., & Xu, E. (2011). Healthcare information systems: Data mining methods in the creation of a clinical recommender system. *Entrepreneurs Information Systems, 5,* 169–181.
24. Hoens, T. R., Blanton, M., Steele, A., & Chawla, N. V. (2013). Reliable medical recommendation systems with patient privacy. *ACM Transactions on Intelligent Systems and Technology, 4,* 1–31.
25. Wiesner, M., & Daniel, P. (2014). Health recommender systems: Concepts requirements, technical basics and challenges. *International Journal of Environmental Research and Public Health, 11*(3), 2580–2607.
26. Signorini, A., Segre, A. M., & Polgreen, P. M. (2011). The use of Twitter to track levels of disease activity and public concern in the U.S. during the influenza A H1N1 pandemic. *PLoSONE, 6*(5), e19467, 1–10. www.plosone.org.
27. Paul, M. J., Dredze, M., & Broniatowski, D. (2014). Twitter improves influenza forecasting. *PLoS Currents.* www.ncbi.nlm.nih.gov.
28. Tripathy, B.K., Chowdhury, R., & Thakur, S. (2016). A classification model to analyze the spread and emerging trends of the Zika virus in Twitter, In *The proceedings of International Conference on Computational Intelligence in Data Mining (ICCIDM 2016).* Advances in Intelligent Systems and Computing (AISC, Vol. 556, pp. 643–650).
29. Freedman, M. L., Monteiro, A. N., Gayther, S. A., et al. (2011). Principles for the post-GWAS functional characterization of cancer risk loci. *Nature Genetics, 43,* 513–518.
30. Zheng, Y., Liu, F., & Hsieh, H.-P. (2013). U-Air: When urban air quality inference meets big data. In *KDD'13, 11–14 August 2013, Chicago, Illinois, USA.*
31. Mei, S., Li, H., Fan, J., Zhu, X., & Dyer C. R. (2013). Inferring air pollution by sniffing social media.
32. Honicky, R.J., Brewer, E. A., Paulos, E., & White, R. M. (2008). N-SMARTS: Networked suite of mobile atmospheric real-time sensors. In *NSDR'08, 18 August 2008, Seattle, Washington, USA* (pp. 25–29).
33. Chen, B. H., Hong, C. J., Pandey, M. R., & Smithd, K. R. (1990). Indoor air pollution in developing countries. *World Health Statistics Quarterly, 43,* 127–138.
34. Davis, B. (2005). Growing pains for metabolomics. *The Scientist, 19*(8), 25–28.

35. Jordan, K. W., Nordenstam, J., Lauwers, G. Y., Rothenberger, D. A., Alavi, K., Garwood, M., et al. (2009). Metabolomic characterization of human rectal adenocarcinoma with intact tissue magnetic resonance spectroscopy. *Diseases of the Colon and Rectum, 52*(3), 520–525.
36. Dettmer, K., Aronov, P. A., & Hammock, B. D. (2007). Mass spectrometry-based metabolomics. *Mass Spectrometry Reviews, 26*(1), 51–78.
37. Zhang, A. H., Qiu, S., Xu, H. Y., Sun, H., & Wang, X. J. (2014). Metabolomics in diabetes. *Clinica Chimica Acta, 429,* 106–110.
38. Donald G., Paul, R., Watkins, B., & Michael, D. (2011). Reily: Metabolomics in toxicology: Preclinical and clinical applications. *Toxicological Sciences, 120*(suppl_1), S146–S170.
39. McKinsey and Company: How big data can revolutionize pharmaceutical R&D. http://www.mckinsey.com/insights/health_systems_and_services/how_big_data_can_revolutionize_pharmaceutical_r_and_d.
40. Medill Reports. (2014). http://news.medill.northwestern.edu/Chicago/news.aspx?id=228875.
41. Xian Sheng, K. (2014). Big data x-learning resources integration and processing in cloud environment. *International Journal of Emerging Technologies and Learning, 9*(5), 22–26.
42. Picciano, A. G. (2012). The evolution of big data and learning analytics in American higher education. *Journal of Asynchronous Learning Networks, 16*(3), 9–20.
43. Romero, C. R., & Ventura, S. (2010). Educational data mining: A review of the state of the art. *IEEE Transactions on Systems, Man and Cybernetics, Part C: Applications and Reviews, 40*(6), 601–618.
44. Wagner, E., & Ice, P. (2012). Data changes everything: Delivering on the promise of learning analytics in higher education. *Educause Review,* 33–42.
45. Niemi, D., & Gitin, E. (2012). Using big data to predict student dropouts technology affordances for research. In *Proceedings from the International Association for Development of the Information Society (IADIS) International Conference on Cognition and Exploratory Learning in Digital Age.*

BWT: An Index Structure to Speed-Up Both Exact and Inexact String Matching

Yangjun Chen and Yujia Wu

Abstract The BWT transformation of a string is originally proposed for string compression, but can also be used to speed up string matchings. In this chapter, we address two issues around this mechanism: (1) how to use BWT to improve the running time of a multiple pattern string matching process; and (2) how to integrate mismatching information into a search of BWT arrays to expedite string matching with k mismatches. For the first problem, we will first construct the BWT array of a target string s, denoted as $BWT(s)$; and then establish a *trie* structure over a set of pattern strings $R = \{r_1, \ldots, r_l\}$, denoted as $T(R)$. By scanning $BWT(s)$ against $T(R)$, the time spent for finding occurrences of r_i's can be significantly reduced. For the second problem, for a given pattern string r, we will precompute its mismatching information (over some different substrings of it, denoted as $M(r)$) and construct a tree structure, called a *mismatching tree*, to record the mismatches between r and s during a search of $BWT(s)$ against r. In this process, the mismatching tree can be effectively utilized to do some kind of useful mismatching information derivation based on $M(r)$ to avoid any possible redundancy. Extensive experiments have been done to compare our methods with the existing ones, which show that for both the problems described above our methods are promising.

1 Introduction

The recent development of next-generation sequencing has changed the way we carry out the molecular biology and genomic studies [1]. It has allowed us to sequence a *DNA* (Deoxyribonucleic acid) sequence at a significantly increased base coverage, as well as at a much faster rate. This requires us considering all the string

Y. Chen (✉) · Y. Wu
Department of Applied Computer Science, University of Winnipeg, Winnipeg, Canada
e-mail: y.chen@uwinnipeg.ca

Y. Wu
e-mail: wyj1128@yahoo.com

© Springer Nature Singapore Pte Ltd. 2018
S. S. Roy et al. (eds.), *Big Data in Engineering Applications*,
Studies in Big Data 44, https://doi.org/10.1007/978-981-10-8476-8_12

221

patterns as a whole, rather than separately check them one by one. Two kinds of string matching need to be handled: *exact matching* and *inexact matching*. By the exact matching, we will find all the occurrences of a pattern string r in a target string s, but by the inexact matching we allow each occurrence having up to k positions different between r and s. The inexact matching is important due to the polymorphisms or mutations among individuals or even sequencing errors, the pattern may disagree in some positions at an occurrence of r in the target s.

The string matching is always an interesting and important research topic in computer science and computer engineering. In the past several decades, a bunch of efficient strategies have been proposed to find all the occurrences of a pattern in a target very fast, such as those discussed in [2–8]. Roughly speaking, all these methods can be classified as illustrated in Fig. 1.

From Fig. 1, we can see that for the exact matching problem we distinguish between two kinds of strategies: the single-pattern oriented and the multi-pattern oriented methods. By the former, each time only one pattern string will be mapped to a target string, and for this we have both on-line methods such as *Knuth-Morris-Pratt* [7], *Boyer-Moore* [6], and *Apostolico-Giancarlo* [9], and off-line (index-based) methods like *suffix* trees [10, 11], *suffix* arrays [12], and *BWT*-transformation (*Burrows-Wheeler* Transformation) [13–15]. However, by the latter, we have only on-line strategies, such as the Aho-Corasick's algorithm proposed in 1975 [16], and its improved versions [17–20], by which an automaton is established over all the patterns and will be searched against a target in one scan.

For the inexact matching problem, we have string matching with k mismatches, k errors, as well as *don't-care* symbols. By the string matching with k mismatches, we will find all the occurrences of a pattern string r in a target string s with each occurrence having up to k positions different between r and s. Different methods for this problem have been proposed, such as the on-line strategies discussed in [2, 4, 21, 22], and the index-based method proposed in [23]. The methods of [4, 21, 22] have the worst-case time complexities bounded by O($kn + m\log m$), where $n = |s|$ and $m = |r|$. By these three methods, the *mismatch information* among substrings of r is used to speed up the working process. The method discussed in [2] is with a slightly better time complexity $O(n\sqrt{k}\log k)$. By this method, the *periodicity*

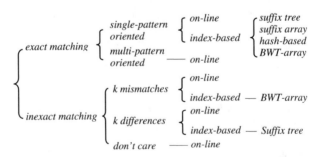

Fig. 1 Classification of methods for string matching

within r is utilized. In [23], a target string s is transformed to a BWT-array (denoted as $BWT(s)$) as an index [13]. In comparison with *suffix trees* [10], $BWT(s)$ uses much less space [13]. However, its time complexity is bounded by $O(mn' + n)$, where n' is the number of leaf nodes of a tree produced during the search of BWT (s). This time requirement can be much worse than the best on-line algorithm for large patterns. The reason for this is that by this method neither mismatch information nor periodicity within r is employed.

The string matching with k errors is quite different from the string matching with k mismatches, by which we will find all the occurrences of a pattern string r in a target string s such that the edit distance [24, 25] between each occurrence and s is $\leq k$. To do such a task, the *dynamic programming paradigm* has to be employed [26], possibly with suffix trees being used as indexes [27, 28]. By the string matching with *don't-care* symbols, we allow *don't-care* to appear in r, in s, or in both of them [29, 30].

In this chapter, we address two issues. One is to construct indexes for the multiple pattern string matching [31], and another one is to construct indexes for the string matching with k mismatches [32]. As discussed above, up to now no effective indexes have been established for these two problems. Specifically, for the first problem, we will

- Construct a trie T over all the pattern sequences, and check T against the BWT-array of s's reverse, denoted as $BWT(\bar{s})$ created as an index for s. This enables us to avoid repeated search of the same part of different pattern strings.
- Change a single-character checking to a multiple-character checking. (That is, each time a set of characters respectively from more than one pattern strings will be checked against a BWT-array in one scan, instead of checking them separately one by one in multiple scans.)

Our experiment shows that it can be more than 40% faster than single-pattern oriented methods when multi-million pattern strings are checked.

For the second problem, two techniques are introduced, which will be combined with a BWT-array scanning as described below:

- An efficient method to calculate the mismatches between $r[i \ldots m]$ and $r[j \ldots m]$ $(i, j \in \{1, \ldots, m\}, i \neq j)$, where $r[i \ldots m]$ represents a substring of r starting from position i and ending at position m. The mismatches between them is stored in an array R such that if $R[p] = q$ then we have $r[i + q - 1] \neq r[j + q - 1]$ and it is their pth mismatch.
- A new tree (forest) structure D to store the mismatches between r and different segments of s. In D, each node v stores an integer i, indicating that there are some positions $i_1, i_2, \ldots, i_l$ in s such that $s[i_q + i - 1] \neq r[i]$ $(q = 1, \ldots, l)$. If v is at the pth level of D, it also shows that it is the pth mismatch between each s $[i_q \ldots i_q + i - 1]$ and r.

By using these two techniques, the time complexity for solving the string matching with k mismatches can be reduced to $O(kn' + n)$. Our experiment shows that $n' \ll n$.

2   Related Work

The string matching problem has always been one of the main focuses in computer science. A huge number of algorithms have been proposed, which can be generally divided into two categories: *exact matching* and *inexact matching*. By the former, all the occurrences of a pattern string r in a target string s will be searched. By the latter, a best alignment between r and s (i.e., a correspondence with the highest score) is searched in terms of a given distance function or a score matrix, which is established to indicate the relevance between different symbols.

- *Exact matching*

The first interesting algorithm for this problem is the famous *Knuth-Morris-Pratt*'s algorithm [7], which scans both r and s from left to right and uses an auxiliary *next-table* (for r) containing the so-called *shift information* (or say, *failure function values*) to indicate how far to shift the pattern from right to left when the current character in r fails to match the current character in s. Its time complexity is bounded by $O(m + n)$, where $m = |r|$ and $n = |s|$. (By the shift information, we mean a largest integer j associated with a position i in r such that $r[1 \ldots j] = r[i - j + 1 \ldots i]$. Thus, if the current character from the target does not match $r[i + 1]$, we will compare $r[j + 1]$ with the character next to the current one at a next step). The *Boyer-Moore*'s approach [6, 33] works a little bit better than the *Knuth-Morris-Pratt*'s. In addition to the next-table, a skip-table *skip* (also for r) is kept, in which each entry $skip[w]$ is a smallest integer j such that $r[m - j] = w$. (Here, we notice that the entries in *skip* are indexed by characters w in the alphabet Σ.) For a large alphabet and small pattern, the expected number of character comparisons is about n/m, and is $O(m + n)$ in the worst case. These two methods have sparked a series of subsequent research on this problem [16, 28, 34, 35]. Especially, the idea of the 'shift information' has also been adopted by Aho and Corasick [16] for the *multiple pattern* matching, by which s is searched for an occurrence of any one of a set of l patterns: $\{r_1, r_2, \ldots, r_l\}$. Their algorithm needs only $O\left(\sum_{i=1}^{l} |r_i| + n\right)$ time. This method has been slightly improved in different ways [17, 18, 36, 37, 38]. In [17], Commentz-Walter combines the Boyer-Moore's technique into the Aho-Corasick's algorithm. In [18], Wu and Mamber extend the Boyer-Moore's algorithm to concurrently search multiple pattern strings. Instead of using *bad* character heuristics to compute shift values, they utilize a character block containing 2 or 3 characters. In addition, hash tables are created to link the blocks and the related patterns. In [36], a concept of *superalphabets* is introduced, in which each (*super*) character corresponds to a set of q-grams (each being a substring from a certain pattern and represented as a bit string, called a *signature*, generated by using a hash function). In this way, a super automaton can be created, in which each transition is labeled with a super character. *s* will also be handled as a sequence of *q-grams* and searched in the same way as the Aho and Corasick's algorithm. The main problem of this method is the *false positive* and a very time-consuming verification process is needed. In [19], Crochemore et al. combine the directed acyclic word graphs into the Aho-Corasick's algorithm. If the total length of all patterns is polynomial with respect to the shortest length m' of a pattern, the average number of comparisons is $O((n/m')\log m')$.

However, all the improved algorithms have the same worst-case time complexity as the Aho-Corasick's.

In situations where a fixed string *s* is to be searched repeatedly, it is worthwhile constructing an index over *s*, such as suffix trees [10, 11], suffix arrays [12], and more recently the BWT-transformation [13, 15, 23, 39]. A suffix tree is in fact a *trie* structure [40] over all the suffixes of *s*; and by using the Weiner's algorithm [11] it can be built in $O(n)$ time. However, in comparison with the BWT-transformation, a suffix tree needs much more space. Especially, for DNA sequences the BWT-transformation works highly efficiently due to the small alphabet Σ of DNA strings. By the BWT, the smaller Σ is, the less space will be occupied by the corresponding indexes. According to a survey done by Li and Homer [41] on sequence alignment algorithms for next-generation sequencing, the average space required for each character is 12–17 bytes for suffix trees while only 0.5–2 bytes for the BWT. The experiments reported in [21] also confirm this distinction. For example, the file size of chromosome 1 of human is 270 Mb. But its suffix tree is of 26 Gb in size while its BWT needs only 390 Mb–1 Gb for different compression rates of auxiliary arrays, completely handleable on PC or laptop machines.

By the hash-table-based algorithms [42], short substrings called 'seeds' will be first extracted from a pattern *r* and a *signature* (a bit string) for each of them will be created. The search of a target string *s* is similar to that of the Brute Force searching, but rather than directly comparing the pattern at successive positions in *s*, their respective signatures are compared. Then, stick each matching seed together to form a complete alignment. Its expected time is $O(m + n)$, but in the worst case, which is extremely unlikely, it takes $O(mn)$ time. The hash technique has also been extensively used in the DNA sequence research [43–47]. However, almost all experiments show that they are generally inferior to the suffix tree and the BWT index in both running time and space requirements.

- *Inexact matching*

By the inexact matching, we will find, for a certain pattern *r* and an integer *k*, all the substrings s' of *s* such that $d(s', r) \leq k$, where *d* is a distance function. In terms of different distance functions, we distinguish between two kinds of inexact matches: string matching with *k* mismatches and string matching with *k* errors. A third kind of inexact matching is that involving Don't Care, or wild-card symbols which match any single symbol, including another Don't Care.

k mismatches When the distance function is the *Hamming* distance, the problem is known as the string matching with *k mismatches* [2, 22]. By the Hamming distance, the number of differences between *r* and the corresponding substring *s'* is counted. There are a lot of algorithms proposed for this problem, such as [2, 22, 24, 48–51]. They are all on-line algorithms. Except those discussed in [2, 22], all the other methods have the worst-case time complexity O(*mn*). The method discussed in [22], however, requires only O(*kn* + *m*log*m*) time, by which the mismatch arrays for *r* is precomputed and exploited to speed up the search of *s*. The method discussed in [2] works slightly better, by which the periodicity within *r* is utilized. Its time complexity is bounded by $O(n\sqrt{k}\log k)$. The algorithm discussed in [23] is index-based, by which *s* is transformed to a BWT-array, used as an index; but its time complexity is bounded by O(*mn'* + *n*), where *n'* is the number of leaf nodes of a tree produced during the search of *BWT*($\bar{s}$). If *m* is large, it can be worse than all those on-line methods discussed in [2, 22, 49, 50]. Another index-based method is based on a brute-force searching of suffix trees [52]. Its time complexity is bounded by $O(m+n+(c\log n^k/k!))$, where *c* is a very large constant. It can also be worse than an on-line algorithm when *n* is large and *k* is larger than a certain constant.

k errors When the distance function is the *Levenshtein* distance, the problem is known as the string matching with *k errors* [24]. By the Levenshtein distance, we have

$$d_{ij} = \min\left\{d_{i-1,j} + w(r_i, \phi), d_{i,j-1} + w\left(\phi, s_j'\right), d_{i-1,j-1} + w\left(r_i, s_j'\right)\right\},$$

where $d_{i,j}$ represents the distance between $r[1 \dots i]$ and $s'[1 \dots j]$, r_i (s_j') the *i*th character in *r* (*j*th character in *s'*), ϕ an empty character, and $w(r_i, s_j')$ the cost to transform r_i into s_j'.

Also, many algorithms have been proposed for this problem [4, 26–28]. They are all some kinds of variants of the *dynamic programming* paradigm [26] with the worst-case time complexity bounded by O(*mn*). However, by the algorithm discussed in [27], the expected time can reach O(*kn*).

don't care As a different kind of inexact matching, the string matching with *Don't-Cares* (or *wild-cards*) has been a third active research topic for decades, by which we may have wild-cards in *r*, in *s*, or in both of them. Due to the wild character's property that it can matches any character, the 'match' relation is no longer transitive, which precludes straightforward adaption of the shift information used by *Knuth-Morris-Pratt* and *Boyer-Moore*. Therefore, all the methods proposed to solve this problem seem not so skillful and need a quadratic time [30]. Using a suffix array as the index, however, the searching time can be reduced to O(log*n*) for some patterns, which contain only a sequence of consecutive Don't Cares [29].

3   BWT Transformation

In this section, we give a brief description of the BWT transformation to provide a discussion background.

3.1   BWT and String Searching

We use s to denote a string that we would like to transform. Assume that s terminates with a special character $\$$, which does not appear elsewhere in s and is alphabetically prior to all other characters. In the case of DNA sequences, we have $\$ < a < c < g < t$. As an example, consider $s = acagaca\$$. We can rotate s consecutively to create eight different strings, and put them in a matrix as illustrated in Fig. 2a.

In Fig. 2a, for ease of explanation, the position of a character in s is represented by its subscript. (That is, we rewrite s as $a_1c_1a_2g_1a_3c_2a_4\$$.) For example, a_2 representing the second appearance of a in s; and c_1 the first appearance of c in s. In the same way, we can check all the other appearances of different characters.

Now we sort the rows of the matrix alphabetically, and get another matrix, as demonstrated in Fig. 2b, which is called the *Burrow-Wheeler Matrix* [13, 14, 39] and denoted as $BWM(s)$. Especially, the last column L of $BWM(s)$, read from top to bottom, is called the *BWT*-transformation (or the *BWT*-array) and denoted as $BWT(s)$. So for $s = acagaca\$$, we have $BWT(s) = acg\$caaa$. The first column is referred to as F.

When ranking the elements x in both F and L in such a way that if x is the ith appearance of a certain character it will be assigned i, the same element will get the same number in the two columns. For example, in F the rank of a_4, denoted as $rk_F(a_4)$, is 1 (showing that a_4 is the first appearance of a in F). Its rank in L, $rk_L(a_4)$ is also 1. We can check all the other elements and find that this property, called the

(a)	(b)	(c) rk_F	F	L	rk_L
$a_1\ c_1\ a_2\ g_1\ a_3\ c_2\ a_4\ \$$	$\$\ a_1\ c_1\ a_2\ g_1\ a_3\ c_2\ a_4$	–	$\$$	a_4	1
$c_1\ a_2\ g_1\ a_3\ c_2\ a_4\ \$\ a_1$	$a_4\ \$\ a_1\ c_1\ a_2\ g_1\ a_3\ c_2$	1	a_4	c_2	1
$a_2\ g_1\ a_3\ c_2\ a_4\ \$\ a_1\ c_1$	$a_3\ c_2\ a_4\ \$\ a_1\ c_1\ a_2\ g_1$	2	a_3	g_1	1
$g_1\ a_3\ c_2\ a_4\ \$\ a_1\ c_1\ a_2$	$a_1\ c_1\ a_2\ g_1\ a_3\ c_2\ a_4\ \$$	3	a_1	$\$$	–
$a_3\ c_2\ a_4\ \$\ a_1\ c_1\ a_2\ g_1$	$a_2\ g_1\ a_3\ c_2\ a_4\ \$\ a_1\ c_1$	4	a_2	c_1	2
$c_2\ a_4\ \$\ a_1\ c_1\ a_2\ g_1\ a_3$	$c_2\ a_4\ \$\ a_1\ c_1\ a_2\ g_1\ a_3$	1	c_2	a_3	2
$a_4\ \$\ a_1\ c_1\ a_2\ g_1\ a_3\ c_2$	$c_1\ a_2\ g_1\ a_3\ c_2\ a_4\ \$\ a_1$	2	c_1	a_1	3
$\$\ a_1\ c_1\ a_2\ g_1\ a_3\ c_2\ a_4$	$g_1\ a_3\ c_2\ a_4\ \$\ a_1\ c_1\ a_2$	1	g_1	a_2	4

Fig. 2   Rotation of a string

rank correspondence, holds for all the elements. That is, for any element a in s, we always have

$$rk_F(a) = rk_L(a) \tag{1}$$

According to this property, a string searching can be very efficiently conducted. To see this, let us consider a pattern string $r = aca$ and try to find all its occurrences in $s = acagaca\$$.

First, we notice that we can store F as $|\Sigma| + 1$ intervals, such as $F_\$ = F[1 \ldots 1]$, $F_A = F[2 \ldots 5]$, $F_C = F[6 \ldots 7]$, $F_G = F[8 \ldots 8]$, and $F_T = \Phi$ for the above example (see Fig. 1c) We can also represent a segment within an F_x with $x \in \Sigma$ as a pair of the form $<x, [\alpha, \beta]>$, where $\alpha \leq \beta$ are two ranks of x. Thus, we have $F_A = F[2 \ldots 5] = <a, [1, 4]>$, $F_C = F[6 \ldots 7] = <c, [1, 2]>$, and $F_G = F[8 \ldots 8] = <g, [1, 1]>$. In addition, we can use L_v to represent a range in L corresponding to a pair v. For example, in Fig. 1c, $L_{<a, [1, 4]>} = L[2 \ldots 5]$, $L_{<c, [1, 2]>} = L[6 \ldots 7]$. $L_{<a, [2, 3]>} = L[3 \ldots 4]$, and so on.

We will also use a procedure *search*(z, v) to search L_v to find the first and the last rank of z (denoted as α' and β', respectively) within L_v, and return $<z, [\alpha', \beta']>$ as the result:

$$search(z, v) = \begin{cases} <z, [\alpha', \beta']>, & \text{if } z \text{ appears in } L_v; \\ \phi, & \text{otherwise.} \end{cases} \tag{2}$$

Then, we work on the characters in r in the reverse order (referred to as a *backward search*). That is, we will search $\bar{r}$ (reverse of r) against $BWT(s)$, as shown below.

Step 1: Check $r[3] = a$ in the pattern string r, and then figure out $F_A = F[2 \ldots 5] = <a, [1, 4]>$.

Step 2: Check $r[2] = c$. Call *search*$(c, L_{<a, [1, 4]>})$. It will search $L_{<a, [1, 4]>} = L[2 \ldots 5]$ to find a range bounded by the first and last rank of c. Concretely, we will find $rk_L(c_2) = 1$ and $rk_L(c_1) = 2$. So, *search*$(c, L_{<a, [1, 4]>})$ returns $<c, [1, 2]>$. It is $F[6 \ldots 7]$.

Step 3: Check $r[3] = a$. Call *search*$(a, L_{<c, [1, 2]>})$. Notice that $L_{<c, [1, 2]>} = L[6 \ldots 7]$. So, *search*$(a, L_{<c, [1, 2]>})$ returns $<a, [2, 3]>$. It is $F[3 \ldots 4]$. Since now we have exhausted all the characters in r and $F[3 \ldots 4]$ contains only two elements, two occurrences of r in s are found. They are a_1 and a_3 in s, respectively.

The above working process can be represented as a sequence of three pairs: $<a, [1, 4]>$, $<c, [1, 2]>$, $<a, [2, 3]>$. In general, for $\bar{r} = C_1 \ldots C_m$, its search against $BWT(s)$ can always be represented as a sequence:

$$<x_1, [\alpha_1, \beta_1]>, \ldots, <x_m, [\alpha_m, \beta_m]>$$

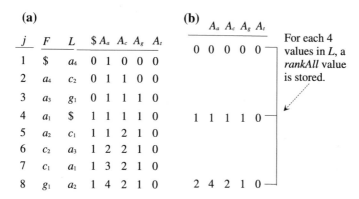

Fig. 3 Illustration for *rankAlls*

where $<x_1, [\alpha_1, \beta_1]> \; = F_{x_1}$, and $<x_i, [\alpha_i, \beta_i]> \; = \; search(x_i, L_{<x_{i-1}, [\alpha_{i-1}, \beta_{i-1}]>})$ for $1 < i \leq m$. We call such a sequence as a *search sequence*. Thus, the time used for this process is bounded by $O\left(\sum_{i=1}^{m} \tau_i\right)$, where τ_i is the time for an execution of $search(x_i, L_{<x_{i-1}, [\alpha_{i-1}, \beta_{i-1}]>})$. However, this time complexity can be reduced to $O(m)$ by using the so-called *rankAll* method [13], by which $|\Sigma|$ arrays each for a character $x \in \Sigma$ are arranged such that $A_x[k]$ (the kth entry in the array for x) is the number of appearances of x within $L[1 \; ... \; k]$ (i.e., the number of x-characters appearing before $L[k + 1]$.) (See Fig. 3a for illustration.)

Now, instead of scanning a certain segment $L[i \; ... \; j]$ ($i \leq j$) to find a subrange for a certain $x \in \Sigma$, we can simply look up the array for x to see whether $A_x[i - 1] = A_x[j]$. If it is the case, then x does not occur in $L[i \; ... \; j]$. Otherwise, $[A_x[i - 1] + 1, A_x[j]]$ should be the range to be found.

For instance, to find the subrange for g within $L[6 \; ... \; 7]$, we will first check whether $A_g[6 - 1] = A_g[7]$. Since $A_g[6 - 1] = A_g[5] = A_g[7] = 1$, we know that g does not appear in $L[6 \; ... \; 7]$. However, since $A_c[2 - 1] \neq A_c[5]$, we immediately get the subrange for c within $L[2 \; ... \; 5]$: $[A_c[2 - 1] + 1, A_c[5]] = [1, 2]$.

The problem of this method is its high space requirement, which can be mitigated by replacing $x[]$ with a compact array A_x for each $x \in \Sigma$, in which, rather than for each $L[i]$ ($i \in \{1, ..., n\}$), only for some entries in L the number of their appearances will be stored. For example, we can divide L into a set of buckets of the same size and only for each bucket a value will be stored in A_x. Obviously, doing so, more search will be required. In practice, the size π of a bucket (referred to as a *compact factor*) can be set to different values. For example, we can set $\pi = 4$, indicating that for each four contiguous elements in L a group of $|\Sigma|$ integers (each in an A_x) will be stored. That is, we will not store all the values in Fig. 3a, but only store $\$[4]$, $a[4]$, $c[4]$, $g[4]$, $t[4]$, and $\$[8]$, $a[8]$, $c[8]$, $g[8]$, $t[8]$ in the corresponding compact arrays, as shown in Fig. 4b. However, each $x[j]$ for $x \in \Sigma$ can be easily derived from A_α by using the following formulas:

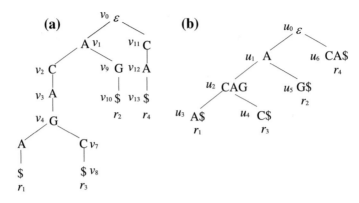

Fig. 4 A trie and its compact version

$$x[j] = A_x[i] + \rho, \tag{3}$$

where $i = \lfloor j/\pi \rfloor$ and ρ is the number of x's appearances within $L[i \cdot \pi + 1 \ldots j]$, and

$$x[j] = A_x[i'] + \rho', \tag{4}$$

where $i' = \lceil j/\pi \rceil$ and ρ' is the number of α's appearances within $L[j + 1 \ldots i' \cdot \pi]$.

Thus, we need two procedures: $sDown(L, j, \pi, x)$ and $sUp(L, j, \pi, x)$ to find ρ and ρ', respectively. In terms of whether $j - i \cdot \pi \leq i' \cdot \pi - j$, we will call $sDown(L, j, \pi, x)$ or $sUp(L, j, \pi, x)$ so that fewer entries in L will be scanned to find $x[j]$.

We notice that the column for $ needn't be stored since it will never be searched. We can also create *rankAlls* only for part of the elements to reduce the space overhead, but at cost of some more searches. See Fig. 3b for illustration.

3.2 Construction of BWT Arrays

A BWT-array can be constructed in terms of a relationship to the *suffix arrays* [13, 14, 39].

As mentioned above, a string $s = a_1 \ldots a_n$ is always ended with $ (i.e., $a_i \in \Sigma$ for $i = 1, \ldots, n - 1$, and $a_n = $). Let $s[i] = a_i$ ($i = 1, 2, \ldots, n$) be the ith character of s, $s[i \ldots j] = a_i \ldots a_j$ a substring and $s[i \ldots n]$ a suffix of s. Suffix array H of s is a permutation of the integers $1, \ldots, n$ such that $H[i]$ is the start position of the ith smallest suffix. The relationship between H and the BWT-array L can be determined by the following formulas:

$$\begin{cases} L[i] = \$, & \text{if } H[i] = 0; \\ L[i] = s[H[i] - 1], & \text{otherwise.} \end{cases} \tag{5}$$

Since a suffix array can be generated in O(n) time [53], L can then be created in a linear time. However, most algorithms for constructing suffix arrays require at least O($n\log n$) bits of working space, which is prohibitively high and amounts to 12 GB for the human genome. Recently, Hon et al. [53] proposed a space-economical algorithm that uses n bits of working space and requires only <1 GB memory at peak time for constructing L of the human genome. We use this for our purpose.

4   Multiple Pattern Matching

In this section, we present our algorithm to search a bunch of pattern strings against a target s. Its main idea is to organize all the reads into a trie T and search T against L to avoid any possible redundancy. First, we present the concept of tries in Sect. 4.1. Then, in Sect. 4.2, we discuss our basic algorithm for the task. We improve this algorithm in Sect. 4.3.

4.1   Tries over Pattern Strings

Let $D = \{s_1, ..., s_n\}$ be a DNA database, where each s_i ($i = 1, ..., n$) is a genome, a very long string $\in \Sigma^*$ ($\Sigma = \{A, T, C, G\}$). Let $R = \{r_1, ..., r_m\}$ be a set of *reads* with each r_j being a short string $\in \Sigma^*$. The problem is to find, for every r_j's ($j = 1, ..., m$), all their occurrences in an s_i ($i = 1, ..., n$) in D.

A simple way to do this is to check each r_j against s_i one by one, for which different string searching methods can be used, such as suffix trees [10, 11], BW-transformation [13], and so on. Each of them needs only a linear time (in the size of s_i) to find all occurrences of r_j in s_i. However, in the case of very large m, which is typical in the new genomic research, one-by-one search of reads against an s_i is no more acceptable in practice and some efforts should be spent on reducing the running time caused by huge m.

Our general idea is to organize all r_j's into a trie structure T and search T against s_i with the BW-transformation being used to check the string matching. For this purpose, we will first attach $ to the end of each s_i ($i = 1, ..., n$) and construct $BWT(s_i)$. Then, attach $ to the end of each r_j ($j = 1, ..., m$) to construct $T = trie$ (R) over R as below.

If $|R| = 0$, $trie(R)$ is, of course, empty. For $|R| = 1$, $trie(R)$ is a single node. If $|R| > 1$, R is split into $|\Sigma| = 5$ (possibly empty) subsets $R_1, R_2, ..., R_5$ so that each R_i ($i \in \{1, ..., 5\}$) contains all those sequences with the same first character $\alpha_i \in \{A, T, C, G\} \cup \{\$\}$. The tries: $trie(R_1), trie(R_2), ..., trie(R_5)$ are constructed in the same way except that at the kth step, the splitting of sets is based on the kth characters in the sequences. They are then connected from their respective roots to a single node to create *trie(R)*.

Example 4.1 As an example, consider a set of four reads:

r_1: ACAGA
r_2: AG
r_3: ACAGC
r_4: CA

For these reads, a trie can be constructed as shown in Fig. 4a. In this trie, v_0 is a virtual root, labeled with an *empty* character ε while any other node v is labeled with a *real* character, denoted as $l(v)$. Therefore, all the characters on a path from the root to a leaf spell a read. For instance, the path from v_0 to v_8 corresponds to the third read r_3 = ACAGC$. Note that each leaf node v is labelled with $ and associated with a *read identifier*, denoted as $\gamma(v)$.

The size of a trie can be significantly reduced by replacing each branchless path segment with a single edge. By a branchless path we mean a path P such that each node on P, except the starting and ending nodes, has only one incoming and one outgoing edge. For example, the trie shown in Fig. 4a can be compacted to a reduced one as shown in Fig. 4b.

4.2 Integrating BWT Search with Trie Search

It is easy to see that exploring a path in a trie T over a set of reads R corresponds to scanning a read $r \in R$. If we explore, at the same time, the L array established over a *reversed* genome sequence $\bar{s}$, we will find all the occurrences of r (without $

(a)	(b) j	F	L	(c)
$ A_4 C_2 A_3 G_1 A_2 C_1 A_1	1	$	A_4	S:
A_1 $ A_4 C_2 A_3 G_1 A_2 C_1	2	A_4	C_2	
A_2 C_1 A_1 $ A_4 C_2 A_3 G_1	3	A_3	G_1	
A_4 C_2 A_3 G_1 A_2 C_1 A_1 $	4	A_1	$	
A_3 G_1 A_2 C_1 A_1 $ A_4 C_2	5	A_2	C_1	
C_1 A_1 $ A_4 C_2 A_3 G_1 A_2	6	C_2	A_3	
C_2 A_3 G_1 A_2 C_1 A_1 $ A_4	7	C_1	A_1	
G_1 A_2 C_1 A_1 $ A_4 C_2 A_3	8	G_1	A_2	$<v_0, 1, 8>$

Fig. 5 Illustration for Step 1

involved) in s. This idea leads to the following algorithm, which is in essence a depth-first search of T by using a stack S to control the process. However, each entry in S is a triplet $<v, a, b>$ with v being a node in T and $a \leq b$, used to indicate a subsegment in $F_{l(v)}[a \ldots b]$. For example, when searching the trie shown in Fig. 5a against the L array shown in Fig. 2a, we may have an entry like $<v_1, 1, 4>$ in S to represent a subsegment $F_A[1 \ldots 4]$ (the first to the fourth entry in F_A) since $l(v_1) = 'A'$. In addition, for technical convenience, we use F_ε to represent the whole F. Then, $F_\varepsilon[a \ldots b]$ represents the segment from the ath to the bth entry in F.

In the algorithm, we first push $<root(T), 1, |s|>$ into stack S (lines 1–2). Then, we go into the main **while-loop** (lines 3–16), in which we will first pop out the top element from S, stored as a triplet $<v, a, b>$ (line 4). Then, for each child v_i of v, we will check whether it is a leaf node. If it is the case, a quadruple $<\gamma(v_i), l(v), a, b>$ will be added to the result $\mathfrak{R}$ (see line 7), which records all the occurrences of a read represented by $\gamma(v_i)$ in s. (In practice, we store compressed suffix arrays [12, 13] and use formulas (1) and (5) to calculate positions of reads in s.) Otherwise, we will determine a segment in L by calculating α' and β' (see lines 8–9). Then, we will use $sDown(L, \alpha' - 1, \pi, x)$ or $sUp(L, \alpha' - 1, \pi, x)$ to find $x[\alpha' - 1]$ as discussed in the previous section. (See line 10.) Next, we will find $x[\beta']$ in a similar way. (See line 11.) If $x[\beta'] > x[\alpha' - 1]$, there are some occurrences of x in $L[\alpha' \ldots \beta']$ and we will push $<v_i, x[\alpha' - 1] + 1, x[\beta']>$ into S, where $x[\alpha' - 1] + 1$ and $x[\beta']$ are the first and last rank of x's appearances within $L[x' \ldots y']$, respectively. (See lines 12–13.) If $x[\beta'] = x[\alpha' - 1]$, x does not occur in $L[\alpha' \ldots \beta']$ at all and nothing will be done in this case. The following example helps for illustration.

ALGORITHM *readSearch(T, LF, π)*

---

begin
1. $v \leftarrow root(T)$; $\mathfrak{R} \leftarrow \varPhi$;
2. $push(S, <v, 1, |s|>)$;
3. **while** S is not empty **do** {
4. $<v, a, b> \leftarrow pop(S)$;
5.   let $v_1, \ldots, v_k$ be the children of v;
6.   **for** $i = k$ **downto** 1 **do** {
7.   **if** v_i is a leaf **then** $\mathfrak{R} \leftarrow \mathfrak{R} \cup \{<\gamma(v_i), l(v), a, b>\}$;
8.   **else**{assume that $F_{l(v)} = <l(v); \alpha, \beta>$;
9. $\alpha' \leftarrow \alpha + a - 1$; $\beta' \leftarrow \alpha + b - 1$; $x \leftarrow l(v_i)$;
10.   find $x[\alpha' - 1]$ by $sDown(L, \alpha'-1, \pi, x)$ or $sUp(L, \alpha'-1, \pi, x)$;
11.   find $x[\beta']$ by $sDown(L, \beta', \pi, x)$ or $sUp(L, \beta', \pi, x)$;
12.   **if** $x[\beta'] > x[\alpha' - 1]$ **then**
13. $push(S, <v_i, x[\alpha' - 1] + 1, x[\beta']>)$;
14.   }
15.   }
16. }
end

Example 4.2 Consider all the reads given in Example 4.1 again. The trie T over these reads are shown in Fig. 4a. In order to find all the occurrences of these reads in s = ACAGACA\$, we will run *readSearch*() on T and the LF of $\bar{s}$ shown in Fig. 5b. (Note that s = $\bar{s}$ for this special string, but the ordering of the subscripts of characters is reversed. In Fig. 5a, we also show the corresponding BWM matrix for ease of understanding.)

In the execution of *readSearch*(), the following steps will be carried out.

Step 1: push $<v_0, 1, 8>$ into S, as illustrated in Fig. 5c.

Step 2: pop out the top element $<v_0, 1, 8>$ from S. Figure out the two children of v_0: v_1 and v_{11}. First, for v_{11}, we will use A_c to find the first and last appearances of l (v_{11}) = 'C' in $L[1 \dots 8]$ and their respective ranks: 1 and 2. Assume that π = 4 (i.e., for each 4 consecutive entries in L a *rankAll* value is stored.) Further assume that for each A_x ($x \in \{a, c, g, t\}$) $A_x[0]$ = 0. The ranks are calculated as follows.

- To find the rank of the first appearance of 'C' in $L[1 \dots 8]$, we will first calculate $C[0]$ by using formula (3) or (4) (i.e., by calling $sDown(L, 0, 4, C)$ or $sUp(L, 0, 4, C)$). Recall that whether (4) or (5) is used depends on whether $j - i \cdot \pi \leq i \cdot \pi - j$, where $i = \lfloor j/\pi \rfloor$ and $i' = \lceil j/\pi \rceil$. For $C[0]$, j = 0. Then, i = i' = 0 and (4) will be used:

$$C[0] = A_c[\lfloor 0/4 \rfloor] + \rho$$

Since $A_c[\lfloor 0/4 \rfloor] = A_c[0] = 0$ and the search of $L[i \cdot \pi \dots j] = L[0 \dots 0]$ finds ρ = 0, $C[0]$ is equal to 0.

- To find the rank of the last appearance of 'C' in $L[1 \dots 8]$, we will calculate $C[8]$ by using (4) for the same reason as above. For $C[8]$, we have j = 8 and i = 2. So we have

Fig. 6 Illustration for stack changes

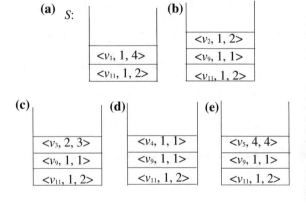

$$C[8] = A_c[\lfloor 8/4 \rfloor] + \rho$$

Since $A_c[\lfloor 8/4 \rfloor] = A_c[2] = 2$, and the search of $L[i \cdot \pi \dots j] = L[8 \dots 8]$ finds $\rho = 0$, we have $C[8] = 2$.

So the ranks of the first and the last appearances of 'C' are $C[0] + 1 = 1$, and $C[8] = 2$, respectively. Push $<v_{11}, 1, 2>$ into S.

Next, for v_1, we will do the same work to find the first and last appearances of $l(v_1) = 'A'$ and their respective ranks: 1 and 4; and push $<v_1, 1, 4>$ into S. Now S contains two entries as shown in Fig. 6a after step 2.

Step 3: pop out the top element $<v_1, 1, 4>$ from S. v_1 has two children v_2 and v_9. Again, for v_9 with $l(v_9) = 'G'$, we will use A_g to find the first and last appearances of G in $L[2 \dots 5]$ (corresponding to $F_A[1 \dots 4]$) and their respective ranks: 1 and 1. In the following, we show the whole working process.

- To find the rank of the first appearance of 'G' in $L[2 \dots 5]$, we will first calculate $G[1]$. We have $j = 1, i = \lfloor j/\pi \rfloor = \lfloor 1/4 \rfloor = 0$ and $i' = \lceil 1/4 \rceil = 1$. Since $j - i \cdot \pi = 0 < i' \cdot \pi - j = 3$, formula (4) will be used:

$$G[1] = A_g[\lfloor 1/4 \rfloor] + \rho$$

Since $A_g[\lfloor 0/4 \rfloor] = A_g[0] = 0$ and search of $L[i \cdot \pi \dots j] = L[0 \dots 0]$ finds $\rho = 0$, $G[1]$ is equal to 0.

- To find the rank of the last appearance of 'G' in $L[2 \dots 5]$, we will calculate $G[5]$ by using (4) based on an analysis similar to above. For $G[5]$, we have $j = 5$ and $i = \lfloor j/\pi \rfloor = 1$. So we have

$$G[5] = A_g[5/4] + \rho$$

Since $A_g[\lfloor 5/4 \rfloor] = A_g[1] = 1$, and search of $L[i \cdot \pi \dots j] = L[4 \dots 5]$ finds $\rho = 0$, we have $G[5] = 1$.

We will push $<v_9, G[1] + 1, G[5]> = <v_9, 1, 1>$ into S.

For v_2 with $l(v_2) = 'C'$, we will find the first and last appearances of C in $L[2 \dots 5]$ and their ranks: 1 and 2. Then, push $<v_2, 1, 2>$ into S. After this step, S will be changed as shown in Fig. 6b.

In the subsequent steps 4, 5, and 6, S will be consecutively changed as shown in Fig. 6c, d, and e, respectively.

In step 7, when we pop the top element $<v_5, 4, 4>$, we meet a node with a single child v_6 labeled with $. In this case, we will store $<\gamma(v_6), l(v_5), 4, 4> = <r_1, A, 4, 4>$ in $\Re$ as part of the result (see line 7 in searchRead()). From this we can find that $rk_L(A_3) = 4$ (note that the same element in both F and L has the same rank), which shows that in $\bar{s}$ the substring of length $|r_1|$ staring from A_3 is an occurrence of r_1. $\square$

4.3 Time Complexity and Correctness Proof

In this subsection, we analyze the time complexity of *readSearch(T, LF, π)* and prove its correctness.

4.3.1 Time Complexity

In the main **while**-loop, each node v in T is accessed only once. If the rankAll arrays are fully stored, only a constant time is needed to determine the range for $l(v)$. So the time complexity of the algorithm is bounded by $O(|T|)$. If only the compact arrays (for the rankAll information) are stored, the running time is increased to $O(|T| \cdot \pi)$, where π is the corresponding compact factor. It is because in this case, for each encountered node in T, $O(\frac{1}{2} \pi)$ entries in L may be checked in the worst case.

4.3.2 Correctness

Proposition 4.1 *Let T be a trie constructed over a collections of reads: $r_1, ..., r_m$, and LF a BWT-mapping established for a reversed genome $\bar{s}$. Let π be the compact factor for the allRank arrays, and $\Re$ the result of readSearch(T, LF, π). Then, for each r_j, if it occurs in s, there is a quadruple $\{<\gamma(v_i), l(v), a, b>\} \in \Re$ such that $\gamma(v_i) = r_j$, $l(v)$ is equal to the last character of r_j, and $F_{l(v)}[a]$, $F_{l(v)}[a + 1], ..., F_{l(v)}[b]$ show all the occurrences of r_j in s.*

Proof We prove the proposition by induction on the height h of T.

Basic step. When $h = 1$. The proposition trivially holds.

Induction hypothesis. Suppose that when the height of T is h, the proposition holds. We consider the case that the height of T is $h + 1$. Let v_0 be the root with $l(v_0) = \varepsilon$. Let $v_1, ..., v_k$ be the children of v_0. Then, $height(T[v_i]) \leq h$ ($i = 1, ..., k$), where $T[v_i]$ stands for the subtree rooted at v_i and $height(T[v_i])$ for the height of $T[v_i]$. Let $l(v_i) = x$ and $F_x = <x; a, b>$. Let $v_{i1}, ..., v_{il}$ be the children of v_i. Assume that α and β be the ranks of the first and last appearances of x in L. According to the induction hypothesis, searching $T[v_{ij}]$ against $L[a' ... b']$, where $a' = a + \alpha - 1$ and $b' = a + \beta - 1$, the algorithm will find all the locations of all those reads with $l(v_i)$ as the first character. This completes the proof. $\square$

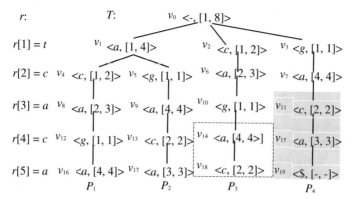

Fig. 7 Search for string matching with 2 mismatches

5 String Matching with k Mismatches

5.1 Basic Working Process

By the string matching with k mismatches, we allow up to k characters in a pattern r to match different characters in a target s. By using the BWT as an index, for finding all such string matches, a tree structure will be generated, in which each path corresponds to a *search sequence* discussed in the previous section. It is due to the possibility that a position in r may be matched to different characters in s and we need to call *search()* multiple times to do this task, leading to a tree representation.

Definition 5.1 (*search tree*) Let r be a pattern string and s be a target string. A search tree T (S-tree for short) is a tree structure to represent the search of r against $BWT(\bar{s})$ (which is equivalent to the search of $\bar{r}$ against $BWT(s)$). In T, each node is a pair of the form $<x, [\alpha, \beta]>$, and there is an edge from v ($=<x, [\alpha, \beta]>$) to u ($=<x', [\alpha', \beta']>$) if $search(x, L_v) = u$.

As an example, consider the case where $r = tcaca$, $s = acagaca$ and $k = 2$. To find all occurrences of r in s with up to two mismatches, a search tree T shown in Fig. 7 will be created.

In Fig. 7, v_0 is a virtual root, representing the whole L, and 'virtually' corresponds to the virtual starting character $r[0] = \text{'-'}$. By exploring paths $P_1 = v_1 \to v_4 \to v_8 \to v_{12} \to v_{16}$ and $P_2 = v_1 \to v_5 \to v_9 \to v_{13} \to v_{16}$, we will find two occurrences of r with 2 mismatches: $s[1 \ldots 5]$ ($=a_1c_1a_2g_1a_3$) and $s[3 \ldots 7]$ ($=a_2g_1a_3c_2a_4$) while by either $P_3 = v_2 \to v_6 \to v_{10} \to v_{14} \to v_{18}$ or $P_4 = v_3 \to v_7 \to v_{11} \to v_{15} \to v_{19}$ no string matching with at most 2 mismatches can be found.

A node $<x, [\alpha, \beta]>$ in such a tree is called a *matching node* if it corresponds to a same character in r. Otherwise, it is called a *mismatching node*. For example, node $v_4 = <c, [1, 2]>$ is a matching node since it corresponds to $r[2] = c$ while $v_1 = <a, [1, 4]>$ is a mismatching node since it corresponds to $r[1] = t$.

For a path P_l, we can store all its mismatching positions in an array B_l of length $k + 1$ such that $B_l[i] = j$ if $P_l[j] \neq r[j]$ and this is the ith mismatch between P_l and r, where $P_l[j]$ is the jth character appearing on P_l. If the number of mismatches, k', say, between P_l and r is less than $k + 1$, then the default value ∞ onwards, i.e.,

$$B_l\left[k' + 1\right] = B_l\left[K' + 2\right] = \cdots = B_l[k+1] = \infty.$$

We call B_l a *mismatch array*. For instance, in Fig. 3, for P_1, we have $B_1 = [1, 4, \infty]$, indicating that at position 1, we have the first mismatch $P_1[1] = a \neq r$ $[1] = t$ and at position 4 we have the second mismatch $P_1[4] = g \neq r[4] = a$. For the same reason, we have $B_2 = [1, 2, \infty]$, $B_3 = [1, 2, 3]$, and $B_4 = [1, 2, 3]$.

These data structures can be easily created by maintaining and manipulating a temporary array B of length $k + 1$ to record the mismatches between the current path P and r. Initially, each entry of B is set to be ∞ and an index variable i pointing to the first entry of B. Each time a mismatch is met, its position is stored in $B[i]$ and then i is increased by 1. Each time r is exhausted or B becomes full (i.e., each entry is set a value not equal to ∞), we will store B as an B_l (and associate it with the leaf node of the corresponding P_l.) Then, 'backtrack' to the lowest ancestor of the current node, which has at least a branch not yet explored, to search a new path. For instance, when we check v_{16}, r is exhausted and the current value of B is $[1, 4, \infty]$. We will store B in B_1 (the array associated with the leaf node v_{16} of P_1) and 'backtrack' to v_1 to explore a new path. At the same time, all those values in B, which are set after v_1, will be reset to ∞, i.e., B will be changed to $[1, \infty, \infty]$.

Now we consider another path P_3. The search along P_3 will stop at v_{10} since when reaching it B becomes full ($B = [1, 2, 3]$). Therefore, the search will not be continued, and v_{14}, v_{18} will not be created.

It is essentially a brute-force search to check all the possible occurrences of r in s. Denote by n' the number of leaf nodes in T. The time used by this process is bounded by $O(mn')$.

In fact, it is the main process discussed in [23]. The only difference is that in [23] a simple heuristics is used, which precomputes, for each position i in r, the number $\sigma(i)$ of consecutive, disjoint substrings in $r[i \ldots m]$, which do not appear in s. For example, in Fig. 3, $\sigma(1) = 2$ since in $r[1 \ldots 5] = tcaca$ both $r[1 \ldots 1] = t$ and $r[2 \ldots 4] = cac$ do not occur in $s = acagaca$. But $\sigma(3) = 0$ since any substring in $r[1 \ldots 3] = aca$ does appear in s. Assume that the number of mismatches between $r[1 \ldots i - 1]$ and $P[1 \ldots i - 1]$ (the current path) is l. Then, if $k - l < \sigma(i)$, we can immediately stop exploring the subtree rooted at $P[i - 1]$ as no satisfactory answers can be found by exploring it.

The time required to establish such a heuristics is $O(n)$ by using $BWT(s)$ [23]. However, the theoretic time complexity of this method is still $O(mn')$. Even in practice, this heuristics is not quite helpful since $\sigma(i)$ delivers only the information related to $r[i \ldots m]$ and the whole s, rather than the information related to $r[i \ldots m]$ and the relevant substrings of s, to which it will be compared. To see this, pay attention to part of the tree marked grey in Fig. 7. Since $\sigma(3) = 0$, the search along P_4 will be continued. But no answer can be found. The heuristics here is in fact useless since it is not about $r[3 \dots 5]$ and $s[5 \dots 7]$, which is to be checked in a next step.

5.2 Mismatch Information

Searching S-trees in an improvement over scanning strings, but it often happens that there are repetitive traversals of similar subtrees due to the multiple appearances of a same pair. However, such repeated appearance of pairs cannot be simply removed since they may be aligned to different positions in r. For example, the first appearance of $<c, [1, 2]>$ (v_4 in Fig. 3) is compared to $r[2]$ while its second appearance (v_2) is to $r[1]$. Hence, we cannot use the result computed for v_4 (when $<c, [1, 2]>$ is first met) as the result for v_2.

However, if we have stored the mismatch information R between substrings of r, like $r[2 \dots 4]$ and $r[1 \dots 3]$, in some way, the mismatches along P_3 can be derived from R and B_1 (the mismatches recorded for P_1), instead of simply exploring P_3 again in a way done for P_1. To do so, for each pair $i, j \in \{1, \dots, m\}$, we need to maintain a data structure R_{ij} containing the positions of the first $k + 1$ mismatches between $r[i \dots m - q + i]$ and $r[j \dots m - q + j]$, where $q = \max\{i, j\}$, such that if $R_{ij}[l] = x (\neq \infty)$ then $r[i + x - 1] \neq r[j + x - 1]$ or one of them does not exist, and it is the lth mismatch between them.

Clearly, this task requires $O(km^2)$ time and space.

For this reason, we will precompute only part of R, instead of R_{ij} for all $i, j \in \{1, \dots, m\}$. Specifically, $R_{12}, \dots, R_{1m}$ for r will be pre-constructed in a way as described in [22], giving the positions of the mismatches between the pattern and itself at various relative shifts. That is, each $R_{1i} (2 \leq i \leq m)$ contains the positions within r of the first $2k + 1$ mismatches between the substring $r[1 \dots m - i]$ and $r[i + 1 \dots m]$, i.e., the overlapping portions of the two copies of pattern r for a relative shift of i. Thus, if $R_{1i}[j] = x$, then $r[x] \neq r[i + x - 1]$ or one of them does not exist, which is the jth mismatch between $r[1 \dots m - i]$ and $r[i + 1 \dots m]$. (See Fig. 8a for illustration.)

In Fig. 8b, we show a pattern $r_1 = tcacg$ and all the possible right-to-left shifts: $r_2 = r[2 \dots 5] = cacg$, $r_3 = r[3 \dots 5] = acg$, and so on. In Fig. 8c, we give R_{12},

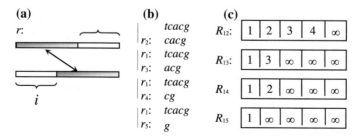

Fig. 8 Illustration for table R

..., R_{15} for r_1. In an R_{1i}, if the number of mismatches, k', say, between $r[1 \ldots m - i]$ and $r[i + 1 \ldots m]$ is less than $2k + 1$, then the default value ∞ onwards, i.e.,

$$R_{1i}[K' + 1] = R_{1i}[k' + 2] = \cdots = R_{1i}[2k + 1] = \infty.$$

We will also use $\delta(R_{1i})$ to represent the number of all those entries in R_{1i}, which are not ∞. Trivially, $R_{11} = [\infty, \ldots, \infty]$.

Using the algorithm of [22], $R_{12}, \ldots, R_{1m}$ can be constructed in O($m\log m$) time, just before the process for the string matching gets started. In addition, we need to keep $2k + 1$, rather than $k + 1$ mismatches in each R_{1i} ($i = 2, \ldots, m$), since for generating an R_{1j}, up to $2k + 1$ mismatches in some R_{1i} with $i < j$ are needed to get an efficient algorithm (see [22] for detailed discussion.)

Each time we meet a node u (compared to a certain $r[j]$), which is the same as an already encountered one v (compared to an $r[i]$), we need to derive dynamically the relevant mismatches, R_{ij}, between $r[i \ldots m - q + i]$ and $r[j \ldots m - q + j]$ from R_{1i} and R_{1j}, as well as r, to compute mismatch information for some new paths (to avoid exploring them by using *search*()). (A node $<x, [\alpha, \beta]>$ is said to be the same as another node $<x', [\alpha', \beta']>$ if $x = x$, $\alpha = \alpha'$ and $\beta = \beta'$.) For this purpose, we design a general algorithm to create R_{ij} efficiently.

- Let ω, ω_1 and ω_2 be three strings. Let A_1 and A_2 be two arrays containing all the positions of mismatches between ω and ω_1, and ω and ω_2, respectively.
- Create a new array A such that if $A[i] = j$ ($\neq \infty$), then $\omega_1[j] \neq \omega_1[j]$, or one of them does not exists. It is the ith mismatch between them.

The algorithm works in a way similar to the *sort-merge-join*, but with a substantial difference in handling a case when an entry in A_1 is checked against an equal entry in A_2. In the algorithm, two index variables p and q are used to scan A_1 and A_2, respectively. The result is stored in A.

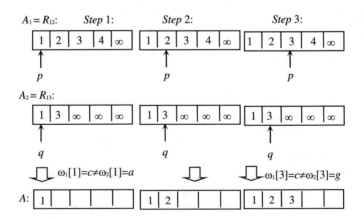

Fig. 9 Illustration for *merge*()

1. $p := 1; q := 1; l := 1;$
2. If $A_2[q] < A_1[p]$, then $\{A[l] := A_2[q]; q := q + 1; l := l + 1;\}$
3. If $A_1[p] < A_2[q]$, then $\{A[l] := A_1[p]; p := p + 1; l := l + 1;\}$
4. If $A_1[p] = A_2[q]$, then $\{$if $\omega_1[p] \neq \omega_2[q]$, then $\{A[l] := q; l := l + 1;\} p := p + 1; q := q + 1;\}$
5. If $p > |A_1|, q > |A_2|$, or both $A_1[p]$ and $A_2[q]$ are ∞, stop (if A_1 (or A_2) has some remaining elements, which are not ∞, first append them to the rear of A, and then stop.)
6. Otherwise, go to (2).

We denote this process as $merge(A_1, A_2, \omega_1, \omega_2)$. As an example, let us consider the case where $A_1 = R_{12} = [1, 2, 3, 4, \infty]$, $A_1 = R_{13} = [1, 3, \infty, \infty, \infty]$, $\omega_1 = r[2 \dots 4] = cacg$ and $\omega_1 = r[3 \dots 5] = acg$, and demonstrate the first three steps of the execution of $merge(A_1, A_2, \omega_1, \omega_2)$ in Fig. 9. The result is $A = [1, 2, 3, 4]$, showing the mismatches between these two substrings.

In step 1: $p = 1, q = 1, l = 1$. We compare $A_1[p] = A_1[1]$ and $A_2[q] = A_2[1]$. Since $A_1[1] = A_2[1] = 1$, we will compare $\omega_1[1]$ and $\omega_2[1]$, and find that $\omega_1[1] = c \neq \omega_2[1] = a$. Thus, $A[1]$ is set to be 1. $p := p + 1 = 2, q := q + 1 = 2, l := l + 1 = 2$.

In step 2: $p = 2, q = 2, l = 2$. we compare $A_1[2]$ and $A_2[2]$. Since $A_1[2] = 2 < A_2[2] = 3$, $A[2]$ is set to be 2. $p := p + 1 = 3, q := 2, l := l + 1 = 3$.

In step 3: $p = 3, q = 2, l = 3$. We compare $A_1[3]$ and $A_2[2]$, and find that $A_1[3] = A_2[2] = 3$. So, we need to compare $\omega_1[3]$ and $\omega_2[3]$. Since $\omega_1[3] = c \neq \omega_2[3] = g$, $A[3]$ is set to be 3. $p := p + 1 = 4, q := 3, l := l + 1 = 4$.

In a next step, we have $p = 4, q = 3, l = 4$. We will compare $A_1[4]$ and $A_2[3]$. Since $A_1[4] = 4 < A_2[3] = \infty$, we set $A[4]$ to 4.

Obviously, the running time of this process is bounded by $O(k)$.

Proposition 5.1 *Let A be the result of $merge(A_1, A_2, \omega_1, \omega_2)$ with $A_1, A_2, \omega_1, \omega_2$ defined as above. Let k be the number of mismatches between ω_1 and ω_2. Then, $A[i]$ must be the position of the ith mismatch between ω_1 and ω_2, or , depending on whether i is $\leq k$.*

Proof Consider $\omega_2[j]$. Position j may satisfy either, neither, or both of the following conditions:

(i) j corresponds to the lth mismatch between ω and ω_2 for some l, i.e., $\omega[j] \neq \omega_2[j]$ and $A_2[l] = j$.
(ii) j corresponds to the fth mismatch between ω and ω_1 for some f, i.e., $\omega[j] \neq \omega_1[j]$ and $A_1[f] = j$.

If (i) holds, but (ii) not, (2) in $merge(A_1, A_2, \omega_1, \omega_2)$ will be executed. Since in this case, we have $\omega[j] \neq \omega_2[j]$ and $\omega[j] = \omega_1[j]$, (2) is correct.

If (ii) holds, but (i) not, (3) will be executed. Since in this case, we have $\omega[j] \neq \omega_1[j]$ and $\omega[j] = \omega_2[j]$, (3) is also correct.

If both (i) and (ii) hold, no conclusion concerning $\omega_1[j]$ and $\omega_2[j]$ can be drawn and we need to compare them. In this case, (4) is executed. If neither (i) nor (ii) is satisfied, we must have $\omega[j] = \omega_2[j]$ and $\omega[j] = \omega_1[j]$. So $\omega_2[j] = \omega_1[j]$, i.e., we have a matching at j.   □

5.3 Main Idea: Mismatch Information Derivation

Now we are ready to present the main idea of our algorithm, which is similar to the generation of an S-tree described in Subsection A. However, each time we meet a node u (compared to a position in r, say, $r[j]$), which is the same as a previous one v (compared to a different position in r, say, $r[i]$), we will not explore $T[u]$ (the subtree rooted at u), but do the following operations to derive the relevant mismatching information:

First, we will create R_{ij} by executing $merge(R_{1i}, R_{1j}, r[i \ldots m - q + i], r[j \ldots m - q + j])$, where $q = \max\{i, j\}$. Then, we will created a set of mismatch arrays for all the sub-paths in $T[u]$, which start at u and end at a leaf node, by doing two steps shown below.

- For each path P_i going through v, figure out a sub-array of B_l, denoted as B_l^i, containing only those values in B_l, which are larger than or equal to i. Moreover, each value in it will be decreased by $i - 1$. (For example, for $B_1 = [1, 4, \infty]$, we have $B_l^1 = [1, 4, \infty]$, $B_l^2 = [3, \infty]$, $B_l^3 = [2, \infty]$, $B_l^4 = [1, \infty]$, and $B_l^5 = [\infty]$.)
- Create the mismatch arrays for all the paths going through u by executing $merge$ $(B_l^i, R_{ij}, P_l[i \ldots m_l], r[j \ldots m])$ for each P_l, where $m_l = |P_i|$.

We denote this process as $mi\text{-}creation(u, v, j, i)$.

As an example, consider v_2 (in Fig. 7, labeled $<c, [1, 2]>$ and compared to r $[1] = t$), which is the same as v_4 (compared to $r[2] = c$). By executing $mi\text{-}creation$ $(v_2, v_4, 1, 2)$, the following operations will be performed, to avoid repeated access of the corresponding subtree (i.e., part of P_3 shown in Fig. 10a):

1. Create R_{21}:
   $R_{12} = [1, 2, 3, 4, \infty]$, $R_{11} = [\infty, \infty, \infty, \infty, \infty]$,

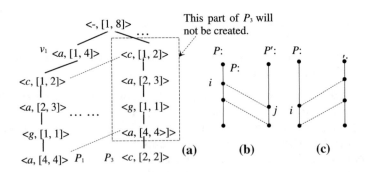

Fig. 10 Illustration for derivation of mismatch information

$R_{21} = merge(R_{12}, R_{11}, r[2 \ldots 5], r[1 \ldots 4]) = [1, 2, 3, 4]$.

2. Create part of mismatch information for P_3:

$B_1 = [1, 4, \infty]$, $B_l^2 = [3, \infty]$. $P_1[2 \ldots 5] = caga$, $r[1 \ldots 4]) = caca$.
$merge(B_l^2, R_{21}, P_1[2 \ldots 5], r[1 \ldots 4]) = [1, 2, 3, 4]$.

In general, we will distinguish between two cases:

(i) $i < j$. This case can be illustrated in Fig. 10b. In this case, the mismatch information for the new paths can be completely derived.

(ii) $i > j$. This case can be illustrated in Fig. 10c. In this case, only part of mismatch information for the new paths can be derived. Thus, after the execution of *merge()*, we have to continue to extend the corresponding paths.

Therefore, among different appearances of a certain node v, we should always use the one compared to $r[i]$ with i being the least to derive as much mismatch information as possible for to be created paths.

Finally, we notice that it is not necessary for us to consider the case $i = j$ since the same node will never appear at the same level more than once. The following lemma is easy to prove.

Lemma 5.1 *In an S-tree T, if two nodes are with the same pair, then they must appear at two different levels.* □

5.4 Algorithm Description

The main idea presented in the previous subsection can be dramatically improved. Instead of keeping a B_l for each P_l, we can maintain a general tree structure, called a *mismatch tree*, to store the mismatch information for all the created paths. First, we define two simple concepts related to S-trees.

Definition 5.2 (*match path*) A sub-path in an S-tree T is called a match path if each node on it is a matching node in T.

Definition 5.3 (*maximal match sub-path*) A maximal match sub-path (*MM-path* for short) in an S-tree T is a match sub-path such that the parent of its first node in T is a mismatching node and its last node is a leaf node or has only mismatching nodes as its children.

For example, edge $v_4 \rightarrow v_8$ in T shown in Fig. 7 is a *MM-path*. Path $v_9 \rightarrow v_{13} \rightarrow v_{17}$ is another one. The node v_{16} alone is also a *MM-path* in T.

Based on the above concepts, we define another important concept, the so-called *mismatch trees*.

Definition 5.4 (*mismatch trees*) A mismatch tree D (*M-tree* for short) for a given S-tree T, is a tree, in which for each mismatching node $<x, [\alpha, \beta]>$ (compared to r $[i]$ for some i) in T we have a node of the form $<x, i>$, and for each *MM-path* a node

of the form $<-, 0>$. There is an edge from u to u' if one of the following two
conditions is satisfied:

- u is of the form $<x, i>$ corresponding to a pair $<x, [\alpha, \beta]>$ (compared to $r[i]$),
  which is the parent of the first node of an MM-path (in T) represented by u'; or
- u is of the form $<-, 0>$ and u' corresponds to a mismatching node which is a
  child of a node on the MM-path represented by u.
  Without causing confusion, we will also call $<-, 0>$ in D a *matching node*,
  and $<x, i>$ a *mismatching node*.

For example, for T shown in Fig. 7, we have its M-tree shown in Fig. 11, in
which u_0 is a virtual root corresponding to the virtual root of the S-tree shown in
Fig. 7. Its value is also set to be $<-, 0>$ since it will be handled as a matching
node. Then, each path in the M-tree corresponds to a B_i. For instance, path $u_0 \to u_1$
$\to u_4 \to u_8 \to u_{12}$ corresponds to $B_1 = [1, 4, \infty]$ if all the matching nodes on the
path are ignored. For the same reason, $u_0 \to u_1 \to u_5 \to u_{19}$ corresponds to
$B_2 = [1, 2, \infty]$.

In addition, we can store all the different nodes v $(=<x, [\alpha, \beta]>)$ in T in a hash
table with each entry associated with a pointer to a node in the corresponding
M-tree D, described as follows.

- If v is a mismatching node compared to $r[i]$ for some $i \in \{1, ..., m\}$, a node
  $u = <x, i>$ will be created in D and a pointer (associated with v, denoted as p
  (v)) to u will be generated.
- If v is a matching node, a node $u = <-, 0>$ will be created in D and $p(v)$ to
  u will be generated. If the parent u' of u itself is $<-, 0>$, u will be merged into
  its parent. That is, v will be linked to u' while u itself will not be generated.

For instance, when $<a, [1, 4]>$ (v_1 in T shown in Fig. 7) is created, it is compared to $r[1] = t$. Since $a \neq t$, we have a mismatch and then $u_1 = <a, 1>$ in the
M-tree D will be generated. At the same time, we will insert $<a, [1, 4]>$ into the
hash table and produce a pointer associated with it to u_1 (see Fig. 11 for

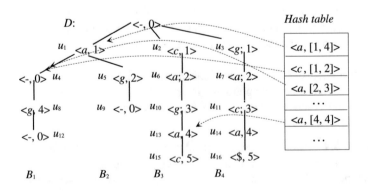

Fig. 11 A mismatch tree

illustration). However, when <c, [1, 2]> (v_4 in T shown in Fig. 7) is created, it is compared to $r[2] = c$ and we have a matching. For this, a node <−, 0> (u_4 in Fig. 7) will be generated, and a link from <c, [1, 2]> to it will be established. But when <a, [2, 3]> (v_8 in T shown in Fig. 7, compared to $r[5] = a$) is met, no node in D will be generated since it is a matching node (in T) and the parent (u_4 in Fig. 11) of the node to be created for it is also <−, 0>. We will simply link it to its parent u_4.

In order to generate D, we will use a stack S to control the process, in which each entry is a quadruple (v, j, κ, u), where

v—a node inserted into the hash table.

j—j is an integer to indicate that v is the jth node on a path in T (counted from the root with the root as the 0th node).

κ—the number of mismatches between the path and $r[0 \ldots j]$ (recall that $r[0] = $ '−').

u—the parent of a node in D to be created for v.

In this way, the *parent/child* link between u and the node to be created for v can be easily established, as described below.

Each time an entry $e = (v, j, \kappa, u)$ with $v = $ <x, [α, β]> is popped out from S, we will check whether $x = r[j]$.

(i)   If $x = r[j]$, we will generate a node $u' = $ <x, j> and link it to u as a child.
(ii) If $x \neq r[j]$, we will check whether u is a node of the form <−, 0>. If it is not the case, generate a node $u' = $ <−, 0>.
      Otherwise, set u' to be u.
(iii) Using *search*() to find all the children of v: $v_1, \ldots, v_l$. Then, push each $(v_i, j + 1, \kappa', u')$ into S with κ' being κ or $\kappa + 1$, depending on whether $y_i = r[j + 1]$, where $v_i = $ <y_i, [$α_i$, $β_i$]> .

Note that in this process it is not necessary to keep T, but insert all the nodes (of T) in the hash table as discussed above.

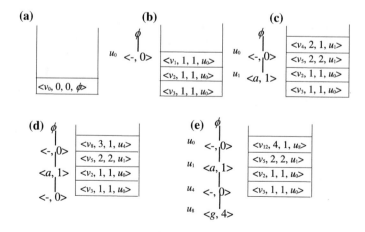

Fig. 12 Illustration for stack changes

Example 5.1 In this example, we run the above process on $r = tcaca$ and $L = BWT\ (\tilde{s})$ shown in Fig. 3c with $k = 2$, and show its first 5 steps. The tree created is shown in Fig. 12.

Step 1: Create the *root*, $v_0 = <-, [1, 8]>$. Push $(v_0, 0, 0, \phi)$ into S, where ϕ is used to represent the parent of the root D. See Fig. 12a.

Step 2: Pop out the top element $(v_0, 0, 0, \phi)$ from S. Create the root u_0 of D, which is set to be a child of ϕ. Push $<v_3, 1, 1, u_0>$, $<v_2, 1, 1, u_0>$, $<v_1, 1, 1, u_0>$ into S, where v_3, v_2, and v_1 are three children of v_0. See Fig. 12b.

Step 3: Pop out $(v_1, 1, 1, u_0)$ from S. $v_1 = <a, [1. 4]>$. Since $r[1] = t \neq a$, a mismatching node $u_1 = <a, 1>$ will be created and set to be a child of u_0. Then, push $(v_4, 2, 1, u_1)$ into S, where v_4 is the child of v_1. See Fig. 12c.

Step 4: Pop out $(v_4, 2, 1, u_1)$ from S. $v_4 = <c, [1, 2]>$. Since $r[2] = c$, we will check whether u_1 is a matching node. It is the case. So, a matching node $u_4 = <-, 0>$ will be created and set to be a child of u_1. Then, push $(v_8, 3, 1, u_4)$ into S, where v_8 is the child of v_4. See Fig. 12d.

Step 5: Pop out $(v_8, 3, 1, u_4)$ from S. $v_8 = <a, [2, 3]>$. $r[3] = a$. However, no new node is created since u_4 is a matching node. Push $(v_{12}, 4, 1, u_4)$ into S, where v_{12} is the child of v_8. See Fig. 12e.   □

From the above sample trace, we can see that D can be easily generated. In the following, we will discuss how to extend this process to a general algorithm for our task.

As with the basic process, each time a node $v = <x, [\alpha, \beta]>$ (compared to $r[j]$) is encountered, which is the same as a previous one $v' = <x', [\alpha', \beta']>$ (compared to $r[i]$), we will not create a subtree in T in a way as for v', but create a new node u for v in D and then go along $p(v')$ (the link associated with v') to find the corresponding nodes u' in D and search $D[u']$ in the breadth-first manner to generate a subtree rooted at u in D by simulating the merge operation discussed in Subsection B. In other words, $D[u]$ (to be created) corresponds to the mismatch arrays for all the paths going though v in T, which will not be actually produced. See Fig. 13 for illustration.

For this purpose, we introduce a third kind of nodes of the form $<-, \infty>$ into D to represent symbol ∞ in mismatching arrays. Such a node is always the last node of a path in D.

To search $D[u']$ breadth-first, a queue data structure Q is used to control the search of $D[u']$ and at the same time generate $D[u]$. In Q, each entry e is a triplet (w, γ, h) with w being a node in $D[u']$, γ an entry in R_{ij}, and h is the number of

Fig. 13 Illustration for generation of subtrees in T

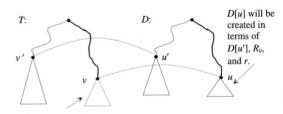

$T[v]$ will not be actually explored.

mismatching nodes on the path from the root to the node to be created in $D[u]$. Initially, put $(u', R_{ij}[1], h')$ into Q, where h' is the number of mismatching nodes on the path from the root to u. In the process, when e is dequeued from Q (taken out from the front), we will make the following operations (simulating the steps in *merge*()):

1. Let $e = (w, R_{ij}[l], h)$. Assume that $w = \langle z, f \rangle$ and $R_{ij}[l] = val$.

   - If $\langle z, f \rangle$ is equal to $\langle -, 0 \rangle$, then create a copy of $\langle -, 0 \rangle$ added to $D[u]$. Let $u_1, ..., u_g$ be the children of w. We will enqueue (append at the end) $(u_1, R_{ij}[l], h), ..., (u_g, R_{ij}[l], h)$ into Q in turn.
   - If is a mismatching node, do (2), (3), or (4).
   - If $\langle z, f \rangle$ is equal to $\langle -, \infty \rangle$, do (5).

2. If $f < i + val - 1$, add $\langle z, j + f - i + 1 \rangle$ to $D[u]$. If $h < k + 1$, enqueue $(u_1, R_{ij}[l], h + 1), ..., (u_g, R_{ij}[l], h + 1)$ into Q.

3. If $f > i + val - 1$ (and $f \neq \infty$), we will scan R_{ij} starting from $R_{ij}[l]$ until we meet the largest $l' \leq k - h + l$ such that $f > i + R_{ij}[l'] - 1$. For each $R_{ij}[q]$ ($l \leq q \leq l'$), we create a new node $\langle r[i + R_{ij}[q] - 1], j + R_{ij}[q] - 1 \rangle$ added to $D[u]$. If $l' < k - h + l$, add $\langle -, \infty \rangle$ to $D[u]$, and enqueue $\langle w, R_{ij}[l' + 1], h + l' - l + 1 \rangle$ into Q.

4. If $f = i + val - 1$, we will distinguish between two subcases: $z \neq r[j + val - 1]$ and $z = r[j + val - 1]$. If $z \neq r[j + val - 1]$, we have a mismatch and a copy of w will be generated and added to $D[u]$. If $h < k + 1$, enqueue $(u_1, R_{ij}[l + 1], h + 1), ..., (u_g, R_{ij}[l + 1], h + 1)$ into Q. If $z = r[j + val - 1]$, create a node $\langle -, 0 \rangle$ added to $D[u]$. (If its parent is also $\langle -, 0 \rangle$, it will be merged into its parent.) Also enqueue $\langle u_1, R_{ij}[l + 1], h \rangle, ..., \langle u_g, R_{ij}[l + 1], h \rangle$ into Q.

5. If $w = \langle -, \infty \rangle$, scan R_{ij} starting from $R_{ij}[l]$ until we find the largest $l' \leq k - h + l$ such that $R_{ij}[l] \neq \infty$. For each $R_{ij}[q]$ ($l \leq q \leq l'$), we create a new node $\langle r[i + R_{ij}[q] - 1], j + R_{ij}[q] - 1 \rangle$ added to $D[u]$. If $l' < k - h + l$, add $\langle -, \infty \rangle$ to $D[u]$, and enqueue $\langle w, R_{ij}[l' + 1], h + l' - l + 1 \rangle$ into Q.

In the above process, (2) corresponds to step 3 in *merge*(), (3) to step 4 in *merge*(), and (4) to step 5 in *merge*().

In (2), we handle the case when $f < i + val - 1$. In this case, we must have $r[f] = r[j + f - i]$. Then, by the following simple inference:

$$P[f] \neq r[f], r[f] = r[j + f - i] \Rightarrow P[f] \neq r[j + f - i],$$

we know that a mismatching node should be added to $D[u]$. Here, P stands for a path starting from v' in T corresponding to a path starting from u' in D, and $P[f]$ for the fth node on P. See Fig. 11a for illustration.

In (3), we handle the case that $f > i + val - 1$. In this case, we have, for each $i' \in \{ i + val - 1, ..., f \}$ with $R_{ij}[q] = i'$ ($l \leq q \leq l'$),

$$p[i'] = r[i'], r[i'] \neq r[j + i' - i] \Rightarrow P[i'] \neq r[j + i' - i].$$

Thus, for each $R_{ij}[q]$ ($l \leq q \leq l'$), a mismatching node will be created and added to $D[u]$.

In the above description, we ignored the technical details on how $D[u]$ is constructed for simplicity. However, in the presence of $D[u']$, it is easy to do such a task by manipulating links between nodes and their respetive parents.

Denote the above process by *node-creation*$(w, \gamma, i, j, R_{ij})$. We have the following proposition.

Proposition 5.2 *node-creation*$(w, \gamma, i, j, R_{ij})$ *create nodes in* $D[u]$ *correctly.*

Proof The correctness of *node-creation*$(w, \gamma, i, j, R_{ij})$ can be derived from Proposition 1.$\square$

Again, if $i > j$, $D[u]$ needs to be extended, which can be done in a way similar to the extension of mismatch arrays as discussed in Subsection C.

As an example, consider Figs. 7 and 11 once again. When we meet $<g, [1, 1]>$ (v_5 in T, compared to $r[2]$) for a second time, we will not generate $T[v_5]$ in Fig. 3, but $D[u_5]$ in Fig. 11. Comparing T and D, we can clearly see the efficiency of this improvement. In D, an MM-path in T is collapsed into a single node of the form $<-, 0>$.

The following is the formal description of the working process.

ALGORITHM $A(L, r, k)$

begin

1. create *root* of T; push(S, (*root*, 0, 0, ϕ));
2. **while** S is not empty **do** {
3. $(v, j, \kappa, u) := \text{pop}(S)$; let $v = <x, \alpha, \beta>$;
4.   **if** v is same as an existing v' (compared to $r[i]$) **then** {
5. $q := \max\{i, j\}$;
6. $R_{ij} := merge(R_{1i}, R_{1j}, r[i .. m - q + i], r[j .. m - q + j])$;
7.   *enqueuer*$(Q, (p(v'), R_{ij}[1]))$;
8.   **while** Q is not empty **do** {
9. $(w, \gamma) := dequeuer(Q)$; *node-creation*$(w, \gamma, i, j, R_{ij})$;}}
10. **else** {
11.   **if** $x \neq r[j]$ **then** create $u' = <x, j>$ and make it a child of u;
12.   **else if** u is $<-, 0>$ **then** $u' := u$
13.   **else** create $u' = <-, 0>$ and make it a child of u;
14. $p(v) := u'$;   (*associate with v a pointer to u'.*)
15.   **if** $j < |r|$ and $\kappa \leq k$ **then** {
16.   **for** each $y \in \Sigma$ within L_v **do** {
17. $w := search(y, L_v)$;
18.   **if** $w \neq \phi$ **then** {
19.   **if** $y = r[j + 1]$ **then** push(S, ($w, j + 1, \kappa, u'$));
20.   **if** $y \neq r[j + 1]$ and $\kappa < k$ **then** {push(S, ($w, j + 1, \kappa + 1, u'$));
21. }}}}

end

If we ignore lines 3–9 in the above algorithm, it is almost a depth-first search of a tree. Each time an entry (v, j, κ, u) is popped out from S (see line 4), it will be checked whether v is the same as a previous one v' (compared to $r[i]$). (See line 4.) If it is not the case, a node u' for v will be created in D (see lines 11–14). Then, all the children of v will be found by using the procedure $search()$ (see line 17) and pushed into S (see lines 18, and 19.) Otherwise, we will first create R_{ij} by executing $merge(R_{1i}, R_{1j}, r[i \ldots m - q + i], r[j \ldots m - q + j])$, where $q = max\{i, j\}$. (see lines 5–6.) Then, we create a subtree in D by executing a series of node-creation operations (see lines 8–9.)

Concerning the correctness of the algorithm, we have the following proposition.

Proposition 5.3
Let L be a BWT-array for the reverse $\tilde{s}$ of a target string s, and r a pattern. Algorithm A(L, r, k) will generate a mismatching tree D, in which each root-to-leaf path represents an occurrence of r in s having up to k positions different between r and s.

Proof In the execution of $A(L, r, k)$, two data structures will be generated: a hash table and a mismatching tree D, in which some subtrees in D are derived by using the mismatching information over r. Replacing each matching node in D with the corresponding maximum matching path and each mismatching node $<x, i>$ with the corresponding pair $<x, [\alpha, \beta]>$ (compared to $r[i]$), we will get an S-tree, in which each path corresponds to a *search sequence* discussed in Section III. Thus, in D each root-to-leaf path represents an occurrence of r in s having up to k positions different between r and s. □

The time complexity of the algorithm mainly consists of three parts: the cost for generating the mismatching information over r which is bounded by O($m\log m$); the cost for generating the M-tree and maintaining the hash table, which is bounded by O(kn'), where n' is the number of the M-tree's leaf nodes; and the cost for checking the characters in s against the characters in r, which is bounded by O(n). So, the total running time is bounded by O($kn' + n + m\log m$).

6 Experiments

In this section, we report the test results. For all the experiments on both the multiple pattern string matching and the string matching with k matches, we use the same data sets summarized in Table 1.

Table 1 Characteristics of genomes

Genomes	Genome sizes (bp)
Rat (Rnor_6.0)	2,909,701,677
Zebra fish (GRCz10)	1,464,443,456
Rat chr1 (Rnor_6.0)	290,094,217
C. elegans (WBcel235)	103,022,290
C. merlae (ASM9120v1)	16,728,967

To store *BWT*, $(\bar{s})$ we use 2 bits to represent a character $\in \{a, c, g, t\}$ and store 4 *rankAll* values (respectively in A_a, A_c, A_g, and A_t) for every 4 elements (in *L*) with each taking 32 bits.

All the tested methods are implemented in C++, compiled by GNU make utility with optimization of level 2. In addition, all of our experiments are performed on a 64-bit Ubuntu operating system, run on a single core of a 2.40 GHz Intel Xeon E5-2630 processor with 32 GB RAM.

6.1 Experiment on Multiple Pattern String Matching

In this experiment, we have tested altogether five different methods:

- *Burrows Wheeler Transformation* (BWT for short),
- *Suffix tree based* (Suffix for short),
- *Hash table based* (Hash for short),
- *Trie-BWT* (tBWT for short, discussed in this paper),
- *Improved Trie-BWT* (itBWT for short, discussed in this paper).

Among them, the codes for the suffix tree based and hash based methods are taken from the *gsuffix* package [54] while all the other three algorithms are implemented by ourselves.

6.1.1 Tests on Synthetic Data Sets

All the synthetic data are created by simulating reads from the five genomes shown in Table 1, with varying lengths and amounts. It is done by using the *wgsim* program included in the *SAMtools* package [45] with default model for single reads simulation.

Over such data, the impact of five factors on the searching time are tested: number *n* of reads, length *l* of *reads* (pattern strings), size *s* of genomes, compact factors f_1 of *rankAll*s (see Sect. 3.1) and compression factors f_2 of suffix arrays [12], which are used to find locations of reads (in a reference genome) in terms of formula (5) (see Sect. 3.2).

- *Tests with varying amount of reads*

In this experiment, we vary the amount *n* of reads with $n = 5, 10, 15, ..., 50$ millions while the reads are 50 bps or 100 bps in length extracted randomly from *Rat chr1* and *C. merlae* genomes. For this test, the compact factors f_1 of *rankAll*s are set to be 32, 64, 128, 256, and the compression factors f_2 of suffix arrays are set to 8, 16, 32, 64, respectively. These two factors are increasingly set up as the amount of reads gets increased.

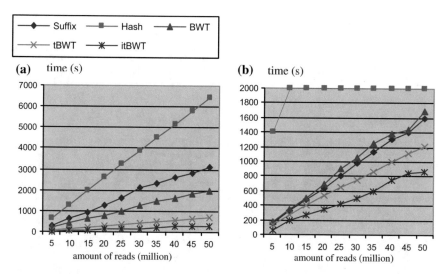

Fig. 14 Test results on varying amount of reads

In Fig. 14a, b, we report the test results of searching the Rat chr1 for matching reads of 50 and 100 bps, respectively. From these two figures, it can be clearly seen that the hash based method has the worst performance while ours works best. For short reads (of length 50 bps) the suffix-based is better than the BWT, but for long reads (of length 100 bps) they are comparable. The poor performance of the hash-based is due to its inefficient brute-force searching of genomes while for both the BWT and the suffix-based it is due to the huge amount of reads and each time only one read is checked. In the opposite, for both our methods tBWT and itBWT, the use of tries enables us to avoid repeated checkings for similar reads.

In these two figures, the time for constructing tries over reads is not included. It is because in the biological research a trie can be used repeatedly against different genomes, as well as often updated genomes. However, even with the time for constructing tries involved, our methods are still superior since the tries can be established very fast as demonstrated in Table 2, in which we show the times for constructing tries over different amounts of reads.

The difference between tBWT and itBWT is due to the different number of BWT array accesses as shown in Table 3. By an access of a BWT array, we will scan a segment in the array to find the first and last appearance of a certain character from a read (by tBWT) or a set of characters from more than one read (by itBWT).

Table 2 Time for trie construction over reads of length 100 BPS

No. of reads	30M	35M	40M	45M	50M
Time for Trie Con. (s)	51	63	82	95	110

Table 3 No. of BWT array accesses

No. of reads	30M	35M	40M	45M	50M
tBWT	47856K	55531K	63120K	70631K	78062K
itBWT	19105K	22177K	25261K	28227K	31204K

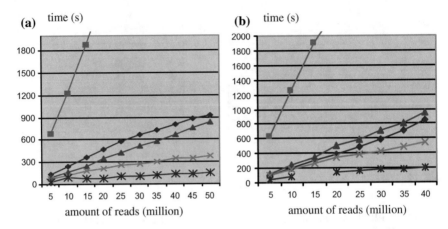

Fig. 15 Test results on varying amount of reads

Figure 15a, b show respectively the results for reads of length 50 bps and 100 bps over the *C. merolae* genome. Again, our methods outperform the other three methods.

- *Tests with varying length of reads*

In this experiment, we test the impact of the read length on performance. For this, we fix all the other four factors but vary length l of simulated reads with $l = 35, 50, 75, 100, 125, ..., 200$. The results in Fig. 16a shows the difference among five methods, in which each tested set has 20 million reads simulated from the Rat chr1 genome with $f_1 = 128$ and $f_2 = 16$. In Fig. 16b, the results show the case that each set has 50 million reads. Figure 17a, b show the results of the same data settings but on *C. merlae* genome.

Again, in this test, the hash based performs worst while the suffix tree and the BWT method are comparable. Both our algorithms uniformly outperform the others when searching on short reads (shorter than 100 bps). It is because shorter reads tend to have multiple occurrences in genomes, which makes the trie used in tBWT and itBWT more beneficial. However, for long reads, the suffix tree beats the BWT since on one hand long reads have fewer repeats in a genome, and on the other hand higher possibility that variations occurred in long reads may result in earlier termination of a searching process. In practice, short reads are more often than long reads.

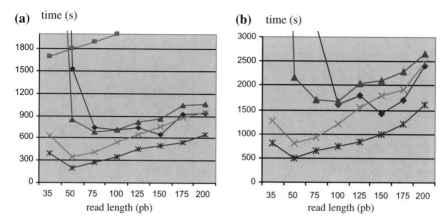

Fig. 16 Test results on varying length of reads

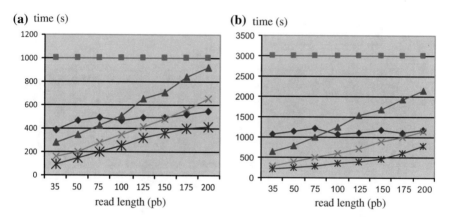

Fig. 17 Test results on varying length of reads

- *Tests with varying sizes of genome*

    To examine the impacts of varying sizes of genomes, we have made four tests with each testing a certain set of reads against different genomes shown in Table 1. To be consistent with foregoing experiments, factors except sizes of genomes remain the same for each test with $f_1 = 128$ and $f_2 = 16$. In Fig. 18a, b, we show the searching time on each genome for 20 million and 50 million reads of 50 bps, respectively. Figures 19a, b demonstrate the results of 20 million and 50 million reads but with each read being of 100 bps.

    These figures show that, in general, as the size of a genome increases the time of read aligning for all the tested algorithms become longer. We also notice that the larger the size of a genome, the bigger the gaps between our methods and the other algorithms. The hash-based is always much slower than the others. For the suffix

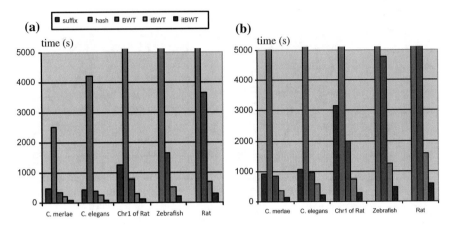

Fig. 18 Test results on varying sizes of genomes

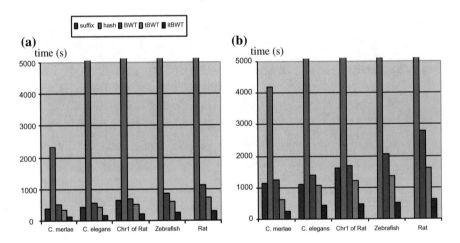

Fig. 19 Test results on varying sizes of genomes

tree, we only show the matching time for the first three genomes. It is because the testing computer cannot meet its huge memory requirement for indexing the Zebra fish and Rat genomes (which is the main reason why people use the BWT, instead of the suffix tree, in practice.) Details for the 50 bp reads in Figs. 17 and 18 show that the tBWT and the itBWT are at least 30% faster than the BWT and the suffix tree, which happened on the *C. elegans* genome. For the Rat genome, our algorithms are even more than six times faster than the others.

Now let us have a look at Fig. 18a, b. Although our methods do not perform as good as for the 50 bp reads due to the increment of length of reads, they still gain at least 22% improvement on speed and nearly 50% acceleration in the best case, compared with the BWT.

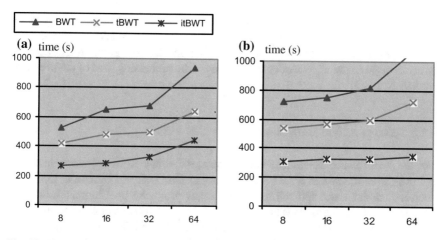

Fig. 20 Test results on varying compact and compression factors

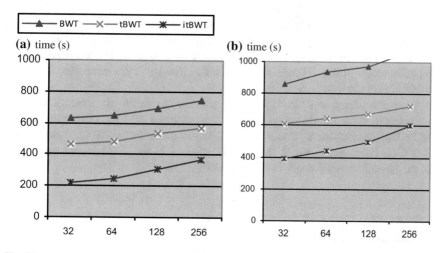

Fig. 21 Test results on varying compact and compression factors

- *Tests with varying compact and compression factors*

In the experiments, we focus only on the BWT method, since there are no compressions in both the suffix tree and the hash-based method. The following test results are all for 20 million reads with 100 bps in length. We first show the impact of f_1 on performance with $f_2 = 16, 64$ in Fig. 20a and b, respectively. Then we show the effect when f_2 is set to 64, 256 in Fig. 21a, b.

From these figures, we can see that the performance of all three methods degrade as f_1 and f_2 increase. Another noticeable point is that both the itBWT and the tBWT are not so sensitive to the high compression rate. Although doubling f_1 or f_2 will slow down their speed, they become faster compared to the BWT. For example, in

Fig. 19, the time used by the BWT grows 80% by increasing f_1 from 8 to 64, whereas the growth of time used by the tBWT is only 50%. In addition, the factor f_1 has smaller impact on the itBWT than the BWT and the tBWT, since the extra data structure used in the itBWT effectively reduced the processing time of the trie nodes by half or more.

6.1.2 Tests on Real Data Sets

For the performance assessment on real data, we obtain RNA-sequence data from the project conducted in an RNA laboratory at University of Manitoba [55]. This project includes over 500 million single reads produced by Illumina from a rat sample. Length of these reads are between 36 bps and 100 bps after trimming using Trimmomatic [56]. The reads in the project are divided into 9 samples with different amount ranging between 20 million and 75 million. Two tests have been conducted. In the first test, we mapped the 9 samples back to rat genome of ENSEMBL release 79 [57]. We were not able to test the suffix tree due to its huge index size. The hash-based method was ignored as well since its running time was too high in comparison with the BWT. In order to balance between searching speed and memory usage of the BWT index, we set $f_1 = 128$, $f_2 = 16$ and repeated the experiment 20 times. Figure 22a shows the average time consumed for each algorithm on the 9 samples.

Since the source of RNA-sequence data is the transcripts, the expressed part of the genome, we did a second test, in which we mapped the 9 samples again directly to the Rat *transcriptome*. This is the assembly of all transcripts in the Rat genome. This time more reads, which failed to be aligned in the first test, are able to be exactly matched. This result is showed in Fig. 22b.

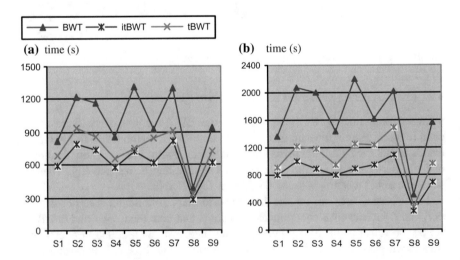

Fig. 22 Test results on real data

From Fig. 22a, b, we can see that the test results for real data set are consistent with the simulated data. Our algorithms are faster than the BWT on all 9 samples. Counting the whole data set together, itBWT is more than 40% faster compared with the BWT. Although the performance would be dropped by taking tries' construction time into consideration, we are still able to save 35% time using itBWT.

6.2   Experiment on String Matching with k Mismatches

In this experiment, we have tested altogether four different methods:

- *BWT-based* [13] (BWT for short),
- *Amir's method* [2] (Amir for short),
- *Cole's method* [52] (Cole for short),
- *Algorithm A discussed in this paper* (A() for short)

By the BWT-based method, an *S*-tree will be created as described in Section IV, but with $\sigma(i)$ being used to cut off branches, where $\sigma(i)$ is the number of consecutive, disjoint substrings in $r[i \ldots m]$ not appearing in *s*. By the Amir's algorithm, a pattern *r* is divided into several periodic stretches separated by 2 *k* aperiodic substrings, called breaks, as illustrated in Fig. 23. Then, for each break b_i, located at a certain position *i*, find all those substrings s_j (located at different positions *j*) in *s* such that $b_i = s_j$, and then mark each of them. After that, discard any position that is marked less than *k* times. In a next step, verify every surviving position in *s*.

By the Cole's, a suffix tree for a target is constructed. (The code for constructing suffix trees is taken from the *gsuffix* package: http://gsuffix.Sourceforge.net/).

For the test, five reference genomes shown in Table 1 are used. Similar to the first experiment, all the simulating reads are taken from these five genomes, with varying lengths and amounts. Concretely, we take 5000 reads with length varying from 100 to 300 bps.

In Fig. 24a, b, we report the average time of testing the Rat (Rnor_6.0) for matching 100 reads of length 100 to 300 bps. From this figure, we can see that Algorithm A() outperforms all the other three methods. But the Amir's method is better than the other two methods. The BWT-based and the Cole's method are comparable. However, for small *k*, the Cole's is a little bit better than the BWT-based method while for large *k* their performances are reversed.

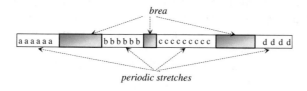

Fig. 23 Illustration for periodic stretches and breaks

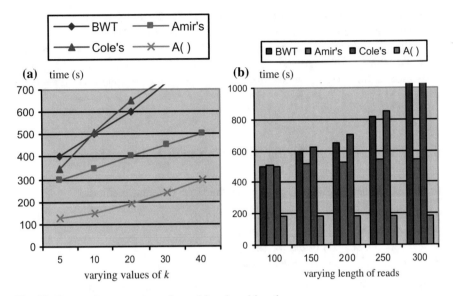

Fig. 24 Test results on varying values of k and read length

k/Length-of-read	5/50	10/100	20/150	30/200
No. of leaf nodes	2K	0.7M	16.5M	102M

Table 4 Number of leaf nodes of S-trees

To show why $A()$ has the best running time, we give the number n' of leaf nodes in the M-trees created by $A()$ for some tests in Table 4, which demonstrates that n' can be much smaller than n. Thus, the time complexity $O(kn')$ of $A()$ should be a significant improvement over $O(n\sqrt{k}\log k)$—the time complexity of Amir's.

In this test (and also in the subsequent tests), the time for constructing $BWT(\bar{s})$ is not included as it is completely independent of r. Once it is created, it can be repeatedly used.

In Fig. 24b, we show the impact of read lengths. For this test, k is set to 25. It can be seen that only the BWT-based and the Cole's are sensitive to the length of reads. For the BWT-based, more time is required to construct S-trees for longer reads while for the Cole's longer paths in a suffix tree will be searched as the lengths of reads increase. For the other two methods: $A()$ and the Amir's, the lengths of reads only impact the time for the read pre-processing, but it is completely overshadowed by the time spent on searching genomes. For the Amir's, the time for recognizing breaks is linear in $|r|$ [2] while for $A()$ the time for generating the mismatch information is bounded by $O(|r|\log|r|)$. No significant difference between them can be measured.

In Fig. 25a, b, we report the test results of searching the Zebra fish (GRCz10).

Again, similar to Fig. 24a, the performance of Algorithm $A()$ is best, and the Amir's is still better than both the BWT-based and the Cole's.

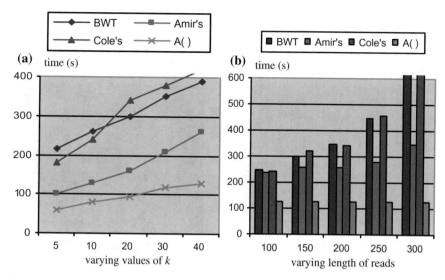

Fig. 25 Test results on varying values of k and read length

Table 5 Number of leaf nodes of S-trees

k/Length-of-read	5/50	10/100	20/150	30/200
No. of leaf nodes	0.7K	0.30M	9.2M	89M

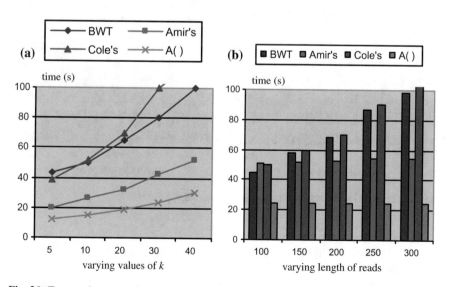

Fig. 26 Test results on varying values of k and read length

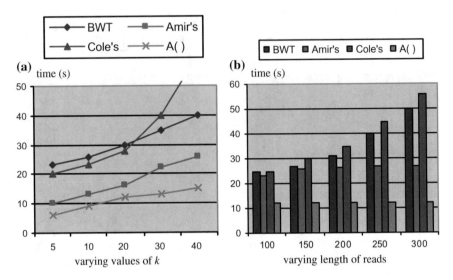

Fig. 27 Test results on varying values of k and read length

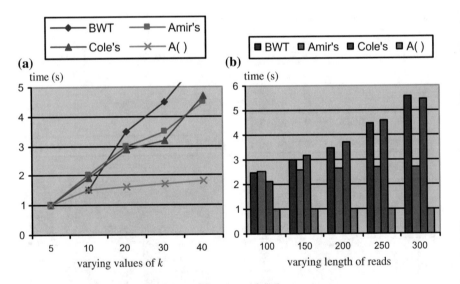

Fig. 28 Test results on varying values of k and read length

In Table 5, we show the number n'.

Figure 25b shares the same features as Fig. 24b. It also shows that only the BWT-based and the Cole's are sensitive to the length of reads.

In Figs. 26, 27, and 28, we show the tests on Rat chr1 (Rnor_6.0), *C. elegans* (WBcel235), and *C. merlae* (ASM9120v1), respectively.

From these figures, the most important feature we can observe is that as the size of genomes becomes smaller, the difference between the Amir's and Cole's diminishes. But the BWT-based and $A(\ )$ remain the worst and the best, respectively. Although $A(\ )$ is impacted by the number of leaf nodes of an S-tree, the impact factor is small in comparison with the size of the whole S-tree, which dominates the time complexity of the BWT-based method. Also, the big difference between $A(\ )$ and Amir's shows that using M-trees the cost for creating mismatch information of r's occurrences in s can be significantly reduced.

7 Conclusion and Future Work

In this chapter, two new methods have been discussed. One is to search a large volume of pattern strings against a single long target string, aiming at efficient next-generation sequencing in DNA databases. The main idea behind it is to combine the search of tries constructed over the patterns and the search of the BWT indexes over the target. Extensive experiments have been conducted, which show that our method improves the running time of the traditional methods by an order of magnitude or more.

The second one is to do the string matching with k mismatches. Its main idea is to transform the reverse $\bar{s}$ of target string s to BWT $(\bar{s})$ and use the mismatch information over a pattern string r to speed up the computation. Its time complexity is bounded by $O(kn' + n + m\log m)$, where $m = |r|$, $n = |s|$, and n' is the number of leaf nodes of a tree structure produced during the search of a $BWT(s)$. Our experiments show that it has a better running time than any existing on-line and index-based algorithms.

As a future work, we will use the BWT to solve another important problem, the string matching with k errors. It seems to be more challenging than the k mismatches since the Levenshtein distance is more difficult to handle than the Hamming distance.

References

1. Li, R., et al. (2008). SOAP: short oligonucleotide alignment program. *Bioinformatics, 24,* 713–714.
2. Amir, A., Lewenstein, M., & Porat, E. (2004). Faster algorithms for string matching with k mismatches. *Journal of Algorithms, 50*(2), 257–275.
3. Aoe, J.-I. (1989). An efficient implementation of static string pattern matching machines. *IEEE Transactions on Software Engineering, 15*(8), 1010–1016.
4. Baeza-Yates, R. A., Perleberg, C. H. Fast and practical approximate string matching. In A. Apostolico, M. Crocchemore, Z. Galil, & U. Manber (Eds.), *Combinatorial pattern matching, lecture notes in computer science* (Vol. 644, pp. 185–192). Berlin: Springer.

5. Baeza-Yates, R. A., & Régnier, M. Fast algorithms for two-dimensional and multiple pattern matching. In *Proceedings of the SWAT '90 the Second Scandinavian Workshop on Algorithm Theory* (pp. 332–347). Bergen, Sweden: Springer.
6. Boyer, R. S., & Moore, J. S. (1977). A fast string searching algorithm. *Communication of the ACM, 20*(10), 762–772.
7. Knuth, D. E., Morris, J. H., & Pratt, V. R. (1977). Fast pattern matching in strings. *SIAM Journal on Computing, 6*(2), 323–350.
8. Landau, G. M., & Vishkin, U. (1985). Efficient string matching in the presence of errors. In *Proceedings of the 26th Annual IEEE Symposium on Foundations of Computer Science* (pp. 126–136).
9. Apostolico, A., & Giancarlo, R. (1986). The Boyer-Moore-Galil string searching strategies revisited. *SIAM Journal on Computing, 15*(1), 98–105.
10. McCreight, E. M. (1976). A space-economical suffix tree construction algorithm. *Journal of the ACM, 23*(2), 262–272.
11. Weiner, P. (1973). Linear pattern matching algorithm. In *Proceedings of the 14th IEEE Symposium on Switching and Automata Theory* (pp. 1–11).
12. Manber, U., & Myers, E. W. (1990). Suffix arrays: a new method for on-line string searches. In *Proceedings of the 1st Annual ACM-SIAM Symposium on Discrete Algorithms* (pp. 319–327). Philadelphia, PA: SIAM.
13. Burrows, M., & Wheeler, D. J. (1994). A block-sorting lossless data compression algorithm.
14. Ferragina, P., & Manzini, G. (2000). Opportunistic data structures with applications. In *Proceedings of the 41st Annual Symposium on Foundations of Computer Science* (pp. 390–398). IEEE.
15. Langmead, B. (2014, September). Introduction to the Burrows-Wheeler transform. www.youtube.com/watch?v=4n7NPk5lwbI.
16. Aho, A. V., & Corasick, M. J. (1975). Efficient string matching: An aid to bibliographic search. *Communication of the ACM, 23*(1), 333–340.
17. Commentz-Walter, B. (1979). A string matching algorithm fast on the average. In *Proceedings of the 6th Colloquium on Automata, Languages and Programming, 16–20 July 1979*, pp. 118–132.
18. Wu, S., & Manber, U. (1994). *A fast algorithm for multi-pattern searching.* Technical Report TR-94-17, Department of Computer Science, Chung-Cheng University.
19. Crochemore, M., et al. (1999). Fast practical multi-pattern matching. *Information Processing Letters, 71,* 107–113.
20. Dandass, Y. S., Burgess, S. C., Lawrence, M., & Bridges, S. M. (2008). Accelerating string set matching in FPGA hardware for bioinformatics research. *BMC Bioinformatics, 9,* 197.
21. Colussi, L., Galil, Z., & Giancarlo, R. (1990). On the exact complexity of string matching. In *Proceedings of the 31st Annual IEEE Symposium of Foundation of Computer Science* (Vol. 1, pp. 135–144).
22. Landau, G. M., & Vishkin, U. (1986). Efficient string matching with k mismatches. *Theoretical Computer Science, 43,* 239–249.
23. Li, H., & Durbin, R. (2009). Fast and accurate short read alignment with Burrows-Wheeler transform. *Bioinformatics, 25*(14), 1754–1760.
24. Baeza-Yates, R. A., & Gonnet, G. H. (1992). A new approach in text searching. *Communication of the ACM, 35*(10), 74–82.
25. Ehrenfeucht, A., & Haussler, D. A new distance metric on strings computable in linear time. *Discrete Applied Mathematics, 20,* 191–203.
26. Eddy, S. R. (2004). What is dynamic programming? *Nature Biotechnology, 22,* 909–910. https://doi.org/10.1038/nbt0704-909.
27. Chang, W. L., & Lampe, J. Theoretical and empirical comparisons of approximate string matching algorithms. In A. Apostolico, M. Crocchemore, Z. Galil, & U. Manber (Eds.), *Combinatorial pattern matching. Lecture notes in computer science* (Vol. 644, pp. 175–184). Berlin: Springer.

28. Ukkonen, E. Approximate string-matching with q-grams and maximal matches. *Theoretical Computer Science, 92*, 191–211.
29. Manber, U., & Baeza-Yates, R. A. (1991). An algorithm for string matching with a sequence of don't cares. *Information Processing Letters, 37*, 133–136.
30. Pinter, R. Y. (1985). Efficient string matching with don't' care patterns. In A. Apostolico & Z. Galil (Eds.), *Combinatorial algorithms on words*. NATO ASI Series (Vol. F12, pp. 11–29). Berlin: Springer.
31. Chen, Y., Wu, Y., & Xie, J. (2016). An efficient algorithm for read matching in DNA databases. In *Proceedings of the International Conference on DBKDA'2016, Lisbon, Portugal, 26–30 June 2016* (pp. 23–34).
32. Chen, Y., & Wu, Y. (2017). Mismatching trees and BWT arrays: A new way for string matching with k-mismatches. In *ICDE2017, 19–22 April 2017* (pp. 339–410). San Diego, USA: IEEE.
33. Galil, Z. (1977). On improving the worst case running time of the Boyer-Moore string searching algorithm. *Communication of the ACM, 22*(9), 505–508.
34. Lecroq, T. (1992). A variation on the Boyer-Moore algorithm. *Theoretical Computer Science, 92*(1), 119–144.
35. Tarhio, J., & Ukkonen, E. Boyer-Moore approach to approximate string matching. In J. R. Gilbert & R. Karlssion (Eds.), *SWAT 90, Proceedings of the 2nd Scandinavian Workshop on Algorithm Theory, Lecture Notes in Computer Science* (Vol. 447, pp. 348–359). Berlin: Springer.
36. Salmela, L., Tarhio, J., & Kytojoki, J. (2006). Multi-pattern string matching with q-grams. *ACM Journal of Experimental Algorithmics, 11*.
37. Jiang, H., & Wong, W. H. (2008). SeqMap: Mapping massive amount of oligonucleotides to the genome. *Bioinformatics, 24*, 2395–2396.
38. Kim, J. Y., & Yaylor, J. S. (1992). Fast multiple keyword searching. In *Proceedings of the Third Annual Symposium on Combinatorial Pattern Matching, 29 April–01 May 1992* (pp. 41–51). Springer.
39. Li, H., & Durbin, R. (2010). Fast and accurate long-read alignment with Burrows-Wheeler transform. *Bioinformatics, 26*(5), 589–595.
40. Knuth, D. E. (1975). *The art of computer programming* (Vol. 3). Massachusetts: Addison-Wesley Publish Com.
41. Li, H., & Homer. (2010). A survey of sequence alignment algorithms for next-generation sequencing. *Briefings in Bioinformatics, 11*(5), 473–483. https://doi.org/10.1093/bib/bbq015.
42. Karp, R. L., & Rabin, M. O. (1987). Efficient randomized pattern-matching algorithms. *IBM Journal of Research and Development, 31*(2), 249–260.
43. Harrison, M. C. (1971). Implementation of the substring test by hashing. *Communication of the ACM, 14*(12), 777–779.
44. Li, H., et al. (2008). Mapping short DNA sequencing reads and calling variants using mapping quality scores. *Genome Research, 18*, 1851–1858.
45. Li, H. (2014). wgsim: a small tool for simulating sequence reads from a reference genome. https://github.com/lh3/wgsim/.
46. Schatz, M. (2009). Cloudburst: Highly sensitive read mapping with mapreduce. *Bioinformatics, 25*, 1363–1369.
47. Lin, H., et al. (2008). ZOOM! Zillions of oligos mapped. *Bioinformatics, 24*, 2431–2437.
48. Baeza-Yates, R. A., & Gonnet, G. H. (1989). A new approach to text searching. In N. J. Belkin & C. J. van Rijsbergen (Eds.), *SIGIR 89, Proceedings of the 12th Annual International ACM Conference on Research and Development in Information Retrieval* (pp. 168–175).
49. Smith, A. D., et al. (2008). Using quality scores and longer reads improves accuracy of Solexa read mapping. *BMC Bioinformatics, 9*, 128.
50. Tarhio, J., & Ukkonen, E. Approximate Boyer-Moore string matching. *SIAM Journal on Computing, 22*(2), 243–260.

51. Nicolas, M., & Rajasekarian, S. (2013). On string matching with k mismatches. https://arxiv.org/pdf/1307.1406.
52. Cole, R., Gottlieb, L., & Lewenstein, M. (2004). Dictionary matching and indexing with errors and don't cares. In *STOC'04* (pp. 91–100).
53. Hon, W., et al. (2007). A space and time efficient algorithm for constructing compressed suffix arrays. *Alrothmica, 48,* 23–36.
54. Bauer, S., Schulz, M. H., & Robinson, P. N. (2014). gsuffix:http://gsuffixSourceforge.net/.
55. Lab website. (2014). http://home.cc.umanitoba.ca/~xiej/.
56. Bolger, A. M., Lohse, M., & Usadel, B. (2014). Trimmomatic: bolger: A flexible trimmer for Illumina Sequence Data. Bioinformatics, btu170.
57. Cunningham, F., et al. (2015). Nucleic Acids Research 2015, 43, Database issue: D662-D669.

Traffic Condition Monitoring Using Social Media Analytics

Taiwo Adetiloye and Anjali Awasthi

Abstract Scientist and practitioner seek innovations that analyze traffic big data for reducing congestion. In this chapter, we propose a framework for traffic condition monitoring using social media data analytics. This involves sentiment analysis and cluster classification utilizing the big data volume readily available through Twitter microblogging service. Firstly, we examine some key aspects of big data technology for traffic, transportation and information engineering systems. Secondly, we consider Parts of Speech tagging utilizing the simplified Phrase-Search and Forward-Position-Intersect algorithms. Then, we use the k-nearest neighbor classifier to obtain the unigram and bigram; followed by application of Naïve Bayes Algorithm to perform the sentiment analysis. Finally, we use the Jaccard Similarity and the Term Frequency-Inverse Document Frequency for cluster classification of traffic tweets data. The preliminary results show that the proposed methodology, comparatively tested for accuracy and precision with another approach employing Latent Dirichlet Allocation is sufficient for predicting traffic flow in order to effectively improve the road traffic condition.

1 Social Media Analytics for Traffic Condition Monitoring

Perhaps the emergence of big data technology could not have been more disruptive anywhere else than in transportation and traffic engineering systems. This is considering that daily traffic flow of human transportation holds vast big data yet to be fully harnessed for real time estimation and prediction. Lu et al. [1] observed that such rapid development of urban "informatization", in the era of big data, offers several details entrenched in some spatio-temporal characteristics, historical correlations and multistate patterns. Undoubtedly, big data have increasingly been used

T. Adetiloye (✉) · A. Awasthi
Concordia Institute for Information and Systems Engineering (CIISE), Montreal, Canada
e-mail: t_adeti@encs.concordia.ca

A. Awasthi
e-mail: anjali.awasthi@concordia.ca

© Springer Nature Singapore Pte Ltd. 2018
S. S. Roy et al. (eds.), *Big Data in Engineering Applications*,
Studies in Big Data 44, https://doi.org/10.1007/978-981-10-8476-8_13

for discovering subtle population patterns and heterogeneities that are not possible with small-scale data [2]. For these reasons amongst others academia, governments, federal and state agencies, industries, and other organizations continue to seek innovations to manage and analyze big data; providing them the prospect of increasing the accuracy of predictions, improving the management and security of transportation infrastructures while enabling informed decision-making to gain better insight into their transportation and traffic engineering phenomena [3].

The practical significance of real-time traffic flow state identification and prediction using big data lies in the ability to identify and predict traffic flow state efficiently, timely and precisely [1]. Various articles [3–5] have employed big data resources to examine traffic demand estimation, traffic flow prediction and performance as well as integration, and validation with existing models. A noteworthy aspect is that the rapidly increasing (big data) volume of leading social media microblogging services such as Twitter (twitter.com) can be pragmatically challenging, and nearly impossible to manually analyze [6]. Nevertheless, the huge volume of data derived from Twitter makes it ideal for machine learning.

Few years ago, researchers developed sentiment and cluster analysis to monitor twitter messages, identify followers and followings, find word resemblances and examine the nature of the comments i.e. positive, negative or neural. Such promising twitter analytic tools appear to be sufficient in solving the aforementioned traffic flow problems. Our objective in this study is tweet mining of the twitter UK traffic delays and to perform sentiment analysis and cluster classification for traffic congestion prediction. The proposed methodology is based on tweet crawling, preprocessing steps, feature extraction and social network generation and cluster.

1.1 Traffic Twitter Sentiment Analysis

Following the launch of twitter in 2006, sentiment analysis has been applied to various areas of interests e.g. extracting adverse drug reactions from tweets [7], news coverage of the nuclear power issues [8], and in the tourism sector for capturing sentiment from integrated resort tweets [6]. Terabytes of twitter data could be from traffic road users expressing their opinions on traffic jam, road accidents and other information which constitute general traffic news update. The question, of course, is how to determine traffic flow state based on the weight as measured by the opinion contained in a twitter message (called "tweet")—a short message that a sender post on twitter that cannot be longer that maximum 140 characters? According to Abidin et al. [9], certain special characters including @, RT, and # symbols used in a tweet creates a collective snapshot of what people are saying about a given topic. An in-depth process of computationally identifying and automatically extracting opinions from a writer's piece of text to determine whether the attitude or emotions towards a topic is positive, negative or neutral is known as sentiment analysis [10, 11]. The technique of sentiment analysis is generally expected to yield a high accuracy rate of roughly 70–80% in training-test data matching tasks [12], while objectively seeking useful insights from a large quantity of aggregated data instead of achieving perfect classification of all data points [6]. Sentiment mining using corpus based and dictionary based methods for semantic orientation of the opinion words in tweets has been presented by Kumar and Sebastian [13].

In drawing the relevance of twitter sentiment analysis to traffic flow state prediction, He et al. [14] consider improving long-term traffic prediction with tweet semantics; and, then, analyze the correlation between traffic volume and tweet counts with various granularities. Finally, an optimization framework to extract traffic indicators based on tweet semantics using a transformation matrix, while integrating them into the traffic prediction using linear regression is proposed. Real-time traffic improvement by semantic mining of social networks has been captured by Grosenick [15]. Abidin et al. [9] introduce the use of Twitter API to retrieve traffic data serving as input to Kalman Filter models for route calculations and updates while fine-tuning the output for new, accurate arrival estimation.

1.2 Traffic Twitter Cluster Classification

Tweets could have a hashtag which consist of any word that starts with "#" symbol. Hashtags help to search messages containing a particular tag. Also of interest is the Part of Speech (POS) tagging in tweets, which has been applied by Elsafoury [16] to monitor urban traffic status. The main idea of POS tagging, also known as word-category disambiguation, is to mark up a word in a corpus and to assign it to a corresponding POS based on its definition and its context. The former is an example of exact term search while the latter, POS, can be considered a typical example of full-text search, which is usually thorough in its search process but can be more challenging to perform when compared to the exact text search. One instance of such text search is classification of tweets into positive and negative sentiments using multinomial Naïve Bayes' unigram with mutual information based on n-grams and POS that has been presented by Go et al. [11]. It outperforms other classifier approaches under consideration. In between the exact and full-text search is the phrase text search for searching a particular word phrase. For instance, an exact term search might be required to search the term "delay" in a tweet stream. This would bring out only tweets containing the term "delay". On the other hand, a phrase term search could be a phrase like "Traffic delay" in which there are more details of the search term. Phrase text search is often more useful when performing cluster classification than the other text search methods. It is noteworthy that using a particular search operation is based on measuring the relevance of the query to efficiently match the terms appropriately. Azam et al. [17] present the functional clustering details of their tweets mining approach which has the following steps:

(1) *Tweet crawling*: It is the process of retrieving tweets from twitter server using Twitter Application Program Interface (API). The crawled tweets are stored on local machine for further processing.
(2) *Tweets pre-processing and tokenization*: It involves the filtering of the crawled tweets of non-entirely textual items like emoticons, URL, special character, stop words etc. A common tokenization method known as the n-gram technique can then be applied to tokenize the tweets into bag-of-works ($n = 1$, known as a unigram is recommended for such tweets tokenization by Broder et al. [18]).
(3) *Feature extraction and social network generation*: It is the process of extracting important features from the preprocessed and tokenized tweets while transforming the feature sets into a social network generation comprising a term tweet matrix A of order $m \times n$, where m is the number of candidate terms and n is the number of tweets. The resulting matrix A is used to compute the weight $w(t_{i,j})$ using the following two equations:

$$w(t_{i,j}) = tf(t_{i,j}) \times idf(t_i) \tag{1}$$

$$idf(t_i) = \log \frac{|D|}{\{d_j : t_i \in d_j\}} + 1 \tag{2}$$

where $tf(t_{i,j})$ is the number of times t_i occurs in jth tweet.
$|D|$ is the total number of tweets and $\{d_j : t_i \in d_j\}$ represents the number of tweets with term, t_i. The objective is to normalize matrix A such that the tweet vectors' length equals to 1.

(4) *Social network clustering*: After generating the social network for the complete set of tweets, Markov clustering is used to achieve the social network clustering by crystallizing the network into various cluster each representing individual events. The Markov clustering algorithm (introduced by van Dongen [19]) is a fast and scalable unsupervised cluster algorithm for graphs (also known as *networks*). It serves as an iterative method for interleaving of the matrix expansion and inflation steps based on simulation of (stochastic) flow in graphs.

More details on the abovementioned steps can be found in Azam et al. [17]. For traffic flow prediction using big data analysis and visualization, McHugh [20] considered among other approaches the use of traffic tweets to test the effectiveness of geographical location of vehicles to determine the location of an incident. A useful method that analyzes traffic tweets in order to generate real-time city traffic insights and predictions for traffic management and city planning has been introduced by Tejaswin et al. [21].

2 Using Tweet Traffic Data for Traffic Condition Monitoring

The logs of twitter traffic data for the sentiment analysis and cluster classification were obtained using twitterR package. The tweets were connected to the Twitter API and OAuth authentication was performed using the ROAuth package all in RStudio. The plyr and stringr packages are used to crawl a number of tweets into RStudio while ensuring they are clean of unwanted symbols. More details of this twitter text mining technique can be found in Rais [22]. Detail documentation of the widely used twitter data mining statistical program can be found in cran.r-project.org [23]. We perform a phrase search based on the phrase using a POS tag: *Uk traffic delay*. This is made possible with a simplified phrase search algorithm derived from Eckert [24], with the original simplified version by Manning et al. [25], given by the following:

Algorithm 1
Phrase-Search(index , phrase)
1. $t \leftarrow$ Terms(phrase)
2. $k \leftarrow 1$
3. answer $\leftarrow$ Index-Get(index, t)
4. $t \leftarrow$ next(t)
5. while $t \neq$ NIL and answer $\neq$ {}
6. do nextTweet $\leftarrow$ Index-Get(index, t)
7. answer $\leftarrow$ Forward-Positional-Intersect(answer, nextTweet, k)
8. $k \leftarrow k + 1$
9. $t \leftarrow$ next(t)
10. return answer

In order to apply the above algorithm for our problem, a positional *index* containing a list of a data mined tweets with a list of positions is used to indicate the search phrase. The *Terms* is taking to be a split-normalization tokenizer that splits the *phrase* into list of tokens, normalizing them and assigning its outputs to k as a bag of words. We consider the weighted k-nearest neighbor classifier [26] which assigns a weight $1/k$ to the outputs. This is done by finding the vector of non-negative weights that is asymptotically optimal while minimizing the misclassification error rate, R_R [26]. Essentially, the asymptotic expansion is needed to ensure strong consistency in the search. This is subject to a regularity class distribution condition:

$$R_R\left(C_n^{wnn}\right) - R_R\left(C^{Bayes}\right) = (B_1 s_n^2 + B_2 t_n^2)\{1 + o(1)\}, \qquad (3)$$

Let C_n^{wnn} be the weighted nearest classifier with weights $\{w_{ni}\}_{i=1}^n$ where B_1 and B_2 are constants determined by:

$$
\begin{aligned}
B_1 &= \int_S \frac{\bar{f}(x_o)}{4\|\dot{\eta}(x_o)\|} dVol^{d-1}(x_o) \\
B_2 &= \int_S \frac{\bar{f}(x_o)}{\|\dot{\eta}(x_o)\|} dVol^{d-1}(x_o),
\end{aligned}
\qquad (4)
$$

Vol^{d-1} denotes the natural $(d-1)$ dimensional volume with measure inherent in $S \in \mathbb{R}^d$ while $\bar{f}(x_o)$ denotes the first derivative of the initial point x_o; $s_n^2 = \sum_{i=1}^n w_{ni}^2$ and $t_n = n^{-2/d} \sum_{i=1}^n w_{ni}\{i^{1+\frac{2}{d}} - (1-i)^{1+\frac{2}{d}}\}$ represent variance and squared bias contributions. C^{Bayes} denotes the Bayes classifiers, minimizing the risk over R. Both are given by:

$$
\begin{aligned}
C_n^{wnn}(x) &= \begin{cases} 1, & if \quad w_{ni\,i=1}^n \geq 1/2 \\ 2, & otherwise \end{cases} \\
C^{Bayes}(x) &= \begin{cases} 1, & if \quad \eta(x) \geq 1/2 \\ 2, & otherwise \end{cases}
\end{aligned}
\qquad (5)
$$

Therefore, there is the interpretation that for the point $x \in \mathbb{R}^d$, $\eta(x)$ belongs to class $C(x)$ with value of 1 in the sense of the weighted nearest neighbor classifier if $w_{ni\,i=1}^n \geq \frac{1}{2}$; and in the sense of the bayesian classifier, if the regression function $\eta(x) = P(Y = 1|X = x) \geq \frac{1}{2}$ and; otherwise, both have a value of 2. Further interpretation of the asymptotic behavior towards optimal classification can be found in Samworth [26]. Subsequently, provided that a single term t from the index is not empty based on the resulting *answer* form the positional *index*, we can iterate over the number of incoming tweets while adapting the document list Forward-Position-Intersect algorithm [24, 25] as follows:

Algorithm 2
Forward-Positional-Intersect(p_1 , p_2 , k)
1. answer ← {}
2. while p_1 ≠ NIL and p_2 ≠ NIL
3. do if tweetId($p1$) = tweetId($p2$)
4.   then:
5. pp_1 ← positions(p_1)
6. pp_2 ← positions(p_2)
7.   while pp_1 ≠ NIL and pp_2 ≠ NIL
8.   do if pos(pp_2) − pos(pp_1) = k
9.   then Add(answer, tweetId(p_1), pos(pp_1))
10. pp_1← next(pp_1)
11. pp_2 ← next(pp_2)
12.   elseif pos(pp_2) − pos(pp_1) > k
13.   then: pp_1 ← next(pp_1)
14.   else: pp_2 ← next(pp_2)
15. $p1$ ←next($p1$)
16. p_2 ←next(p_2)
17.   elseif tweetId(p_2) > tweetId(p_2)
18.   then p_2 ←next(p_2)
19.   else p_1 ←next(p_1)
20.   return answer

Re-defining the variables in Eckert [24] let p_1, p_2, pp_1 and pp_2 be the pointers to tweet lists and let p_1 and p_2 reference the tweet lists of the two terms to be intersected while pp_1 and pp_2 reference the inner position lists for each tweet with *tweetId* and *pos* dereferencing the pointers to their actual value in the list. Let *positions* extract the inner position list from an entry in the tweet list. *Add* adds a list identifier and a position to the resulting tweet list. The tweet lists represents the tweets logs of traffic information saved into file.

For our sentiment analysis, we consider the approach of Hu and Liu [27] lexicon of opinion words (LOWs). With our earlier derivations, we posit that the index of sentiments word would require correct interpretation of the word context in relevance to the topic of traffic delay and congestion by scoring the opinion contained in the traffic tweets based on the contextual polarity: positive, negative and neutral. The first method of the improved Naïve Bayes Algorithm (INB-1) by Kang et al. [28] was helpful in computing the score for the crawled filtered traffic tweets based on the following conditional probability:

$$Class(t_i) = \arg \max R_1\left(p_{ij}\right)P\left(c_j\right) \prod_{i=1}^{d} P\left(p_i|c_j\right) \qquad (6)$$

$$R_1\left(p_{ij}\right) = \frac{\sum_{p_{ij} \in L_j}^{|L|} C(p_{ij})}{\sum_{p_{ij} \in L}^{|L|} C(p_{ij})} \qquad (7)$$

where $Class(t_i)$ denotes the function that determines whether a traffic tweet (t_i) is positive, negative or neural. The probability of class c_j is calculated by $P\left(c_j\right)$ while $P\left(p_i|c_j\right)$ computes the probability that p_i belongs to c_j. $R_1\left(p_{ij}\right)$ denotes the ratio of number of patterns. $C(p_{ij})$ present in the class j of LOWs when the number of patterns $|L|$ is counted over number of patterns $C(p_{ij})$ present in the class j of LOWs when the number of patterns $|L|$ is uncounted. The pattern essentially an n-gram, dwells on the form of $n-1$ Markov model, representing contiguous sequence of n items from a corpus widely known as shingles. We used the Jaccard index to know the extent of similarity between sample sets of shingles irrespective of the ordering. This is given by:

$$J(C_1, C_2) = \frac{|C_1 \cap C_2|}{|C_1 U C_2|} \qquad (8)$$

$J(C_1, C_2)$ denotes the similarity between set C_1 and C_2. It follows that when item C_1 and C_2 are unrelated then $J(C_1, C_2) = 1$; otherwise $0 \leq J(C_1, C_2) \leq 1$. The cluster formation provide enough evidence to support the interrelations between traffic incidents with regards to the trending causatives of traffic congestions. Furthermore, we employ the term-frequency-inverse-document-frequency, $tdidf$ [29] to classify each term in the traffic congestion clusters based on the frequency of occurrence. This is performed by invoking the TF log-normalization with the smooth $tdidf$ weight-schemes as follows:

$$tf(t, d) = 1 + \log\left(f_{t,d}\right) \qquad (9)$$

$$idf(t, D) = \log\frac{N}{n_t} \qquad (10)$$

Such that tweet document term weight is given by:

$$tdidf(t, d, D) = tf(t, d) \cdot idf(t, D) \qquad (11)$$

With $N = |D|$ denoting the total number of document in the corpus; $n_t = 1 + |\{d \in D: t \in d\}|$ representing number of times term t appears in document d which belongs to D in the corpus. Notice that the addition of 1 to $|\{d \in D: t \in d\}|$ ensure that infinity value $idf(t, D)$ is avoided.

3 Experimental Evaluation

3.1 Discussion of Results

A sample of 121 tweets were retrieved based on the phrase search UK traffic delay. The data was cleaned of irrelevant symbols. After tweets crawling, preprocessing, tokenization and feature extraction, we obtained the sentiment analysis results as presented in Table 1.

In the time period of obtaining the traffic delay tweets, it was observed that possible severity of 22 were negative sentiments; most likely attributed to serious accidents on the road way (12 negative sentiments). Other relevant phrases are generated in the sentiment analysis such as "serious accidents", "long delays", "looking good", "serious delays" etc. The Jaccard index or similarity and *tdidf* is

Table 1 Traffic twitter sentiment analysis

Phrase	Negative	Neural	Positive	Total
Possible severity	22			22
Serious accident	12			12
Latest	9	3		12
Long delay	1	6		7
Looking good			6	6
Serious delays		6		6
Huge		5		5
Broken down	5			5
Heavy		5		5
Updates		4		4
Emergency	4			4
Blocked		4		4
Main work		4		4
Delays		3		3
Uninjured			3	3
Travel heavy		3		3
Accident	3			3
Update		3		3
Nightmare	2			2
Shocking	2			2
Severe accident	2			2
Bridge congestion delay	2			2
Severe	2			2
Total	69	43	9	121

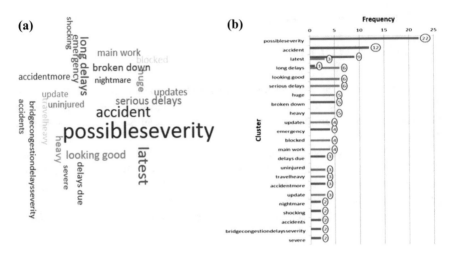

Fig. 1 **a** Traffic delay trending events. **b** Cluster classification index

used to generate the relevant traffic trending events contributing to the cluster classification index as shown in Fig. 1a, b.

3.2 Classification Accuracy

The sentiment classification accuracy of our model is measured in order to determine the performance following the split of the traffic tweet dataset into the training sets (70%) for which the true values are known; validation set (15%) for tuning the classifier during training; and testing set (15%) with unknown values associated with the traffic congestion situation. This is based on the following measures:

$$\text{Accuracy}, a = \frac{\sum TP + \sum TN}{TP_o} \tag{12}$$

$$\text{Precision}, P_r = \frac{TN}{TP + FN} \tag{13}$$

Let TP be the true positive rate denoting the number of the traffic tweets that were correctly identified. TN is the true negative rate denoting the number of traffic tweets correctly rejected; FN be the false negative rate denoting the number of traffic tweets incorrectly rejected; FP be the false positive rate denoting the number of traffic tweets incorrectly accepted; TP_o be the total count of traffic tweets which belongs to a set; P_r be the precision which represents the fraction of the tweets

Table 2 Sentiment classification accuracy

Traffic tweet data sets	Sentiment classification	TP rate	FP rate	Precision
Training	Positive	0.990	0.031	0.882
	Neural	0.987	0.008	0.964
	Negative	0.912	0.049	0.920
Validation	Positive	0.976	0.028	0.905
	Neural	0.989	0.006	0.977
	Negative	0.966	0.045	0.933
Testing	Positive	0.908	0.034	0.855
	Neural	0.955	0.042	0.977
	Negative	0.963	0.055	0.961

relevant to the search query; a be the (overall) accuracy which determines the number of correct queries as per the total number of queries. The results show an average accuracy and average precision of 0.95 and 0.91, respectively. Table 2 summarizes the performance of the classifiers for each class under consideration with regards to some clusters associated with the traffic congestion delay.

In the training set, the TP rate yields highest value of 0.990 for the positive sentiment traffic tweet classification with a least value of 0.908 in the testing set for the positive sentiment. The classifier of neural opinion has the least FP of 0.006 in the validation set while its highest value of 0.055 emerges in the testing set for the negative sentiments. The precision yields highest value of 0.977 in the neural sentiment found in the validation and testing set while its least value is in the positive sentiment classification contained in the testing set. We envisage that correctly classifying the traffic congestion based on the twitter sentiments would depend on the location of the user, internet accessibility and tweets time-proximity to the real time the traffic congestion persists with respect to the incident time leading to it.

3.3 Model Validation

To validate the model, the performance of Latent Dirichlet Allocation (LDA) is compared with the model employing the Naïve Bayes and Jaccard similarity with n-gram (JCn-g). The LDA is a typical example of a topic model that can be used for clustering data points; for instance, Azam et al. [17] applied it for clustering of tweets. It is also considered a generative probabilistic model that allows documents to be represented as random mixtures over latent topics characterized by a distribution over words [30]. Table 3 presents the comparative evaluation of JCn-g using unigram and bigram with LDA.

Table 3 Comparative evaluation of JCn-g with LDA

Performance metrics	JCn-g ($n = 1$)	JCn-g ($n = 2$)	LDA
Accuracy	0.871	0.882	0.880
Precision	0.742	0.753	0.762

As observed JCn-g with bigram yields the best accuracy while LDA yields the most precise result. This can be attributed to the fact that LDA not only serves as a generative probabilistic model but also combines it topics interpretability with prior Dirichlet distribution form. Figure 2 presents the cluster generative probabilistic models for the JCn-g and LDA respectively. It shows the data compression of JCn-g ($n = 2$) and LDA as well as the better similarity between them to buttress our earlier statement. In fact, it can be seen that the green and black tweet clusters are approximately within the same dimensional vector space in the JCn-g ($n = 2$) and LDA. The best precision observe in LDA becomes obvious from the yellow tweets cluster data points which share same vector space with the JCn-g ($n = 1$).

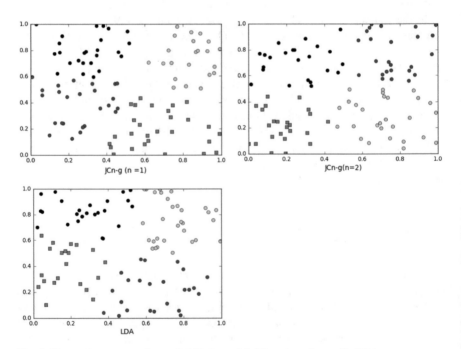

Fig. 2 Tweet cluster generative probabilistic model: JCn-g ($n = 1$, $n = 2$), LDA

4 Conclusions and Future Work

Exploring traffic condition using social media data, which can be readily obtained from Twitter, continues to influence traffic information and transportation engineering management decision makers. Applying the proposed data mining techniques on different strata of the UK traffic delay tweets yielded interesting results on traffic congestion, incidents and control.

The validation of JCn-g using LDA shows that the JCn-g with bigram has better accuracy than LDA; however, LDA maintained its high precision over the JCn-g with unigram and bigram. Precious works have suggested that LDA combines its topics interpretability with prior Dirichlet distribution form.

Future work should seek to improve the precision of our cluster classification algorithm. It should seek to improve our preliminary results with a view to seeing if a hybrid approach of the JCn-g with LDA can be more feasible. Also, investigating the reliability for seamless integration with well-known traffic management software system tools should be explored.

References

1. Lu, H-P., Sun, Z., & Qu, W. (2015). Big data-driven based real-time traffic flow state identification and prediction. *Discrete Dynamics in Nature and Society, 2015*, Article ID 284906, 1–11.
2. Villars, R. L., Olofson, C. W., & Eastwood, M. (2011). Big data: What it is and why you should care. IDC.
3. Vlahogianni, E. I, Park, B. B., & van Lint, J. W. C. (2015). Big data in transportation and traffic engineering. *Transportation Research Part C: Emerging Technologies, 58*(Part B), 1–161.
4. Stopher, P. R., & Greaves, S. P. (2007). Household travel surveys: Where are we going? *Transportation Research Part A: Policy and Practice, 41*(5), 367–381.
5. Wang, X., & Li, Z. (2016). Traffic and transportation smart with cloud computing on big data. *International Journal of Computer Science and Applications, 13*(1), 1–16.
6. Philander, K., & Zhong, Y. (2016). Twitter sentiment analysis: Capturing sentiment from integrated resort tweets. *International Journal of Hospitality Management, 55*, 16–24.
7. Korkontzelos, I., Nikfarjam, A., Shardlow, M., Sarker, A., Ananiadou, S., & Gonzalez, G. H. (2016). Analysis of the effect of sentiment analysis on extracting adverse drug reactions from tweets and forum posts. *Journal of Biomedical Informatics, 62*, 148–158.
8. Burscher, B., Vliegenthart, R., & de Vreese, C. H. (2016). Frames beyond words: Applying cluster and sentiment analysis to news coverage of the nuclear power issue. *Social Science Computer Review, 34*(5), 530–545.
9. Abidin, A. F., Kolberg, M., & Hussain, A. (2015). Integrating Twitter traffic information with Kalman filter models for public transportation vehicle arrival time prediction. In M. Trovati, R. Hill, A. Anjum, S. Y. Zhu & L. Liu (Eds.), *Big-data analytics and cloud computing* (pp. 67–82).
10. Pak, A., & Paroubek, P. (2010). Twitter as a corpus for sentiment analysis and opinion mining. In *Proceedings of the Seventh conference of International language Resources and Evaluation (LREC' 10)*.

11. Go, A., Huang, L., & Bhayani, R. (2009). Twitter sentiment analysis. Stanford University, Stanford California, USA, CS224N - Final Year Project.
12. Wang, J., Gu, Q., & Wang, G. (2013). Potentila power and problems in sentiment mining of social media. *International Journal of Strategic Decision Science, 4*(2), 16–26.
13. Kumar, A., & Sebastian, T. M. (2012). Sentiment analysis on Twitter. *International Journal of Computer Science Issues, 9*(4:3), 372–378.
14. He, J., Shen, W., Divakaruni, P., Wynter, L., & Lawrence, R. (2013). Improving traffic prediction with tweet semantics. In *Proceedings of the Twenty-Third International Joint Conference on Artificial Intelligence*, Beijing, China.
15. Grosenick, S. (2012). Real-time traffic prediction improvement through semantic mining of social networks. Unpublished master thesis. University of Washington, Washington.
16. Elsafoury, F. A. (2013). *Monitoring urban traffic status using twitter messages* (pp. 1–46).
17. Azam, N., Abulaish, M., & Haldar, N. A.-H. (2015). Twitter data mining for events classification and analysis. In *Second International Conference on Soft Computing and Machine Intelligence*.
18. Broder, A. Z., Glassman, S. C., Manasse, M. S., & Zweig, G. (1997). Syntactic clustering of the web. *Computer Networks and ISDN Systems, 29*(8), 1157–1166.
19. van Dongen, S. (2000). *Graph clustering by flow simulation*. Utrecht, Netherlands: University of Utrecht.
20. McHugh, D. (2014). *Traffic prediction and analysis using a big data and visualisation approach*. Ireland: Blanchardstown.
21. Tejaswin, P., Kumar, R., & Gupta, S. (2015). Tweeting traffic: Analyzing Twitter for generating real-time city traffic insights and predictions. In *CODS-IKDD '15*, Bangalore, India.
22. Rais, K. (2014). Twitter analysis.
23. cran.r-project.org. https://cran.r-project.org/web/packages/.
24. Eckert, K. (2008). Simplified phrase search algorithm.
25. Manning, C. D., Raghavan, P., & Schütze, H. (2008). *Introduction to information retrieval*. Cambridge University Press.
26. Samworth, R. J. (2012). Optimal weighted nearest neighbour classifiers. *Annals of Statistics, 40*(5), 2733–2763.
27. Hu, M., & Liu, B. (2004). Mining opinion features in customer review. In *Proceedings of the 19th National Conference on Artificial Intelligence, AAAAI'04*.
28. Kang, H., Yoo, S. J., & Han, D. (2012). Senti-lexicon and improved Naïve Bayes algorithms for sentiment analysis. *Expert Systems with Applications, 39*, 6000–6010.
29. Spärck Jones, K. (1972). A statistical interpretation of term specificity and its application in retrieval. *Journal of Documentation, 28*(1), 11–21.
30. Blei, D. M., Ng, A. Y., & Jordan, M. I. (2003). Latent dirichlet allocation. *Journal of Machine Learning Research, 3*(4–5), 993–1022.

Modelling of Pile Drivability Using Soft Computing Methods

Wengang Zhang and Anthony T. C. Goh

Abstract Driven piles are commonly used to transfer the loads from the super-structure through weak strata onto stiffer soils or rocks. For driven piles, the impact of the piling hammer induces compression and tension stresses in the piles. Hence, an important design consideration is to check that the strength of the pile is suffi-cient to resist the stresses caused by the impact of the pile hammer. Due to its complexity, pile drivability lacks a precise analytical theory or understanding of the phenomena involved. In situations where measured or numerical data are available, various soft computing methods have shown to offer great promise for mapping the nonlinear interactions between the system's inputs and outputs. In this study, two soft computing methods, the Back propagation neural network (BPNN) and Mul-tivariate adaptive regression splines (MARS) algorithms were used to assess pile drivability in terms of the Maximum compressive stresses, Maximum tensile stresses, and Blow per foot. A database of more than four thousand piles is utilized for model development and comparative performance of the predictions between BPNN and MARS.

Keywords Back propagation neural network · Multivariate adaptive regression splines · Pile drivability · Computational efficiency · Nonlinearity

W. Zhang
Key Laboratory of New Technology for Construction of Cities in Mountain Area,
Chongqing University, Chongqing 400045, China

W. Zhang
School of Civil Engineering, Chongqing University, Chongqing 400045, China

A. T. C. Goh (✉)
School of Civil and Environmental Engineering, Nanyang Technological University,
Singapore 639798, Singapore
e-mail: ctcgoh@ntu.edu.sg

© Springer Nature Singapore Pte Ltd. 2018
S. S. Roy et al. (eds.), *Big Data in Engineering Applications*,
Studies in Big Data 44, https://doi.org/10.1007/978-981-10-8476-8_14

1 Introduction

Driven piles are commonly used to transfer the loads from the superstructure through weak strata onto stiffer soils or rocks. For these piles, the impact of the piling hammer induces compression and tension stresses in the piles. Hence, an important design consideration is to ensure that the strength of the pile is sufficient to resist the stresses introduced by the impact of the pile hammer. One common method of calculating the driving stresses is based on the stress-wave theory [18] which involves the discrete idealization of the hammer-pile-soil system. Considering that the conditions at each site are different, generally a wave equation based computer program is required to generate the pile driving criteria for each individual project. The pile driving criteria include:

- Hammer stroke versus Blow per foot BPF (1/set) for required bearing capacity,
- Maximum compressive stresses versus BPF,
- Maximum tension stress versus BPF.

However, this process can be rather time consuming and requires very specialized knowledge of the wave equation program.

The essence of modeling/numerical mapping is prediction, which is obtained by relating a set of variables in input space to a set of response variables in output space through a model. The analysis of pile drivability involves a large number of design variables and nonlinear responses, particularly with statistically dependent inputs. Thus, the commonly used regression models become computationally impractical. Another limitation is the strong model assumptions made by these regression methods.

An alternative soft computing technique is the artificial neural network (ANN). The ANN structure consists of one or more layers of interconnected neurons or nodes. Each link connecting each neuron has an associated weight. The "learning" paradigm in the commonly used Back-propagation (BP) algorithm [14] involves presenting examples of input and output patterns and subsequently adjusting the connecting weights so as to reduce the errors between the actual and the target output values. The iterative modification of the weights is carried out using the gradient descent approach and training is stopped once the errors have been reduced to some acceptable level. The ability of the trained ANN model to generalize the correct input-output response is performed in the testing phase and involves presenting the trained neural network with a separate set of data that has never been used during the training process.

This paper explores the use of ANN and another soft computing technique known as multivariate adaptive regression splines (MARS) [3] to capture the intrinsic nonlinear and multidimensional relationship associated with pile drivability. Similar with neural networks, no prior information on the form of the numerical function is required for MARS. The main advantages of MARS lie in its capacity to capture the intrinsic complicated data mapping in high-dimensional data patterns and produce simpler, easier-to-interpret models, and its ability to perform analysis on parameter relative importance. Previous applications of the MARS algorithm in civil engineering include predicting the doweled pavement performance, estimating shaft resistance of piles in sand and deformation of asphalt mixtures, analyzing shaking table tests of reinforced soil wall, determining the undrained shear strength of clay, predicting liquefaction-induced lateral spread, assessing the ultimate and serviceability performances of underground caverns, estimating the EPB tunnel induced ground surface settlement, and inverse analysis for braced excavation [1, 7, 8, 12, 13, 15–17, 19–23]. In this paper, the Back propagation neural network (BPNN) and MARS models are developed for pile drivability predictions in relation to the Maximum compressive stresses (MCS), Maximum tensile stresses (MTS), and Blow per foot (BPF). A database of more than four thousand piles is utilized for model development and comparative performance between BPNN and MARS predictions.

2 Methodologies

2.1 Back-Propagation Algorithm

A three-layer, feed-forward neural network topology shown in Fig. 1 is adopted in this study. As shown in Fig. 1, the back-propagation algorithm involves two phases of data flow. In the first phase, the input data are presented forward from the input to output layer and produces an actual output. In the second phase, the error between the target values and actual values are propagated backwards from the output layer to the previous layers and the connection weights are updated to reduce

Fig. 1 Back-propagation neural network architecture used in this study

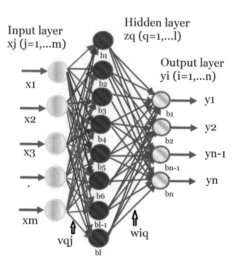

the errors between the actual output values and the target output values. No effort is made to keep track of the characteristics of the input and output variables. The network is first trained using the training data set. The objective of the network training is to map the inputs to the output by determining the optimal connection weights and biases through the back-propagation procedure. The number of hidden neurons is typically determined through a trial-and-error process; normally the smallest number of neurons that yields satisfactory results (judged by the network performance in terms of the coefficient of determination R^2 of the testing data set) is selected. In the present study, a Matlab-based back-propagation algorithm BPNN with the Levenberg-Marquardt (LM) algorithm [2] was adopted for neural network modeling.

2.2 Multivariate Adaptive Regression Splines Algorithm

MARS was first proposed by [3] as a flexible procedure to organize relationships between a set of input variables and the target dependent that are nearly additive or involve interactions with fewer variables. It is a nonparametric statistical method based on a divide and conquer strategy in which the training data sets are partitioned into separate piecewise linear segments (splines) of differing gradients (slope). MARS makes no assumptions about the underlying functional relationships between dependent and independent variables. In general, the splines are connected smoothly together, and these piecewise curves (polynomials), also known as basis functions (BFs), result in a flexible model that can handle both linear and nonlinear behavior. The connection/interface points between the pieces are called knots. Marking the end of one region of data and the beginning of another, the candidate knots are placed at random positions within the range of each input variable.

MARS generates BFs by stepwise searching over all possible univariate candidate knots and across interactions among all variables. An adaptive regression algorithm is adopted for automatically selecting the knot locations. The MARS algorithm involves a forward phase and a backward phase. The forward phase places candidate knots at random positions within the range of each predictor variable to define a pair of BFs. At each step, the model adapts the knot and its corresponding pair of BFs to give the maximum reduction in sum-of-squares residual error. This process of adding BFs continues until the maximum number is reached, which usually results in a very complicated and overfitted model. The backward phase involves deleting the redundant BFs that made the least contributions. An open MARS source code from [10] is adopted in performing the analyses presented in this paper.

Let y be the target dependent responses and $X = (X_1, \ldots, X_P)$ be a matrix of P input variables. Then it is assumed the data are generated based on an unknown "true" model. For a continuous response, this would be

$$y = f(X_1, \ldots, X_P) + e = f(\mathbf{X}) + e \tag{1}$$

in which e is the fitting error. f is the built MARS model, comprising of BFs which are splines piecewise polynomial functions. For simplicity, only the piecewise linear function is expressed and considered in this paper. Piecewise linear functions follow the form $\max(0, x-t)$ with a knot defined at value t. Expression $\max(\cdot)$ means that only the positive part of (.) is used otherwise it is assigned a zero value. Formally,

$$\max(0, x-t) = \begin{cases} x-t, & \text{if } x \geq t \\ 0, & \text{otherwise} \end{cases} \tag{2}$$

The MARS model $f(X)$, which is a linear combination of BFs and their interactions, is expressed as

$$f(X) = \beta_0 + \sum_{m=1}^{M} \beta_m \lambda_m(X) \tag{3}$$

where each λ_m is a BF. It can be a spline function, or interaction BFs produced by multiplying an existing term with a truncated linear function involving a new/different variable (higher orders can be used only when the data warrants it; for simplicity, at most second-order is adopted). The terms β are constant coefficients, estimated using the least-squares method.

Figure 2 shows an example illustration of how the MARS algorithm would make use of piecewise linear spline functions to fit provided data patterns. The MARS mathematical equation is as follows

$$y = -5.0875 - 2.7678 \times BF1 + 0.5540 \times BF2 + 1.1900 \times BF3 \tag{4}$$

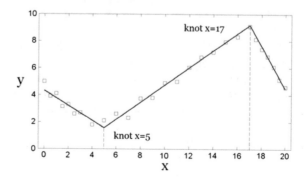

Fig. 2 Knots and linear splines for a simple MARS example

in which BF1 = max(0, $x - 17$), BF2 = max(0, $17 - x$) and BF3 = max(0, $x - 5$) and max is defined as: max(a, b) is equal to a if $a > b$, else b. The knots are located at $x = 5$ and 17. These two knots delimit/cut the x range into three intervals where different linear relationships are identified.

The MARS modeling is a data-driven process. To construct the model in Eq. (3), first the forward phase is performed on the training data starting initially with only the intercept β_0. At each subsequent step, the basis pair that produces the maximum reduction in the training error is added. Considering a current model with M basis functions, the next pair to be added to the model is in the form of

$$\beta_{M+1} \lambda_l(X)\max(0, X_j - t) + \beta_{M+2} \lambda_l(X)\max(0, t - X_j) \qquad (5)$$

with each β being estimated by the least-squares method. This process of adding BFs continues until the model reaches some predetermined maximum number, generally leading to a purposely overfitted model.

The backward phase improves the model by removing the less significant terms until it finds the best sub-model. Model subsets are compared using the less computationally expensive method of Generalized Cross-Validation (GCV). The GCV is the mean-squared residual error divided by a penalty that is dependent on model complexity. For the training data with N observations, GCV is calculated as [9]

$$GCV = \frac{\frac{1}{N} \sum_{i=1}^{N} [y_i - f(x_i)]^2}{[1 - \frac{M + d \times (M-1)/2}{N}]^2} \qquad (6)$$

in which M is the number of BFs, d is a penalty for each basis function included in the developed sub-model, N is the number of data sets, and $f(x_i)$ denotes the MARS predicted values. Thus the numerator is the mean square error of the evaluated model in the training data, penalized by the denominator which accounts for the increasing variance in the case of increasing model complexity. Note that $(M-1)/2$ is the number of hinge function knots. The GCV penalizes not only the number of BFs but also the number of knots. A default value of 3 is assigned to penalizing parameter d and further suggestions on choosing the value of d can be referred to [3]. At each deletion step, a basis function is pruned to minimize Eq. (3), until an adequately fitting model is found.

After the optimal MARS model is determined, by grouping together all the BFs involving one variable and another grouping of BFs involving pairwise interactions, the analysis of variance (ANOVA) decomposition procedure [3] can be used to assess the parameter relative importance based on the contributions from the input variables and the BFs.

Table 1 Summary of performance measures

Performance measure	Definition
Coefficient of determination (R^2)	$R^2 = 1 - \dfrac{\frac{1}{n}\sum_{i=1}^{n}(Y_i - \overline{Y})^2}{\frac{1}{n}\sum_{i=1}^{n}(y_i - \overline{y})^2}$
Coefficient of correlation (r)	$r = \dfrac{\sum_{i=1}^{N}(Y_i - \overline{Y})(y_i - \overline{y})}{\sqrt{\sum_{i=1}^{N}(Y_i - \overline{Y})^2}\sqrt{\sum_{i=1}^{N}(y_i - \overline{y})^2}}$
Relative root mean squared error (RRMSE)	$RRMSE = \dfrac{\sqrt{\frac{1}{N}\sum_{i=1}^{N}(Y_i - y_i)^2}}{\frac{1}{N}\sum_{i=1}^{N}y_i} \times 100$
Performance index (ρ)	$\rho = \dfrac{RRMSE}{1+r}$

$\overline{y}$ is the mean of the target values of y_i; $\overline{Y}$ is the mean of the predicted Y_i; N denotes the number of data points in the used set, training set, testing set or the overall set; Definitions of RRMSE, r and ρ are based on [4].

3 Performance Measures

Table 1 shows the performance measures and the corresponding definitions utilized for prediction comparison of the two surrogate methods.

4 Pile Drivability Data Sets

In this paper, a database containing 4072 piles with a total of seventeen variables is developed from the information on piles already installed for bridges in the State of North Carolina [11]. Seventeen variables including hammer characteristics, hammer cushion material, pile and soil parameters, ultimate pile capacities, and stroke were regarded as inputs to estimate the three dependent responses comprising of the Maximum compressive stresses (MCS), Maximum tensile stresses (MTS), and Blow per foot (BPF). A summary of the input variables and outputs is listed in Table 2.

For purpose of simplifying the analyses considering the extensive number of parameters and large data set, Joen and Rahman [11] divided the data into five categories (Q_1–Q_5) based on the ultimate pile capacity, as detailed in Table 3. In this paper, for each category 70% of the data patterns were randomly selected as the training dataset and the remaining data were used for testing. For details of the entire data set as well as each design variable and responses, the report by Joen and Rahman [11] can be referred to.

Table 2 Summary of input variables and outputs

Inputs and outputs	Parameters and parameter descriptions		
Input variables	Hammer	Hammer weight (kN)	Variable 1 (x1)
		Energy (kN · m)	Variable 2 (x2)
	Hammer cushion material	Area (m^2)	Variable 3 (x3)
		Elastic modulus (GPa)	Variable 4 (x4)
		Thickness (m)	Variable 5 (x5)
		Helmet weight (kN)	Variable 6 (x6)
	Pile information	Length (m)	Variable 7 (x7)
		Penetration (m)	Variable 8 (x8)
		Diameter (m)	Variable 9 (x9)
		Section area (m^2)	Variable 10 (x10)
		Slenderness L/D	Variable 11 (x11)
	Soil information	Quake at toe (m)	Variable 12 (x12)
		Damping at shaft (s/m)	Variable 13 (x13)
		Damping at toe (s/m)	Variable 14 (x14)
		Shaft resistance (%)	Variable 15 (x15)
	Ultimate pile capacity Q_u (kN)		Variable 16 (x16)
	Stroke (m)		Variable 17 (x17)
Outputs	Maximum compressive stress MCS (MPa)		
	Maximum tensile stress MTS (MPa)		
	BPF		

Table 3 Division of data with respect to ultimate pile capacities

Pile type	Q_u range (kN)	Data		
		No. of training data	No. of testing data	Total
Q_1	133.4–355.9	270	90	360
Q_2	360.0–707.3	428	144	572
Q_3	707.4–1112.1	808	249	1057
Q_4	1112.2–1774.8	1296	421	1717
Q_5	1774.9–3113.7	276	90	366
Total		3078	994	4072

5 BPNN Models

For simplicity, only BPNN models with one single hidden layer structure are considered. The optimal BPNN model is selected from models with different hidden neurons since the other main parameters for BPNN algorithms have been fixed as:
logsig transfer function from the input layer to the hidden layer;
tansig transfer function from the hidden layer to the output layer;
maxepoch = 500;
learning rate = 0.01;
min_grad = 1×10^{-15};
decrease factor mu_dec = 0.7;
increase factor mu_inc = 1.03.

5.1 The Optimal BPNN Model

The BPNN with the highest coefficient of determination R^2 value for the testing data sets is considered to be the optimal model. Figure 3 plots the R^2 values of the testing data sets for BPNN models with different neurons (from 5 to 15) in the hidden layer for MCS, MTS and BPF predictions. It can be observed that for the optimal MCS, MTS, and BPF models, the number of the neurons in the hidden layer is 9, 7 and 11, respectively.

5.2 Modeling Results

Figures 4, 5 and 6 show the BPNN predictions for the training and testing data patterns for MCS, MTS, and BPF, respectively. For the MCS predictions, considerably high R^2 (>0.97) are obtained for both the training and testing patterns. Compared with the MCS predictions, the developed BPNN model is slightly less

Fig. 3 R^2 for different neuron numbers for MCS, MTS and BPF models

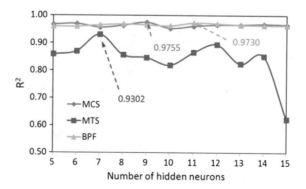

Fig. 4 Prediction of MCS
using BPNN

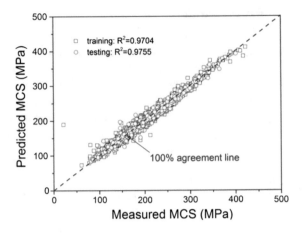

Fig. 5 Prediction of MTS
using BPNN

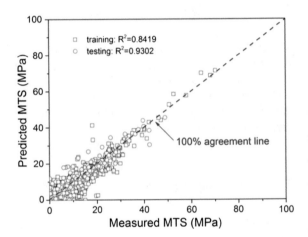

Fig. 6 Prediction of BPF
using BPNN

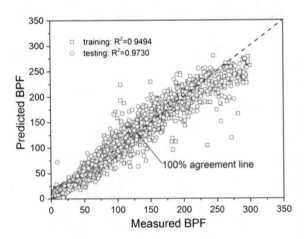

Fig. 7 Prediction of BPF
using BPNN

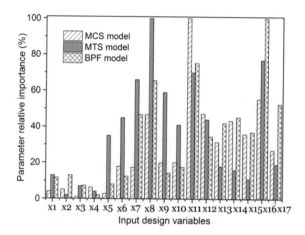

accurate in predicting the MTS mainly as a result of the bias (errors) due to the significantly smaller tensile stress values in comparison to the compressive stresses. For the BPF estimation, high R^2 are also obtained for both the training and testing patterns, with the latter slightly greater than the training sets. In addition, the three optimal BPNN models can serve as reliable tools for prediction of MCS, MTS and BPF.

5.3 Parameter Relative Importance

The parameter relative importance determined by BPNN is based on the method by [5] and discussed by Goh [6]. Figure 7 gives the plot of the relative importance of the input variables for the three BPNN models. It can be observed that MCS is mostly influenced by the input variable x11 (Slenderness) and MTS is mostly influenced by the input variable x8 (Penetration). Interestingly, BPF is primarily influenced by the input variable x16 (Ultimate pile capacity).

5.4 Model Interpretability

For brevity, only the developed BPNN MCS model is expressed in mathematical form through the trained connections weights, the bias, and the transfer functions. The Mathematical expression for MCS obtained by the optimal MCS analysis is shown in the Appendix 1.

6 MARS Models

It is assumed that at most the 2nd order interaction is considered for the prediction of MCS, MTS and BPF using MARS. The number of basis functions changes from $2n$ to n^2 ($n = 17$ in this study, numerical trials indicate that overfitting occurs when the number of BFs exceeds 80).

6.1 The Optimal MARS Model

The MARS model with the highest R^2 value and less BFs for the testing data set is considered to be the optimal. Figure 8 plots the R^2 values of the testing data sets for the MARS models with different BFs (from 34 to 78) in the hidden layer for the MCS, MTS and BPF predictions. It can be observed that for the optimal MCS, MTS, and BPF models, the number of BFs is 52, 36 and 38, respectively.

6.2 Modeling Results

Figures 9, 10 and 11 show the MARS predictions for the training and testing data patterns for MCS, MTS, and BPF, respectively. For the MCS prediction, considerably high R^2 (>0.95) are obtained for both the training and testing patterns. As in the BPNN analysis, the developed MARS model is less accurate in predicting MTS compared with the MCS predictions, mainly due to the bias brought about by the smaller tensile stress values. For the BPF estimation, high R^2 (>0.90) are also obtained for both the training and testing patterns, with the latter slightly greater than the training sets. Consequently, the three optimal MARS models can serve as reliable tools for prediction of MCS, MTS and BPF.

Fig. 8 R^2 for different number of BFs for MCS, MTS and BPF models

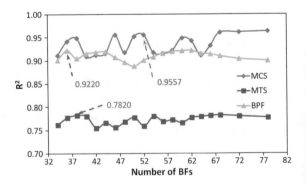

Fig. 9 Prediction of MCS using MARS

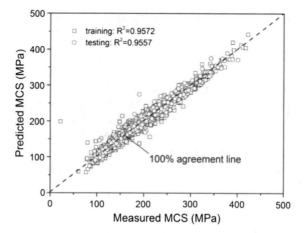

Fig. 10 Prediction of MTS using MARS

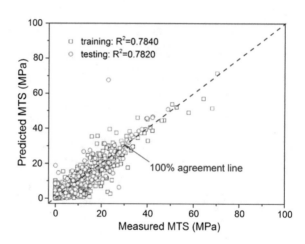

Fig. 11 Prediction of BPF using MARS

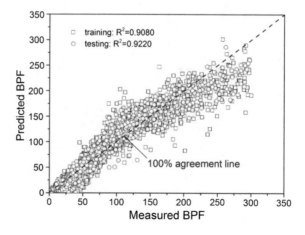

6.3 Parameter Relative Importance

Table 4 displays the ANOVA decomposition of the built MARS models for MCS, MTS and BPF respectively. For each model, the ANOVA functions are listed. The GCV column provides an indication on the significance of the corresponding ANOVA function, by listing the GCV value for a model with all BFs corresponding to that particular ANOVA function removed. It is this GCV score that is used to assess whether the ANOVA function is making a significant contribution to the model, or whether it just marginally improves the global GCV score. The #basis column gives the number of BFs comprising the ANOVA function and the variable (s) column lists the input variables associated with this ANOVA function.

Table 4 ANOVA decomposition of MARS model for MCS, MTS and BPF

Function	MCS			MTS			BPF		
	GCV	# Basis	Variable (s)	GCV	# Basis	Variable (s)	GCV	# Basis	Variable (s)
1	28.82	1	1	1.047	2	5	39.657	2	1
2	8.346	2	6	575.191	1	6	9.750	2	2
3	7.073	1	8	109.688	2	7	1.760	2	13
4	10.226	1	12	305.352	1	8	3.005	2	15
5	5.629	3	17	251.585	2	11	8.034	2	16
6	11.184	1	1 3	25.373	1	17	2.976	2	17
7	48.344	2	1 17	0.441	1	1 6	66.894	3	1 3
8	8.048	5	2 4	337.341	2	3 7	0.370	2	1 6
9	11.846	2	3 4	0.893	2	3 17	0.235	2	1 13
10	21.733	2	3 17	5.626	2	5 7	0.231	1	1 16
11	63.062	1	4 15	2.229	1	5 11	43.396	2	2 3
12	8.017	1	6 8	795.122	4	6 7	0.357	1	2 4
13	4.976	3	8 17	92.069	3	6 8	0.403	2	2 16
14				6.797	1	6 9	0.557	4	2 17
15				48.170	4	6 11	0.280	2	3 13
16				1.472	1	6 16	0.705	2	4 15
17				2.593	2	6 17	0.227	1	4 17
18				0.626	1	7 8	0.170	1	5 13
19				11.173	1	7 17	0.191	1	6 15
20				0.447	1	8 16	0.221	2	7 15
21				50.089	2	8 17	0.375	1	13 15
22				0.828	1	11 15	0.984	1	16 17
23				148.475	2	11 17			
24				1.472	2	14 17			
25				0.466	1	15 17			

Fig. 12 Relative importance
of the input variables in
MARS pile drivability models

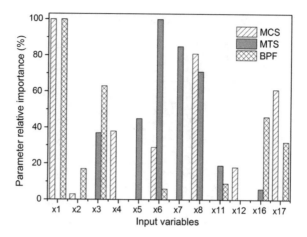

Figure 12 gives the plot of the relative importance of the input variables for the three HP drivability models developed by MARS. It can be observed that both MCS and BPF are mostly influenced by the input variable x1 (hammer weight). Interestingly, MTS is primarily influenced by the input variable x6 (the weight of helmet). It should be noted that since the BPNN and MARS algorithms adopt different methods in assessing the parametric relative importance, it is understandable that the two algorithms give different results.

6.4 Model Interpretability

Table 5 lists the BFs of the MCS model. The MARS model is in the form of

$MCS(MPa) = 169.4 + 0.0095 \times BF1 + 35.6 \times BF2 - 47.5 \times BF3 - 0.46 \times BF4 - 2 \times BF5 + 8847 \times BF6 + 9.2 \times BF7 - 8.2 \times BF8 - 0.0025 \times BF9 + 0.0062 \times BF10 - 3.2 \times BF11 + 470 \times BF12 - 0.0036 \times BF13 - 0.8 \times BF14 - 0.0012 \times BF15 + 0.006 \times BF16 + 9.43 \times BF17 - 6.1 \times BF18 + 0.136 \times BF19 - 0.098 \times BF20 - 0.83 \times BF21 - 0.17 \times BF22 - 540 \times BF23 + 1.34 \times 10^5 \times BF24 + 1.672 \times BF25 - 0.42 \times BF26 + 0.144 \times BF27 - 4.57 \times BF28 - 0.0054 \times BF29 + 0.052 \times BF30 + 87 \times BF31 + 250 \times BF32 - 763 \times BF33 - 16 \times BF34 - 28.1 \times BF35 + 0.217 \times BF36 - 0.2 \times BF37 + 34.5 \times BF38 + 31.3 \times BF39 - 50.2 \times BF40 - 425 \times BF41 + 0.0018 \times BF42 - 0.003 \times BF43 - 7.4 \times BF44 + 341 \times BF45 + 51.4 \times BF46 + 5.67 \times BF47 + 12 \times BF48 + 0.96 \times BF49 + 100.2 \times BF50 - 0.2 \times BF51 + 0.23 \times BF52$

$$(7)$$

Table 5 BFs and corresponding equations of MARS MCS model

BF	Equation	BF	Equation
BF1	max(0, x16−1550)	BF27	max(0, x15 − 15)
BF2	max(0, x17−2.29)	BF28	max(0, 15−x15)
BF3	max(0, 2.29−x17)	BF29	BF28 × max(0, x16−289)
BF4	max(0, x6−7.38)	BF30	BF28 × max(0, 289−x16)
BF5	max(0, 7.38−x6)	BF31	BF2×max(0, x1−29.4)
BF6	max(0, 0.014−x10)	BF32	BF6 × max(0, x6−6.67)
BF7	max(0, x2−30.7)	BF33	BF6 × max(0, 6.67−x6)
BF8	max(0, 30.7−x2)	BF34	BF5 × max(0, 1.81−x17)
BF9	BF1 × max(0, x7−8.00)	BF35	BF3 × max(0, x1−29.4)
BF10	BF1 × max(0, 8.00−x7)	BF36	BF7 × max(0, x11−50)
BF11	max(0, x11−9)	BF37	BF7 × max(0, 50−x11)
BF12	max(0, 9−x11)	BF38	BF28 × max(0, x13−0.59)
BF13	max(0, 1550−x16) × max(0, x8−3.05)	BF39	BF28 × max(0, 0.59−x13)
BF14	max(0, 1550−x16) × max(0, 3.05−x8)	BF40	BF4 × max(0, x5−0.05)
BF15	max(0, 1550−x16) × max(0, x6−9.34)	BF41	BF4 × max(0, 0.05−x5)
BF16	max(0, 1550−x16) × max(0, 9.34−x6)	BF42	max(0, 1550−x16) × max(0, x11−24)
BF17	BF6×max(0, x16−1067.5)	BF43	max(0, 1550−x16) × max(0, 24−x11)
BF18	BF6 × max(0, 1068−x16)	BF44	BF7 × max(0, 0.18−x3)
BF19	BF11 × max(0, x4−3.24)	BF45	max(0, x3−0.26)
BF20	BF11 × max(0, 3.24−x4)	BF46	max(0, 0.26−x3)
BF21	BF11 × max(0, x1−29.4)	BF47	BF5 × max(0, x4−1.97)
BF22	BF11 × max(0, 29.4−x1)	BF48	BF5 × max(0, 1.97−x4)
BF23	BF6 × max(0, x7−3.05)	BF49	BF5 × max(0, 44.7−x2)
BF24	BF6 × max(0, 3.05−x7)	BF50	BF45 × max(0, 30−x11)
BF25	BF7 × max(0, x17−2.90)	BF51	BF11 × max(0, x2−54.2)
BF26	BF7 × max(0, 2.90−x17)	BF52	BF11 × max(0, 54.2−x2)

7 Discussions

Comparisons of R^2, r, RRMSE and ρ, as well as the built model interpretability between MARS and BPNN are shown in Table 6. It can be observed that generally BPNN models are slightly more accurate than MARS. However, in terms of the model interpretability, MARS outperforms BPNN through easy-to-interpret model. Thus, both these two methods can actually be used for cross-validation.

Table 6 Comparison of performance measures for BPNN and MARS

Performance measures		MCS	BPNN MTS	BPF	MCS	MARS MTS	BPF
R^2	Training	0.9704	0.8419	0.9494	0.9572	0.7840	0.9080
	Testing	0.9755	0.9302	0.9730	0.9557	0.7820	0.9220
r	Training	0.9852	0.8931	0.9742	0.9782	0.8855	0.9534
	Testing	0.9874	0.9414	0.9762	0.9784	0.8945	0.9604
RRMSE (%)	Training	4.2382	80.828	18.487	5.0764	83.604	24.731
	Testing	3.7621	62.364	18.543	4.8102	73.222	23.464
ρ (%)	Training	2.1357	42.688	9.3624	2.5663	44.342	12.672
	Testing	1.8945	32.468	9.3884	2.4321	38.651	11.975
Model interpretability		Poor, as shown in Appendix 1			Good, Eq. (7) and Table 5		

8 Summary and Conclusions

A database containing 4072 pile data sets with a total of 17 variables is adopted to develop the BPNN and MARS models for drivability predictions. Performance measures indicate that both the BPNN and MARS models for the analyses of pile drivability provide similar predictions and can thus be used for predicting pile drivability as cross-validation. In addition, the MARS algorithm builds flexible models using simpler linear regression and data-driven stepwise searching, adding and pruning. The developed MARS models are much easier to be interpreted.

Acknowledgements The authors are grateful to the support by the National Natural Science Foundation of China (No. 51608071) and the Advanced Interdisciplinary Special Cultivation program (No. 106112017CDJQJ208850).

Appendix 1

Calculation of BPNN Output MCS Model

The transfer functions used MCS are 'logsig' transfer function for hidden layer to output layer and 'tansig' transfer function for output layer to target. The calculation process of BPNN output for MCS is elaborated in detail as follows:

From connection weights for a trained NN, it is possible to develop a mathematical equation relating input parameters and the single output parameter Y using

$$Y = f_{sig}\left\{ b_0 + \sum_{k=1}^{h}\left[w_k f_{sig}\left(b_{hk} + \sum_{i=1}^{m} w_{ik}X_i \right)\right]\right\} \qquad (8)$$

in which b_0 is the bias at the output layer, ω_k is the weight connection between neuron k of the hidden layer and the single output neuron, b_{hk} is the bias at neuron k of the hidden layer ($k = 1, h$), ω_{ik} is the weight connection between input variable i ($i = 1, m$) and neuron k of the hidden layer, xxx is the input parameter i, and f_{sig} is the sigmoid (logsig & *tansig*) transfer function.

Using the connection weights of the trained neural network, the following steps can be followed to mathematically express the BPNN model:

Step 1: Normalize the input values for x_1, x_2,... and x_{17} linearly using $X_{norm} = 2(x_{actual} - x_{min})/(x_{max} - x_{min}) - 1$

Let the actual $x_1 = X_{1a}$ and the normalized $x_1 = X_1$

$$X_1 = -1 + 2 \times (X_{1a} - 7.8)/(31.1 - 7.8) \tag{9}$$

Let the actual $x_2 = X_{2a}$ and the normalized $x_2 = X_2$

$$X_2 = -1 + 2 \times (X_{2a} - 23.9)/(102.3 - 23.9) \tag{10}$$

Let the actual $x_3 = X_{3a}$ and the normalized $x_3 = X_3$

$$X_3 = -1 + 2 \times (X_{3a} - 0.15)/(0.27 - 0.15) \tag{11}$$

Let the actual $x_4 = X_{4a}$ and the normalized $x_4 = X_4$

$$X_4 = -1 + 2 \times (X_{4a} - 1.21)/(3.72 - 1.21) \tag{12}$$

Let the actual $x_5 = X_{5a}$ and the normalized $x_5 = X_5$

$$X_5 = -1 + 2 \times (X_{5a} - 0.0)/(0.18 - 0.0) \tag{13}$$

Let the actual $x_6 = X_{6a}$ and the normalized $x_6 = X_6$

$$X_6 = -1 + 2 \times (X_{6a} - 4.0)/(34.5 - 4.0) \tag{14}$$

Let the actual $x_7 = X_{7a}$ and the normalized $x_7 = X_7$

$$X_7 = -1 + 2 \times (X_{7a} - 3.0)/(30.5 - 3.0) \tag{15}$$

Let the actual $x_8 = X_{8a}$ and the normalized $x_8 = X_8$

$$X_8 = -1 + 2 \times (X_{8a} - 3.0)/(30.5 - 3.0) \tag{16}$$

Let the actual $x_9 = X_{9a}$ and the normalized $x_9 = X_9$

$$X_9 = -1 + 2 \times (X_{9a} - 0.30)/(0.36 - 0.30) \tag{17}$$

Let the actual $x_{10} = X_{10a}$ and the normalized $x_{10} = X_{10}$

$$X_{10} = -1 + 2 \times (X_{10a} - 0.010)/(0.014 - 0.010) \tag{18}$$

Let the actual $x_{11} = X_{11a}$ and the normalized $x_{11} = X_{11}$

$$X_{11} = -1 + 2 \times (X_{11a} - 8.4)/(100.1 - 8.4) \tag{19}$$

Let the actual $x_{12} = X_{12a}$ and the normalized $x_{12} = X_{12}$

$$X_{12} = -1 + 2 \times (X_{12a} - 0.0025)/(0.0084 - 0.0025) \tag{20}$$

Let the actual $x_{13} = X_{13a}$ and the normalized $x_{13} = X_{13}$

$$X_{13} = -1 + 2 \times (X_{13a} - 0.16)/(0.83 - 0.16) \tag{21}$$

Let the actual $x_{14} = X_{14a}$ and the normalized $x_{14} = X_{14}$

$$X_{14} = -1 + 2 \times (X_{14a} - 0.20)/(0.66 - 0.20) \tag{22}$$

Let the actual $x_{15} = X_{15a}$ and the normalized $x_{15} = X_{15}$

$$X_{15} = -1 + 2 \times (X_{15a} - 10)/(95 - 10) \tag{23}$$

Let the actual $x_{16} = X_{16a}$ and the normalized $x_{16} = X_{16}$

$$X_{16} = -1 + 2 \times (X_{16a} - 137.9)/(2891.2 - 137.9) \tag{24}$$

Let the actual $x_{17} = X_{17a}$ and the normalized $x_{17} = X_{17a}$

$$X_{17} = -1 + 2 \times (X_{17a} - 1.02)/(3.46 - 1.02) \tag{25}$$

Step 2: Calculate the normalized value (Y_1) using the following expressions:

$$
\begin{aligned}
A_1 = {} & -21.3261 + 6.2318 logsig(X_1) - 3.1654 logsig(X_2) + 17.1602 logsig(X_3) \\
& - 1.459 logsig(X_4) - 4.3521 logsig(X_5) + 13.216 logsig(X_6) - 9.768 logsig(X_7) \\
& + 3.5715 logsig(X_8) - 0.1209 logsig(X_9) + 0.2208 logsig(X_{10}) + 15.9897 logsig(X_{11}) \\
& - 9.4443 logsig(X_{12}) - 1.9824 logsig(X_{13}) - 5.5972 logsig(X_{14}) - 6.3521 logsig(X_{15}) \\
& + 5.4767 logsig(X_{16}) - 0.7102 logsig(X_{17})
\end{aligned}
$$

$$\tag{26}$$

$$
\begin{aligned}
A_2 = {} & -8.4258 - 15.0031 logsig(X_1) + 11.7647 logsig(X_2) + 1.1075 logsig(X_3) \\
& - 5.1013 logsig(X_4) + 6.9054 logsig(X_5) - 10.1146 logsig(X_6) - 5.4258 logsig(X_7) \\
& - 23.0086 logsig(X_8) + 0.5226 logsig(X_9) + 0.4659 logsig(X_{10}) + 4.8115 logsig(X_{11}) \\
& + 8.377 logsig(X_{12}) + 18.6713 logsig(X_{13}) - 13.0335 logsig(X_{14}) + 15.2353 logsig(X_{15}) \\
& - 12.7608 logsig(X_{16}) + 2.239 logsig(X_{17})
\end{aligned}
$$

$$\tag{27}$$

$$A_3 = -7.9671 - 21.6681logsig(X_1) + 6.0201logsig(X_2) + 6.4033logsig(X_3)$$
$$-0.5677logsig(X_4) + 21.532logsig(X_5) + 11.305logsig(X_6) - 21.4426logsig(X_7)$$
$$+24.4447logsig(X_8) - 1.7981logsig(X_9) - 1.526logsig(X_{10}) - 6.618logsig(X_{11})$$
$$-32.874logsig(X_{12}) + 3.4611logsig(X_{13}) - 5.9862logsig(X_{14}) + 9.1232logsig(X_{15})$$
$$-15.876logsig(X_{16}) - 1.1918logsig(X_{17})$$

$$(28)$$

$$A_4 = -0.8699 + 3.7546logsig(X_1) - 2.2402logsig(X_2) - 1.2905logsig(X_3)$$
$$+0.2448logsig(X_4) - 0.1977logsig(X_5) + 0.0614logsig(X_6) + 1.206logsig(X_7)$$
$$-0.6279logsig(X_8) - 2.19logsig(X_9) + 2.1303logsig(X_{10}) - 0.3518logsig(X_{11})$$
$$-0.4643logsig(X_{12}) + 1.0234logsig(X_{13}) - 0.1317logsig(X_{14}) + 0.1105logsig(X_{15})$$
$$-0.2714logsig(X_{16}) - 0.3666logsig(X_{17})$$

$$(29)$$

$$A_5 = -1.8394 + 0.5777logsig(X_1) + 0.0039logsig(X_2) - 0.1447logsig(X_3)$$
$$-0.0038logsig(X_4) + 0.1173logsig(X_5) - 0.2946logsig(X_6) - 1.2601logsig(X_7)$$
$$+3.2074logsig(X_8) - 0.496logsig(X_9) + 0.4693logsig(X_{10}) - 2.415logsig(X_{11})$$
$$+0.2184logsig(X_{12}) + 0.5232logsig(X_{13}) + 0.1453logsig(X_{14}) + 0.1121logsig(X_{15})$$
$$-0.1928logsig(X_{16}) + 0.2347logsig(X_{17})$$

$$(30)$$

$$A_6 = -10.0517 - 1.6316logsig(X_1) - 9.722logsig(X_2) - 0.7598logsig(X_3)$$
$$-0.6052logsig(X_4) + 6.2292logsig(X_5) - 15.005logsig(X_6) + 13.487logsig(X_7)$$
$$-14.773logsig(X_8) + 0.2152logsig(X_9) + 0.1029logsig(X_{10}) - 7.4019logsig(X_{11})$$
$$-9.094logsig(X_{12}) + 0.7162logsig(X_{13}) + 3.5733logsig(X_{14}) + 5.7949logsig(X_{15})$$
$$-9.2971logsig(X_{16}) + 0.3107logsig(X_{17})$$

$$(31)$$

$$A_7 = -19.748 + 17.4566logsig(X_1) - 6.3092logsig(X_2) - 11.431logsig(X_3)$$
$$+2.612logsig(X_4) - 10.9304logsig(X_5) + 3.7079logsig(X_6) - 6.078logsig(X_7)$$
$$-11.0792logsig(X_8) - 0.0797logsig(X_9) + 2.5652logsig(X_{10}) - 13.752logsig(X_{11})$$
$$+21.45logsig(X_{12}) + 5.628logsig(X_{13}) + 4.9272logsig(X_{14}) + 0.3388logsig(X_{15})$$
$$-10.0783logsig(X_{16}) + 11.6775logsig(X_{17})$$

$$(32)$$

$$A_8 = -0.8092 + 0.8928logsig(X_1) + 0.2205logsig(X_2) - 0.1265logsig(X_3)$$
$$-0.0339logsig(X_4) + 0.2392logsig(X_5) - 0.5454logsig(X_6) - 2.6061logsig(X_7)$$
$$+7.1711logsig(X_8) - 1.018logsig(X_9) + 0.8755logsig(X_{10}) - 5.613logsig(X_{11})$$
$$+0.8703logsig(X_{12}) + 0.9022logsig(X_{13}) + 0.3572logsig(X_{14}) + 0.202logsig(X_{15})$$
$$-0.3748logsig(X_{16}) + 0.2867logsig(X_{17})$$

$$(33)$$

$$A_9 = -1.3878 - 1.4052logsig(X_1) + 0.6705logsig(X_2) - 0.5285logsig(X_3)$$
$$- 0.0814logsig(X_4) + 0.668logsig(X_5) + 1.052logsig(X_6) - 4.0946logsig(X_7)$$
$$+ 3.5151logsig(X_8) - 0.4901logsig(X_9) + 1.104logsig(X_{10}) + 0.3529logsig(X_{11})$$
$$0.9261logsig(X_{12}) + 1.442logsig(X_{13}) - 0.0338logsig(X_{14}) + 0.2956logsig(X_{15})$$
$$- 0.365logsig(X_{16}) - 1.0274logsig(X_{17})$$

$$(34)$$

$$B_1 = -1.9078 \times tanh(A_1) \tag{35}$$

$$B_2 = -0.2020 \times tanh(A_2) \tag{36}$$

$$B_3 = 0.5773 \times tanh(A_3) \tag{37}$$

$$B_4 = -1.8211 \times tanh(A_4) \tag{38}$$

$$B_5 = 59.5399 \times tanh(A_5) \tag{39}$$

$$B_6 = -1.8844 \times tanh(A_6) \tag{40}$$

$$B_7 = -17.4645 \times tanh(A_7) \tag{41}$$

$$B_8 = -17.3515 \times tanh(A_8) \tag{42}$$

$$B_9 = -2.1634 \times tanh(A_9) \tag{43}$$

$$C_1 = -1.2957 + B_1 + B_2 + B_3 + B_4 + B_5 + B_6 + B_7 + B_8 + B_9 + B_{10} + B_{11} \tag{44}$$

$$Y_1 = C_1 \tag{45}$$

Step 3: De-normalize the output to obtain MCS

$$MCS = 21.93 + (422.18 - 21.93) \times (Y_1 + 1)/2 \tag{46}$$

Note: $logsig(x) = 1/(1 + exp(-x))$ while $tanh(x) = 2/(1 + exp(-2x)) - 1$

References

1. Attoh-Okine, N. O., Cooger, K., & Mensah, S. (2009). Multivariate adaptive regression (MARS) and hinged hyperplanes (HHP) for doweled pavement performance modeling. *Journal of Construction and Building Materials, 23,* 3020–3023.
2. Demuth, H., & Beale, M. (2003). *Neural network toolbox for MATLAB-user guide version 4.1.* The Math Works Inc.
3. Friedman, J. H. (1991). Multivariate adaptive regression splines. *The Annals of Statistics, 19,* 1–141.

4. Gandomi, A.H., Roke, D.A. (2013). Intelligent formulation of structural engineering systems. In *Seventh MIT Conference on Computational Fluid and Solid Mechanics- Focus: Multiphysics & Multiscale*, 12–14 June, Cambridge, USA.
5. Garson, G. D. (1991). Interpreting neural-network connection weights. *AI Expert, 6*(7), 47–51.
6. Goh, A. T. C. (1994). Seismic liquefaction potential assessed by neural networks. *Journal of Geotechnical Engineering, ASCE, 120*(9), 1467–1480.
7. Goh, A. T. C., & Zhang, W. G. (2014). An improvement to MLR model for predicting liquefaction-induced lateral spread using multivariate adaptive regression splines. *Engineering Geology, 170*, 1–10.
8. Goh, A. T. C., Zhang, W. G., Zhang, Y. M., Xiao, Y., & Xiang, Y. Z. (2016). Determination of EPB tunnel-related maximum surface settlement: A Multivariate adaptive regression splines approach. *Bulletin of Engineering Geology and the Environment.* https://doi.org/10.1007/s10064-016-0937-8.
9. Hastie, T., Tibshirani, R., Friedman, J. (2009). *The elements of statistical learning: Data mining, inference and prediction*, 2nd ed., Springer.
10. Jekabsons, G. (2010). *VariReg: A software tool for regression modelling using various modeling methods*. Riga Technical University. http://www.cs.rtu.lv/jekabsons/.
11. Jeon, J. K., Rahman, M. S. (2008). *Fuzzy neural network models for geotechnical problems*. Research Project FHWA/ NC/ 2006–52. North Carolina State University, Raleigh, N.C.
12. Lashkari, A. (2012). Prediction of the shaft resistance of non-displacement piles in sand. *International Journal for Numerical and Analytical Methods in Geomechanics, 37*, 904–931.
13. Mirzahosseini, M., Aghaeifar, A., Alavi, A., Gandomi, A., & Seyednour, R. (2011). Permanent deformation analysis of asphalt mixtures using soft computing techniques. *Expert Systems with Applications, 38*(5), 6081–6100.
14. Rumelhart, D. E., Hinton, G. E., Williams, R. J. (1986. Learning internal representation by error propagation. In Parallel distributed processing, Rumelhart DE, McClelland JL (eds). MIT Press, (pp. 318–362) Cambridge, vol. 1.
15. Samui, P. (2011). Determination of ultimate capacity of driven piles in cohesionless soil: A multivariate adaptive regression spline approach. *International Journal for Numerical and Analytical Methods in Geomechanics, 36*, 1434–1439.
16. Samui, P., Das, S., & Kim, D. (2011). Uplift capacity of suction caisson in clay using multivariate adaptive regression splines. *Ocean Engineering, 38*(17–18), 2123–2127.
17. Samui, P., & Karup, P. (2011). Multivariate adaptive regression splines and least square support vector machine for prediction of undrained shear strength of clay. *Applied Metaheuristic Computing, 3*(2), 33–42.
18. Smith, E. A. L. (1960). Pile driving analysis by the wave equation. *Journal of the Engineering Mechanics Division ASCE, 86*, 35–61.
19. Zarnani, S., El-Emam, M., & Bathurst, R. J. (2011). Comparison of numerical and analytical solutions for reinforced soil wall shaking table tests. *Geomechanics & Engineering, 3*(4), 291–321.
20. Zhang, W. G., & Goh, A. T. C. (2013). Multivariate adaptive regression splines for analysis of geotechnical engineering systems. *Computers and Geotechnics, 48*, 82–95.
21. Zhang, W. G., & Goh, A. T. C. (2014). Multivariate adaptive regression splines model for reliability assessment of serviceability limit state of twin caverns. *Geomechanics and Engineering, 7*(4), 431–458.
22. Zhang, W. G., & Goh, A. T. C. (2017). Reliability assessment of ultimate limit state of twin cavern. *Geomechanics and Geoengineering, 12*(1), 48–59.
23. Zhang, W. G., Zhang, Y. M., & Goh, A. T. C. (2017). Multivariate adaptive regression splines for inverse analysis of soil and wall properties in braced excavation. *Tunneling and Underground Space Technology, 64*, 24–33.

Author Biographies

Zhang Wengang is Professor of Geotechnical Research Institute in the School of Civil Engineering, Chongqing University (CQU), China. He received his Ph.D. degree from the School of Civil and Environmental Engineering at Nanyang Technological University (NTU) Singapore and his B.Eng and M.Eng degrees from Hohai University (HHU) China. Before his career in CQU, he worked as a postdoctoral Research Fellow in NTU for more than two years. He received the Hulme Best Paper Award from Tunneling and Underground Construction Society Singapore (TUCSS) in 2013. Dr Zhang was invited in 2017 to act as the Lead Guest Editor for the Journal Geoscience Frontiers in view of his previously well-received paper in that journal. He joined CQU in 2016 as the Hundredth Young Professor and later in the same year was given the prestigious China Thousand Youth Talents plan award. His current research interests include: Big data geotechnical analysis, Braced excavation and deformation analysis and Rainfall induced slope deformation and instability.

Goh Anthony Teck Chee is Associate Professor in the School of Civil and Environmental Engineering (CEE) at Nanyang Technological University (NTU) Singapore. Prof Goh received both his Ph.D. and B.Eng. in Monash University Australia. Dr Goh is a registered Professional Engineer in Singapore. His teaching, research and professional practice have covered many aspects of geotechnical engineering including soft computing, finite element analysis, earth retaining structures, pile foundations, and slope stability. He is currently a Technical Committee member on Urban Geoengineering for the ISSMGE. Since 1993, Dr Goh has been at the forefront of research into the application of artificial intelligence (AI) methodologies in geotechnical engineering.

Three Different Adaptive Neuro Fuzzy Computing Techniques for Forecasting Long-Period Daily Streamflows

Ozgur Kisi, Jalal Shiri, Sepideh Karimi and Rana Muhammad Adnan

Abstract A modeling study was presented here using three different adaptive neuro-fuzzy (ANFIS) approach algorithms comprising grid partitioning (ANFIS-GP), subtractive clustering (ANFIS-SC) and fuzzy C-Means clustering (ANFIS-FCM) for forecasting long period daily streamflow magnitudes. Long-period data (between 1936 and 2016) from two hydrometric stations in USA were used for training, evaluating and testing the approaches. Five different input combinations were applied based on the autoregressive analysis of the recorded streamflow data. A sensitivity analysis was also carried out to investigate the effect of different model architectures on the obtained outcomes. When using ANFIS-GP, the double-input model gives the best results for different model architectures, while the triple-input models produce the most accurate results using both ANFIS-SC and ANFIS-FCM models, which is due to increasing the model complexity for ANFIS-GP by using more input parameters. Comparing the all three algorithms it was observed that the ANFIS-FCM generally gave the most accurate results among others.

O. Kisi (✉)
Faculty of Natural Sciences and Engineering, Ilia State University, Tbilisi, Georgia
e-mail: ozgur.kisi@iliauni.edu.ge

J. Shiri · S. Karimi
Faculty of Agriculture, Water Engineering Department, University of Tabriz, Tabriz, Iran
e-mail: j_shiri2005@yahoo.com

S. Karimi
e-mail: karimi_sepide@yahoo.com

R. M. Adnan
Faculty of Agricultural and Biosystems Engineering and Technology,
Muhammad Nawaz Sharif University of Agriculture, Multan, Pakistan
e-mail: adnan.ikram@mnsuam.edu.pk

© Springer Nature Singapore Pte Ltd. 2018
S. S. Roy et al. (eds.), *Big Data in Engineering Applications*,
Studies in Big Data 44, https://doi.org/10.1007/978-981-10-8476-8_15

303

1 Introduction

Accurate streamflow forecast is very important in water resources system planning, design, operation and management as well as identifying hydrologic drought spells [6], controlling flood events [28], optimizing hydrologic system [17], determining environmental flow portions [33], modeling surface water-groundwater interactions [10], and modeling suspended sediment load in rivers [18]. Traditionally, conceptual simple models have been developed by numerous researchers to describe the rainfall-runoff process for computing the total amount of surface water flows. Although such models do not require more detailed information on the physical parameters, they can produce acceptable results in some cases [34]. In the contrary, physically-based models of river flow forecast are generally time consuming and complex which need lots of input variables for simulating river flow magnitudes [2]. So, using autoregressive moving average (ARMA) approaches for forecasting the streamflow magnitudes using the previously recorded flow magnitudes have been proposed as alternatives for physically-based models [23].

As a substitute, application of heuristic models e.g. adaptive neuro-fuzzy inference system (ANFIS) in streamflow forecasting has become viable. For instance, Wang et al. [35], Kagoda et al. [16] and Humphrey et al. [14] applied artificial neural networks (ANNs) models for streamflow forecasting. Shiri and Kisi [30] introduced a wavelet-neuro-fuzzy model of streamflow forecasting. Sharma et al. [29] compared neuro-fuzzy model with a physically based watershed model for streamflow forecasting and concluded that the neuro-fuzzy model was equally comparable to physical model especially when rain gauge stations were not adequate. Ballini et al. [3] applied ANFIS for seasonal river flow forecast. Nayak et al. [25] ANFIS modeling approach to model the long-term daily river flow magnitudes in India and reported that ANFIS gave promising results in this case. Vernieuwe et al. [34] compared data-driven Takagi–Sugeno models for rainfall–discharge dynamics modeling. Zounemat Kermani and Teshnelab [39] introduced ANFIS approach as a strong method of daily streamflow prediction when compared to ANN and traditional regression models. El-Shafie et al. [9] proposed ANFIS technique to forecast the inflow for the Nile River at Aswan High Dam on monthly basis. Wang et al. [36] examined different heuristic models for forecasting monthly discharge time series and introduced the ANFIS models as the most accurate technique in this field. Rath et al. [26] applied hierarchical neuro-fuzzy model for real-time flood forecasting. He et al. [11] compared ANN, ANFIS and SVM for forecasting riverflow in a semiarid mountain region and found that the performance of the applied models in river flow forecasting was satisfactory. Yarar [37] introduced a hybrid wavelet-ANFIS model for forecasting monthly streamflow time series. Yilmaz and Muttil [38] utilized different machine learning techniques including ANFIS for runoff estimation. Talei et al. [32] applied Takagi–Sugeno neuro-fuzzy model with online learning for runoff forecasting. Anusree and Varghese [1] compared ANFIS, ANN and MNLR models for daily streamflow forecasting and found the ANFIS as the superior model.

The main aim of this study is to forecast long period daily streamflows using three different adaptive neuro fuzzy techniques, i.e. ANFIS with Grid Partition (ANFIS-GP), ANFIS with subtractive clustering (ANFIS-SC) and ANFIS with fuzzy C-Means clustering (ANFIS-FCM). Two different ANFIS-GP methods were considered in the present study: ANFIS-GP with constant output and ANFIS-GP with linear output. The models were also compared according to their complexity and training durations.

2 Methods

2.1 Adaptive Neuro-fuzzy Inference System (ANFIS)

ANFIS is a merger of an adaptive neural network (ANN) and a fuzzy inference system (FIS), where the parameters of FIS are identified by the ANN learning algorithms. ANFIS is able to estimate real continuous functions on a compact set of parameters with any degree of accuracy [15]. There are two approaches for FIS, namely, Mamdani and Assilian [24] and Takagi and Sugeno [31]. The differences between the two approaches corresponds to the consequent part where Mamdani's method uses fuzzy membership functions, while linear or constant functions are utilized in Sugeno's method.

2.1.1 ANFIS Architecture

Let's assume a FIS having two input variables of x and y and one output variable f. The first-order Sugeno fuzzy model, a typical rule set with two fuzzy If-Then rules would read:

$$\text{Rule 1: If } x \text{ is } A_1 \text{ and } y \text{ is } B_1, \text{ then } f_1 = p_1 x + q_1 y + r_1 \tag{1}$$

$$\text{Rule 2: If } x \text{ is } A_2 \text{ and } y \text{ is } B_2, \text{ then } f_2 = p_2 x + q_2 y + r_2 \tag{2}$$

where A_1, A_2 and B_1, B_2 are the membership functions (MFs) of the inputs x and y, respectively and p_1, q_1, r_1 and p_2, q_2, r_2 are the parameters of the output function. Here, the output f is the weighted mean of the single rule outputs.

The output of the ith node in layer l is shown as $O_{l,i}$. Every node i in Layer 1 is an adaptive node with node $O_{l,i} = \varphi A_i(x)$, for $i = 1, 2$, or $O_{l,i} = \varphi B_{i-2}(y)$, for $i = 3, 4$, where x (or y) is the input to the ith node and A_i (or B_{i-2}) is a linguistic label (such as 'low' or 'high') associated with this node. The MFs for A and B are generally described by generalized bell functions as:

$$\varphi A_i(x) = \frac{1}{1 + \left[(x - c_i)/a_i\right]^{2b_i}} \tag{3}$$

where $\{a_i, b_i, c_i\}$ is the parameter set. Parameters in this layer are called as premise parameters. The outputs of this layer are the membership values of the premise part. Layer 2 includes the nodes labeled Π which multiply incoming signals and sending the product out. For instance,

$$O_{2,i} = w_i = \varphi A_i(x)\varphi B_i(y), \ i = 1, 2. \tag{4}$$

Each node output shows the firing strength of a rule. The nodes labeled N computes the ratio of the ith rule's firing strength to the sum of all rules' firing strengths in Layer 3,

$$O_{3,i} = \overline{w}_i = \frac{w_i}{w_1 + w_2}, \ i = 1, 2. \tag{5}$$

The outputs of this layer are referred to as normalized firing strengths. The nodes of the Layer 4 are adaptive with node functions

$$O_{4,i} = \overline{w}_i f_i = \overline{w}_i(p_i x + q_i y + r_i) \tag{6}$$

where $\overline{w}_i$ is the output of Layer 3, and $\{p_i, q_i, r_i\}$ are the parameter set. Parameters of this layer are called as consequent parameters. The single fixed node of the Layer 5 labeled Σ computes the final output as the summation of all incoming signals

$$O_{5,i} = \sum_{i=1} \overline{w}_i f_i = \frac{\sum_i w_i f_i}{\sum_i w_i} \tag{7}$$

So, an adaptive network which is functionally equivalent to a Sugeno first-order fuzzy inference system is built.

2.1.2 Grid Partition Method

Grid partition (GP) is one of the commonly used methods for producing initial FIS rules for ANFIS building, where the space including input- output parameters is divided into certain partitions called as grids. Each grid expresses a fuzzy surface, and interference areas between grids make a continuous output surface [13, 22]. There is no fixed rule for defining the number of MFs for each variable, so they are identified through a trial and error process. The learning process is begun from zero output and during the learning process, functions and fuzzy rules are trained gradually [7]. Although they are different membership function types which can be applied in modeling various procedures, the literature review shows that triangular MFs are commonly used and the most optimal MF types for practical applications

[27]. Nevertheless, other studies (e.g. [34]) have confirmed that the type of MFs cannot affect the results of simulation majorly, though other studies have demonstrated the major effect of MFs in modeling accuracy (e.g. [19]).

2.1.3 Subtractive Clustering Method

The subtractive clustering method assumes that each data point is a potential cluster center and calculates a measure of the likelihood that each data point would define the cluster center on the basis of the density of surrounding data points. Considering a set of n data points $\{x_1, x_2, \ldots, x_i\}$ in m-dimensional space, it is assumed that all data points within a cubic space have been normalized. In subtractive clustering, each of the data points is considered as a potential cluster center. As a result, the density index D_i corresponding to the data x_i can be expressed as follows:

$$D_i = \sum_{j=1}^{n} \exp\left(- \frac{\|x_i - x_j\|}{(r_a/2)^2} \right) \qquad (8)$$

Here, r_a is a positive quantity called cluster radius. If many data points are adjacent to a data point, hence, that data point has the maximum density. After measuring the density of each data point, data point with the highest density is selected as the first data center clustering [12]. If the effect limited area of the center of the first cluster center is removed, following formula is used to measure the other points density.

$$D_i = D_i - D_{c1} \sum_{j=1}^{n} \exp\left(- \frac{\|x_i - x_{cl}\|^2}{(r_b/2)^2} \right) \qquad (9)$$

Here, x_{c1} and D_{c1} are the selected points and density potential, respectively. r_b is a positive constant. To avoid approaching the cluster centers, the r_b constant value is normally larger than r_a (r_b is considered $1.5r_a$). After measuring the density for each data point, the next cluster center x_{c2} is selected and all the measured density for data points will be recalculated. This process continues until a sufficient number of cluster center produce [2, 20].

2.1.4 Fuzzy C-Means Method

In fuzzy clustering, each pattern might belong to several clusters or segment. One of the most functional clustering algorithms is K-mean algorithm. This unsupervised algorithm in large datasets, exposures with some limitations in the process, may not work properly. To deal with the disadvantages, different clustering algorithms have been proposed. Among them, fuzzy c-means as a proper alternative method is used

[21]. Fuzzy C-means (FCM) was developed by Dunn [8] and Bezdek [4] improved it.

The FCM method blocks a set of N vector x_i, $i = 1,..., n$, into c fuzzy clusters, where each pattern is corresponded to a cluster with a degree specified by a membership grade u_{ij} between 0 and 1. The final object by the FCM algorithm is to find c cluster centers so that the cost function of the dissimilarity measure can be minimized. The aim is minimizing the objective function that is defined as below:

$$MinJ_{FCM} = \sum_{c=1}^{C} \sum_{i=1}^{N} w_{ic}^p \|w_i - v_c\|^2 \quad s.t. \sum_{c=1}^{C} w_{ic} = 1, i = 1, 2, \ldots, N \quad (10)$$

which p $(1 < p)$ is known as fuzzifier portion and N, is the number of data points; C, the number of clusters; w_{ic}, the number of belongings of the ith data point to the cth cluster; v, is the clusters center and x is the number of the input. for calculating the amount of w_{ic} the following formula is used [5]:

$$w_{ic} = \frac{1}{\sum_{L=1}^{C} \left(d_{ic}^2 / d_{ij}^2 \right)^{1/(p-1)}} \quad for \ i = 1, 2, \ldots, N \ and \ c = 1, 2, \ldots, C \quad (11)$$

For beginning of the center vectors, centers are calculated by:

$$v_c = \frac{\sum_{j=1}^{N} w_{jc}^p x_j}{\sum_{j=1}^{N} w_{jc}^p} \quad (12)$$

FCM procession continues until a convergence condition is obtained.

3   Case Study

Daily streamflow data from two stations, Murder Creek near Evergreen (Hydrologic Unit Code 03140304, Latitude 31°25′06″, Longitude 86°59′12″, Drainage area 176.00 square miles, Gage datum 178.29 feet a.s.l.) and Choctawhatchee River Near Newton (Hydrologic Unit Code 03140201, Latitude 31°20′34″, Longitude 85°36′38″, Drainage area 686.00 square miles, Gage datum 138.56 feet a.s.l.), Alabama, USA were used in this research. The location of the study area and stations are illustrated in Fig. 1. The reason of selection of these stations is due to having long data period. Data covering the period of 1936–2016 for the both stations were divided into three parts, training (01/01/1938–12/31/1976, 14245 values), validation (01/01/1977–12/31/1996, 7305 values) and testing (01/01/1997–12/31/2016, 7305).

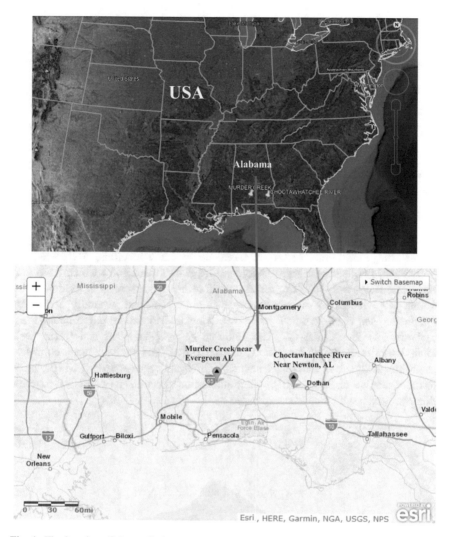

Fig. 1 The location of the studied area

The summary of statistical properties is reported in Table 1 for the used streamflow data. From the table, it is clear that data have highly skewed distributions, skewness coefficient ranging between 8.19 and 20.7. Data of the Choctawhatchee River has more autocorrelations than those of the Murder Creek. This may be due to the high discharge volume of the Choctawhatchee.

Table 1 The statistical properties of the streamflow data sets

Station	Data set	Min	Max	Mean	Sd	Csx	r_1	r_2	r_3
Choctawhatchee River	Training	1.73	584	27.9	34.4	4.52	0.887	0.702	0.564
	Validation	1.23	205	25.8	47.9	20.7	0.789	0.505	0.339
	Testing	0.99	975	22.7	36.7	8.19	0.867	0.650	0.498
Murder Creek	Training	1.11	346	7.93	10.1	10.1	0.668	0.427	0.341
	Validation	1.62	264	8.49	9.68	8.94	0.719	0.399	0.287
	Testing	0.94	241	6.89	9.19	8.45	0.743	0.484	0.357

Min, Max, Mean, Sd, Csx, r_1, r_2 and r_3 show the mean, minimum, maximum, standard deviation, skewness coefficients, lag-1, lag-2 and lag-3 autocorrelations, respectively

4 Application and Results

In this study, the ability of cluster based neuro fuzzy methods, ANFIS-SC and ANFIS-FCM, was investigated in forecasting daily streamflows which have long data period (1936–2016). The results of these methods were compared with the ANFIS-GP which uses all possible rule combinations and generally has higher complexity and computational time when compared to cluster based neuro fuzzy methods. For the ANFIS-GP, two different outputs, constant and linear, were applied to detect the difference with each other. The models were compared according to the four different statistics, root mean square error (RMSE), mean absolute relative error (MARE), determination coefficient (R^2) and Nash-Sutcliffe efficiency (NE) which can be expressed as

$$RMSE = \sqrt{\frac{1}{n} \sum_{i=1}^{n} (Q_{m,i} - Q_{o,i})^2} \tag{13}$$

$$MARE = \frac{1}{n} \sum_{i=1}^{n} \frac{|Q_{m,i} - Q_{o,i}|}{Q_{o,i}} 100 \tag{14}$$

$$NSE = 1 - \frac{\sum_{i=1}^{n} (Q_{m,i} - Q_{o,i})}{\sum_{i=1}^{n} (Q_{o,i} - \overline{Q}_o)} \tag{15}$$

where $Q_{o,i}$ and $Q_{m,i}$ are the observed and estimated streamflows, N is the number of time steps, $\overline{Q}_o$ is the mean of the observed streamflows. First, auto and partial auto correlation analysis were employed and they suggested three previous lags. In the applications, however, five input scenarios comprising five previous lags were used from input(i) to input(iv) comprising Qt-1 to Qt-1, Qt-2, Qt-3, Qt-4 and Qt-5.

The results of the ANFIS-GP models with constant and linear outputs are presented for the Choctawhatchee River in Tables 2 and 3. For the ANFIS-GP models, different number of triangular membership functions were tried and the best one that gave the minimum RMSE in the validation period was selected. It is clear from

Table 2 Results of ANFIS-GP models with constant output—Choctawhatchee River

Data set	Statistics	Input (i)	Input (ii)	Input (iii)	Input (iv)	Input (v)	Mean
Training	RMSE (m³/s)	51.78	46.67	16.99	10.99	11.52	25.79
	MARE (%)	48.52	16.36	7.700	6.180	5.950	15.13
	R^2	49.36	15.99	9.200	6.950	7.200	16.11
	NE	45.06	9.160	7.180	4.930	5.610	12.93
	Duration (s)	48.68	22.05	10.27	7.260	7.570	17.49
Validation	RMSE (m³/s)	31.37	29.44	29.04	29.21	30.18	29.85
	MARE (%)	29.66	16.54	16.96	17.29	18.60	19.81
	R^2	0.594	0.657	0.664	0.655	0.618	0.638
	NE	0.570	0.622	0.632	0.628	0.602	0.611
Testing	RMSE (m³/s)	18.08	16.15	15.70	15.96	16.33	16.44
	MARE (%)	42.28	21.08	21.91	22.48	24.38	26.42
	R^2	0.759	0.811	0.821	0.814	0.804	0.802
	NE	0.758	0.807	0.817	0.811	0.802	0.799

Table 3 Results of ANFIS-GP models with linear output—Choctawhatchee River

Data set	Statistics	Input (i)	Input (ii)	Input (iii)	Input (iv)	Input (v)	Mean
Training	RMSE (m³/s)	15.81	13.53	13.31	13.14	12.88	13.73
	MARE (%)	24.60	15.46	16.44	15.60	15.54	17.53
	R^2	0.789	0.845	0.850	0.854	0.860	0.839
	NE	0.789	0.845	0.850	0.854	0.860	0.839
	Duration (s)	3.680	8.710	26.42	116.6	619.0	154.9
Validation	RMSE (m³/s)	31.37	10,687	22,450	1711	1648	7306
	MARE (%)	29.66	76.32	233.5	23.33	20.14	76.59
	R^2	0.594	0.016	0.054	0.241	0.257	0.233
	NE	0.570	−49,861	−220,032	−1278	−1185	−54,471
Testing	RMSE (m³/s)	18.08	6744	13,108	2042	1930	4768
	MARE (%)	42.28	64.13	203.4	28.39	27.48	73.14
	R^2	0.759	0.010	0.000	0.041	0.034	0.169
	NE	0.758	−33,719	−127,390	−3093	−2762	−33,392

the tables that the ANFIS-GP with constant output has the best accuracy in test period for the 3rd input combination while the 1st input provides the best results for the ANFIS-GP with linear output. First model comprising constant output seems to be superior to the second model. This can also be seen from the mean of the all input combinations. From the mean statistics, it is clear that the training accuracy of ANFIS-GP models with linear output is better compared to the other one. The 2nd model with linear output can approximate better than the 1st model with constant output because it has higher number of parameters and more flexible than the latter one. Assume that we used 2 Gaussian membership functions (each has 2 parameters) and 5 inputs for each model. In this case, the premise parameters of the both models will be $2 * 2^5 = 64$. The 1st model will have 32 rules, each has 1 constant output parameter and totally it will have 32 output parameters while the 2nd model will have 32 rules, each has 6 output parameters and totally $32 * 6 = 192$ output parameters. For this reason, the training duration of the ANFIS-GP models with linear output is also much higher than those of the ANFIS-GP models with constant output especially when the number of inputs is higher than 2.

Tables 4 and 5 report the results of ANFIS-FCM and ANFIS-SC models with respect to RMSE, MARE, R^2 and NE for the Choctawhatchee River. For the ANFIS-FCM models, different number of cluster numbers (vary between 1 and 8)

Table 4 Results of ANFIS-FCM models—Choctawhatchee River

Data set	Statistics	Input (i)	Input (ii)	Input (iii)	Input (iv)	Input (v)	Mean
Training	RMSE (m³/s)	15.58	13.77	13.63	13.70	13.71	14.08
	MARE (%)	16.91	15.17	15.66	15.63	15.84	15.84
	R^2	0.795	0.839	0.843	0.841	0.841	0.832
	NE	0.795	0.839	0.843	0.841	0.841	0.832
	Duration (s)	13.20	20.80	28.23	20.98	25.43	21.73
Validation	RMSE (m³/s)	28.92	24.32	24.75	24.60	24.75	25.47
	MARE (%)	17.05	15.09	15.79	15.83	16.18	15.99
	R^2	0.635	0.759	0.756	0.756	0.750	0.731
	NE	0.635	0.742	0.733	0.736	0.733	0.716
Testing	RMSE (m³/s)	18.00	15.99	15.69	15.70	15.71	16.22
	MARE (%)	19.39	16.89	18.54	17.99	18.51	18.26
	R^2	0.760	0.811	0.817	0.817	0.817	0.804
	NE	0.760	0.811	0.817	0.817	0.817	0.804

Table 5 Results of ANFIS-SC models—Choctawhatchee River

Data set	Statistics	Input (i)	Input (ii)	Input (iii)	Input (iv)	Input (v)	Mean
Training	RMSE (m³/s)	15.64	14.04	13.76	13.77	13.78	14.20
	MARE (%)	19.55	17.74	16.83	16.90	17.04	17.61
	R^2	0.793	0.833	0.840	0.839	0.839	0.829
	NE	0.793	0.833	0.840	0.839	0.839	0.829
	Duration (s)	12.89	16.68	20.64	30.70	36.03	23.39
Validation	RMSE (m³/s)	34.17	33.35	33.44	33.66	33.94	33.71
	MARE (%)	22.71	20.74	18.68	18.54	18.74	19.88
	R^2	0.503	0.534	0.528	0.518	0.506	0.518
	NE	0.490	0.515	0.512	0.505	0.497	0.504
Testing	RMSE (m³/s)	18.71	17.17	16.40	16.62	16.92	17.16
	MARE (%)	29.84	27.41	23.58	23.08	23.44	25.47
	R^2	0.741	0.783	0.802	0.796	0.788	0.782
	NE	0.740	0.781	0.801	0.795	0.788	0.781

which decides the number of rules were tried and the best one that gave the minimum RMSE in the validation period was selected. For the ANFIS-SC models, different number of radii values (vary between 0.1 and 1) which decides the number of membership functions and rules were tried. It is apparent from the tables, the 3rd input combination has the best accuracy for the both methods and after 3 lags input, the accuracy of the models does not considerably increase. Comparison with the ANFIS-GP models indicates that the ANFIS-FCM model slightly performs superior to the ANFIS-GP with constant output and ANFIS-SC models. The training duration of the ANFIS-GP with constant output is less than those of the cluster

Table 6 Results of ANFIS-GP models with constant output—Murder Creek

Data set	Statistics	Input (i)	Input (ii)	Input (iii)	Input (iv)	Input (v)	Mean
Training	RMSE (m³/s)	7.520	7.370	7.350	15.63	13.60	10.30
	MARE (%)	35.33	28.05	25.91	20.75	16.11	25.23
	R^2	0.450	0.471	0.474	0.793	0.843	0.606
	NE	0.450	0.471	0.474	0.793	0.843	0.606
	Duration (s)	3.720	6.950	12.71	26.61	64.08	22.81
Validation	RMSE (m³/s)	6.730	6.540	6.450	32.71	30.42	16.57
	MARE (%)	29.03	25.24	24.51	24.98	18.37	24.43
	R^2	0.519	0.548	0.561	0.546	0.623	0.559
	NE	0.516	0.543	0.556	0.533	0.596	0.549
Testing	RMSE (m³/s)	6.180	5.990	5.980	18.11	18.17	10.89
	MARE (%)	48.31	35.83	31.66	28.80	20.91	33.10
	R^2	0.554	0.583	0.582	0.757	0.757	0.647
	NE	0.548	0.576	0.577	0.757	0.755	0.642

Table 7 Results of ANFIS-GP models with linear output—Murder Creek

Data set	Statistics	Input (i)	Input (ii)	Input (iii)	Input (iv)	Input (v)	Mean
Training	RMSE (m³/s)	7.140	6.880	6.720	15.31	13.15	9.840
	MARE (%)	19.15	18.55	19.19	16.97	16.09	17.99
	R^2	0.503	0.539	0.560	0.802	0.854	0.651
	NE	0.503	0.539	0.560	0.802	0.854	0.651
	Duration (s)	4.100	8.760	26.62	116.2	941.1	219.4
Validation	RMSE (m³/s)	6.640	6.280	13.93	7442	7013	2896
	MARE (%)	20.26	18.89	18.82	29.37	48.98	27.26
	R^2	0.529	0.578	0.133	0.214	0.018	0.295
	NE	0.529	0.578	−1.075	−24,179	−21,472	−9130
Testing	RMSE (m³/s)	5.830	5.480	7.320	4193.5	3420	4768
	MARE (%)	19.83	19.83	22.05	35.92	38.87	73.14
	R^2	0.598	0.646	0.471	0.044	0.027	0.169
	NE	0.598	0.644	0.365	−13,036	−8674	−33,392

based ANFIS-FCM and ANFIS-SC models. The reason of this might be fact that the ANFIS-FCM and ANFIS-SC have linear output comprising more consequent parameters. However, the main advantage of the cluster based neuro fuzzy methods is that their rules are automatically determined based on the selected cluster number or radii value. For example, in case of 8 clusters, we will have only 8 rules for the whole fuzzy model while the ANFIS-GP has 32 rules when the input number is 5.

The statistics of the ANFIS-GP models with constant and linear outputs are compared in Tables 6 and 7 for the Murder Creek. As seen from the tables, the both methods have the best accuracy in the test period for the 2nd input combination. After 2nd input, the accuracy of ANFI-GP model comprising linear output is

Table 8 Results of ANFIS-FCM models—Murder Creek

Data set	Statistics	Input (i)	Input (ii)	Input (iii)	Input (iv)	Input (v)	Mean
Training	RMSE (m³/s)	7.100	6.870	7.000	15.58	14.10	10.13
	MARE (%)	18.73	16.94	18.15	22.58	25.42	20.36
	R^2	0.509	0.540	0.523	0.795	0.832	0.640
	NE	0.509	0.540	0.523	0.795	0.832	0.640
	Duration (s)	13.31	19.90	15.26	13.90	16.03	15.68
Validation	RMSE (m³/s)	6.670	6.290	6.250	27.21	25.41	14.37
	MARE (%)	19.78	17.71	18.84	27.60	32.46	23.28
	R^2	0.525	0.577	0.583	0.683	0.749	0.623
	NE	0.525	0.577	0.582	0.677	0.718	0.616
Testing	RMSE (m³/s)	5.860	5.570	5.620	18.09	16.18	10.26
	MARE (%)	19.63	18.12	19.59	35.66	43.00	27.20
	R^2	0.594	0.634	0.628	0.757	0.806	0.684
	NE	0.594	0.634	0.627	0.757	0.806	0.683

Table 9 Results of ANFIS-SC models—Murder Creek

Data set	Statistics	Input (i)	Input (ii)	Input (iii)	Input (iv)	Input (v)	Mean
Training	RMSE (m³/s)	92.24	126.8	8.690	15.59	14.01	51.46
	MARE (%)	33.02	40.23	8.460	18.70	17.41	23.56
	R^2	0.030	0.453	0.327	0.794	0.834	0.488
	NE	0.030	0.453	0.327	0.794	0.834	0.488
	Duration (s)	0.266	0.543	0.790	30.71	36.03	13.67
Validation	RMSE (m³/s)	122.5	272.2	10.96	33.98	33.37	94.60
	MARE (%)	40.22	73.53	11.81	21.83	20.41	33.56
	R^2	0.000	0.034	0.157	0.510	0.531	0.247
	NE	−0.047	−1.347	0.093	0.496	0.514	−0.058
Testing	RMSE (m³/s)	114.2	165.1	9.620	18.36	16.67	64.80
	MARE (%)	41.93	57.57	9.910	26.28	25.62	32.26
	R^2	0.002	0.178	0.244	0.751	0.796	0.394
	NE	−0.067	0.061	0.215	0.750	0.794	0.351

worsening. It can be said that increasing input number increases the complexity of the model and this results in worse streamflow forecasts. Similar to the Chocta-whatchee River, the training durations of the ANFIS-GP with linear output is higher

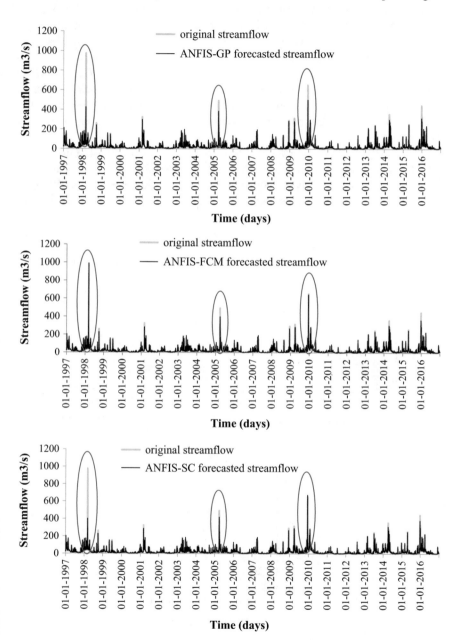

Fig. 2 The time variation of the observed and forecasted streamflows by using the optimal ANFIS-GP-constant, ANFIS-FCM and ANFIS-SC models—Choctawhatchee River

than those of the models with constant output especially for the inputs higher than 2. Tables 8 and 9 present the training, validation and testing results of ANFIS-FCM and ANFIS-SC models for the Murder Creek. ANFIS-FCM model has the best accuracy in 2nd input combination while the 3rd input combination provides the best accuracy for ANFIS-SC model. Comparison of the Tables 6, 7 and 8 clearly shows that the ANFIS-GP model with linear output slightly performs superior to the ANFIS-GP with constant output and ANFIS-FCM models. ANFIS-SC model has the worst accuracy even though it has the least training duration. Comparison of two stations obviously indicates that the accuracy of the applied models is better for the Choctawhatchee River compared to Murder Creek. The main reason of this might be the fact that the data of first station has lower autocorrelations than the latter one.

The time variation of observed and forecasted streamflows by using the optimal ANFIS-GP-constant, ANFIS-FCM and ANFIS-SC models can be seen from Fig. 2 for the Choctawhatchee River. From the figures, it is clear that the ANFIS-FCM model catches the high streamflow values better than the other models. The

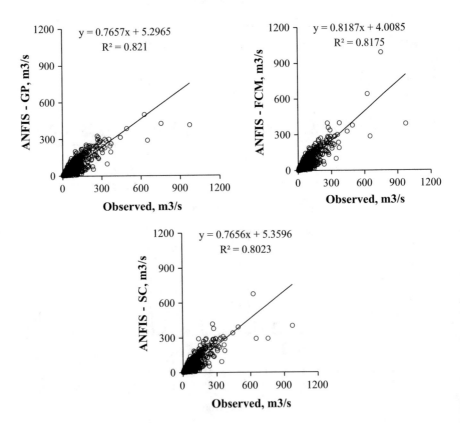

Fig. 3 The scatterplots of the observed and forecasted streamflows by using the optimal ANFIS-GP-constant, ANFIS-FCM and ANFIS-SC models—Choctawhatchee River

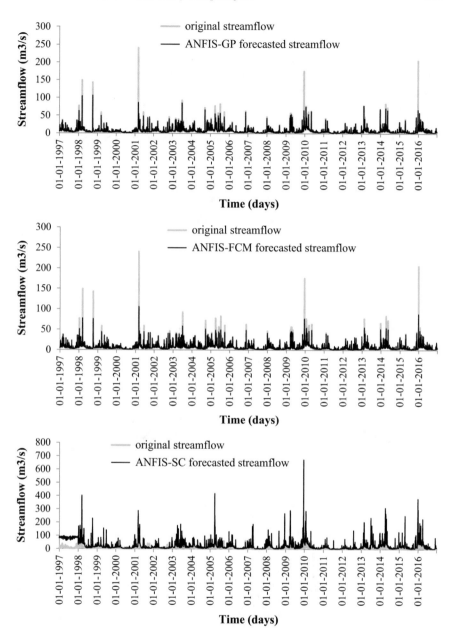

Fig. 4 The time variation of the observed and forecasted streamflows by using the optimal ANFIS-GP-linear, ANFIS-FCM and ANFIS-SC models—Murder Creek

ANFIS-SC also seems to be better than ANFIS-GP model. Figure 3 makes the scatterplot comparison of the applied models. ANFIS-GP model has slightly higher R^2 than the ANFIS-FCM. However, the a and b coefficients of the fit line equation (assume that the fit line is $y = ax + b$) are respectively closer to the 1 and 0 (exact line is $y = x$) for the ANFIS-FCM compared to the ANFIS-GP model. Figure 4 illustrates the time variation graphs of the observed and forecasted streamflows by using the optimal ANFIS-GP-linear, ANFIS-FCM and ANFIS-SC models for the Murder Creek. The ANFIS-GP-linear and ANFIS-FCM models considerably underestimate peak discharges while the ANFIS-SC overestimates. The scatter diagrams of the three methods are given in Fig. 5. As seen from the figure, the ANFIS-GP model has slightly higher R^2 than the ANFIS-FCM while the slope coefficient of the latter model closer to the 1 compared to the first model. The ANFIS-SC seems to be insufficient in forecasting daily streamflows of Murder Creek.

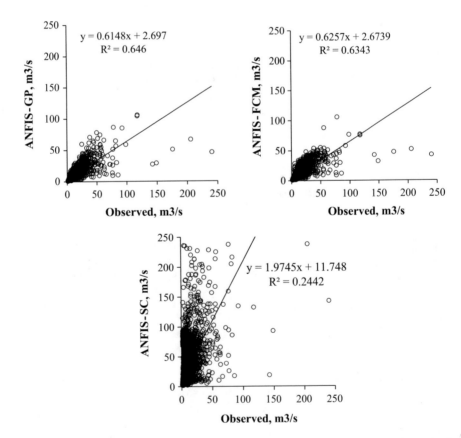

Fig. 5 The scatterplots of the observed and forecasted streamflows by using the optimal ANFIS-GP-linear, ANFIS-FCM and ANFIS-SC models—Murder Creek

5 Conclusion

Long period streamflow data from two hydrometric stations in USA were used in the present research to forecast streamflow magnitudes in daily forecast horizon. Adaptive neuro-fuzzy inference system (ANFIS) with three different running algorithms, namely, ANFIS grid partitioning (ANFIS-GP), ANFIS sub clustering (ANFIS-SC) and ANFIS fuzzy C means (ANFIS-FCM) were then trained, validated and tested using these data. Five input combinations were tried by also considering the auto- and partial-auto-correlation functions of the streamflow records during the study period to see the effect of 5 time lags on the predictions. Using different models and input combinations it was revealed that the best input combination (which can be used to feed the predictive models) is somewhat model-specific, where introducing more input parameters (beyond the double-input combination) has deteriorated the ANFIS-GP accuracy. This might be linked to the model complexity by using more inputs and might dictate a risk of redundancy when using inputs roughly based on linear measures (e.g. auto correlation). Nevertheless, the models architectures had monotonous influence on the predictive models performance that showed the necessity of performing sensitivity analysis on the models architectures. This might be crucially important when using short period data, where the time domain is limited and general trend of data which can affect the predictions are not involved in model training. It was seen that the ANFIS-GP model with linear output produce complex model structure especially in case of high number of inputs compared to ANFIS-GP with constant output.

In this study, high number of membership functions were not tried for that ANFIS-GP model because its parameters exponentially increase when the number of MFs was increased. In future studies, the effect of MFs numbers may be investigated by using high speed computers.

References

1. Anusree, K., & Varghese, K. O. (2016). Streamflow prediction of Karuvannur River Basin using ANFIS, ANN and MNLR models. *Procedia Technology, 24,* 101–108.
2. Aqil, M., Kita, I., Yano, A., et al. (2007). A comparative study of artificial neural networks and neuro-fuzzy in continuous modeling of the daily and hourly behavior of runoff. *Journal of Hydrology, 337,* 22–34.
3. Ballini, R., Soares, S., & Andrade, M. G. (1999). Seasonal streamflow forecasting via a neural fuzzy system. In: 14th Triennial World Congress, Beijing, P.R. China (pp. 5249–5254).
4. Bezdek, J. C. (1981). *Pattern recognition with fuzzy objective function algorithms.* New York: Plenum.
5. Bezdek, J. C., Ehrlich, R., & Full, W. (1984). FCM: The fuzzy C-means clustering algorithm. *Computers & Geosciences, 10*(2–3), 191–203.
6. Chemeda Edossa, D., & Singh Babel, M. (2011). Application of ANN-based streamflow forecasting model for agricultural water management in the Awash River Basin, Ethiopia. *Water Resources Management, 25,* 1759–1773.

7. Cobaner, M. (2011). Evapotranspiration estimation by two different neuro-fuzzy inference systems. *Journal of Hydrology, 398*(3–4), 299–302.
8. Dunn, J. C. (1973). A fuzzy relative of the ISODATA process and its use in detecting compact well-separated clusters. *Journal of Cybernetics, 3*(3), 32–57.
9. El-Shafie, A., Taha, M. R., & Noureldin, A. (2007). A neuro-fuzzy model for inflow forecasting of the Nile River at Aswan high dam. *Water Resources Management, 21,* 533–556.
10. Gunduz, O., & Aral, M. M. (2005). River networks and groundwater flow: A simultaneous solution of a coupled system. *Journal of Hydrology, 301*(1–4), 216–234.
11. He, Z., wen, X., Liu, H., & Du, J. (2013). A comparative study of artificial neural network, adaptive neuro fuzzy inference system and support vector machine for forecasting river flow in the semiarid mountain region. *Journal of Hydrology, 509,* 379–386.
12. Hiremath, S. M., Patra, S. K., & Mishra, A. K. (2012). ANFIS with subtractive clustering-based extended data rate prediction for cognitive radio. In *Proceeding of the 5th International Conference on Computers and Devices for Communication (CODEC).* https://doi.org/10.1109/codec.2012.6509239.
13. Hu, Y. C. (2007). Sugeno fuzzy integral for finding fuzzy if–Then classification rules. *Applied Mathematics and Computation, 185,* 72–83.
14. Humphrey, G. B., Gibbs, M. S., Dandy, G. C., & Maier, H. R. (2016). A hybrid approach to monthly streamflow forecasting: Integrating hydrological model outputs into a Bayesian artificial neural network. *Journal of Hydrology, 540,* 623–640.
15. Jang, J. S. R., Sun, C. T., & Mizutani, E. (1997). *Neurofuzzy and soft computing: A computational approach to learning and machine intelligence.* New Jersey: Prentice-Hall.
16. Kagoda, P. A., Ndiritu, J., Ntuli, C., & Mwaka, B. (2010). Application of radial basis function neural networks to short-term streamflow forecasting. *Physics and Chemistry of the Earth, 35* (13–14), 571–581.
17. Kisi, O. (2008). River flow forecasting and estimation using different artificial neural network techniques. *Hydrology Research, 39*(1), 27–40.
18. Kisi, O., Hossein zadeh Dalir, A., Cimen, M., & Shiri, J. (2012). Suspended sediment modeling using genetic programming and soft computing techniques. *Journal of Hydrology, 450–451,* 48–58.
19. Kisi, O., Shiri, J., & Tombul, M. (2013). Modeling rainfall-runoff process using soft computing techniques. *Computers & Geosciences, 51,* 108–117.
20. Kisi, O., Karimi, S., Shiri, J., Makarynskyy, O., & Yoon, H. (2014). Forecasting sea water levels at Mukho Station, South Korea using soft computing techniques. *The International Journal of Ocean and Climate Systems, 5*(4), 175–188.
21. Kisi, O., & Zounemat-Kermani, M. (2016). Suspended sediment modeling using neuro-fuzzy embedded fuzzy c-means clustering technique. *Water Resources Management, 30*(11), 3979–3994.
22. Lin, C. T., Lin, C. J., & Lee, C. S. G. (1995). Fuzzy adaptive learning control network with on-line neural learning. *Fuzzy Sets Systems, 71,* 25–45.
23. Maier, H. R., & Dandy, G. C. (1996). Use of artificial neural networks for prediction of water quality parameters. *Water Resources Research, 32*(4), 1013–1022.
24. Mamdani, E. H., & Assilian, S. (1975). An experiment in linguistic synthesis with a fuzzy logic controller. *International Journal of Man-Machine Studies, 7*(1), 1–13.
25. Nayak, P. C., Sudheer, K. P., Rangan, D. M., & Ramasastri, K. S. (2004). A neuro-fuzzy computing technique for modeling hydrological time series. *Journal of Hydrology, 291,* 52–66.
26. Rath, S., Nayak, P. C., & Chatterjee, C. (2013). Hierarchical neuro-fuzzy model for real-time flood forecasting. *International Journal of River Basin Management, 11*(3), 253–268.
27. Russel, S. O., & Campbell, P. F. (1996). Reservoir operating rules with fuzzy programming. *Journal of Water Resources Planning and Management, 122*(3), 165–170.
28. Sarlak, N. (2008). Annual streamflow modelling with asymmetric distribution function. *Hydrological Processes, 22,* 3403–3409.

29. Sharma, S., Srivastava, P., Fang, X., & Kalin, L. (2015). Performance comparison of Adoptive Neuro Fuzzy Inference System (ANFIS) with Loading Simulation Program C++ (LSPC) model for streamflow simulation in El Niño Southern Oscillation (ENSO)-affected watershed. *Expert Systems with Applications, 42*(4), 2213–2223.
30. Shiri, J., & Kisi, O. (2010). Short-term and long-term streamflow forecasting using a wavelet and neuro-fuzzy conjunction model. *Journal of Hydrology, 394,* 486–493.
31. Takagi, T., & Sugeno, M. (1985). Fuzzy identification of systems and its application to modeling and control. *IEEE Transactions on Systems, Man, and Cybernetics, 15*(1), 116–132.
32. Talei, A., Chua, L. H., Queck, C., & Jansson, P. E. (2013). Runoff forecasting using a Takagi-Sugeno neuro-fuzzy model with online learning. *Journal of Hydrology, 488,* 17–32.
33. Tennant, D. L. (1976). Instream flow regimes for fish, wildlife, recreation and related environmental resources. *Fisheries, 1,* 6–10.
34. Vernieuwe, H., Georgieva, O., De Baets, B., Pauwels, V. R. N., Verhoest, N. E. C., & De Troch, F. P. (2005). Comparison of data-driven Takagi-Sugeno models of rainfall-discharge dynamics. *Journal of Hydrology, 302*(1–4), 173–186.
35. Wang, W., Van Gelder, P., Vrijling, J. K., & Ma, J. (2006). Forecasting daily streamflow using hybrid ANN models. *Journal of Hydrology, 324,* 383–399.
36. Wang, W., Chau, K. W., Cheng, C. T., & Qiu, L. (2009). A comparison of performance of several artificial intelligence methods for forecasting monthly discharge time series. *Journal of Hydrology, 374,* 294–306.
37. Yarar, A. (2014). A hybrid wavelet and neuro-fuzzy model for forecasting the monthly streamflow data. *Water Resources Management, 28,* 553–565.
38. Yilmaz, A. G., & Muttil, N. (2014). Runoff estimation by machine learning methods and application to the Euphrates Basin in Turkey. *Journal of Hydrologic Engineering, 19*(5), 1015–1025.
39. Zounemat Kermani, M., & Teshnelab, M. (2008). Using adaptive neuro-fuzzy inference system for hydrological time series prediction. *Applied Soft Computing, 8,* 928–936.

Prediction of Compressive Strength of Geopolymers Using Multi-objective Feature Selection

Lasyamayee Garanayak, Sarat Kumar Das and Ranajeet Mohanty

Abstract To reduce the carbon dioxide emission to the environment, production of geopolymer is one of the effective binding materials to act as a substitute of cement. The strength of the geopolymer depends upon different factors such as chemical constituents, curing temperature, curing time, super plasticizer etc. In this paper, prediction models for compressive strength of geopolymer is presented using recently developed artificial intelligence techniques; multi-objective feature selection (MOFS), functional network (FN), multivariate adaptive regression spline (MARS) and multi gene genetic programming (MGGP). The MOFS is also used to find the subset of influential parameters responsible for the compressive strength of geopolymers. MOFS has been applied with artificial neural network (ANN) and non-dominated sorting genetic algorithm (NSGA II). The parameters considered for development of prediction models are curing time, NaOH concentration, $Ca(OH)_2$ content, superplasticizer content, types of mold, types of geopolymer and H_2O/Na_2O molar ratio. The developed AI models were compared in terms of different statistical parameters such as average absolute error, root mean square error correlation coefficient, Nash-Sutcliff coefficient of efficiency.

Keywords Geopolymer · Compressive strength · Multi-objective feature selection · Artificial neural network · NSGA II · Multivariate adaptive regression spline · Genetic programming · Functional network

L. Garanayak (✉) · R. Mohanty
Civil Engineering Department, National Institute of Technology, Rourkela, India
e-mail: lizaoec@gmail.com

R. Mohanty
e-mail: ranajeetmohanty@gmail.com

S. K. Das
Civil Engineering Department, Indian Institute of Technology (ISM), Dhanbad, India
e-mail: saratdas@rediffmail.com; sarat@nitrkl.ac.in

© Springer Nature Singapore Pte Ltd. 2018
S. S. Roy et al. (eds.), *Big Data in Engineering Applications*,
Studies in Big Data 44, https://doi.org/10.1007/978-981-10-8476-8_16

1 Introduction

Climatic change, one of the biggest global issues, primarily caused by elevated concentrations of carbon dioxide that increased from 280 to 370 ppm mainly due to industry resources [17]. Every ton of cement consumes 1.5 tons of raw materials i.e. limestone and sand [26] and 0.94 tons of carbon dioxide [22]. The world is attaining an uphill task in terms of sustainable development by using the waste industrial byproducts as an alternate resource of binder material infrastructure development and producing green environment without consumption of natural resources with low-energy, low-CO_2 binders [15]. After lime, ordinary Portland cement and its variants, geopolymer or Alkali-activated material (AAM) in general is considered as third generation cement. In stable geopolymer, source material should be highly amorphous consisting with sufficient reactive glass content and should consist of low water demand, which able to release the aluminum easily [28].

Strength of the geopolymer depends on different factors such as—solid solution ratio, curing temperature, curing time, chemical concentration, molar ratio of alkali solution, type of alkali solution, type of primary materials composed of silica and aluminium, type of admixtures and additives [23, 26]. There is a complex relationship between the compressive strength of the geopolymer with the above-discussed factors, particularly for different types of geopolymer. Hence, in order to achieve a desired compressive strength, it needs a trial and error approach to fix the above parameters, which is cumbersome and time-consuming.

Now a day, artificial intelligence (AI) techniques are found to be more efficient in the development of prediction models compared to traditional statistical methods [16, 30, 31]. Nazari and Torgal [19] used artificial neural network (ANN) for predicting the compressive strength of different types of geopolymers. They used the database of Pacheco-Torgal et al. [23–25] which contained different types of geopolymers obtained from waste materials based on different compositions containing aluminosilicate as an elementary source. First, set of dataset contained 180 data samples where the basic material used was tungsten mine waste [23], thermally treated at a temperature of 950 °C for 24 h. For mortar test authors used crushed sand as fine aggregates having a specific gravity of 2.7, fineness modules of 2.8 and 0.9% water adsorption for 24 h. Sodium hydroxide (NaOH) solution was used as an activator by dissolving NaOH flakes in distilled water. The solution and extra water were added to the dry mix of sand, tungsten mine waste and calcium hydroxide (Ca(OH)$_2$) maintaining 4% as water to dry solid ratio. The compressive strength of 50 mm cube cured under ambient condition was per ASTM C109 [1], were determined. Output data collected was the compressive strength of the cube, which was the average of three specimens. The second set of dataset belongs to Pacheco-Torgal et al. [25], which contains total 144 data samples. Its geopolymer is also based on metakaoline, which was subjected to thermal treatment at 650 °C temperature. Fine aggregate of specific gravity 3, 1% water absorption, fineness modulus of 2.8 were used for mortar preparation. Different concentration of NaOH like 10M, 12M, 14M and 16M were developed by mixing of NaOH flakes with distilled water and then mixed with sodium silicate solution of 1:2.5 as mass ratio. Mortar was added with admixtures like superplasticizer content of 1, 2, and 3%. The $Ca(OH)_2$ was used as a replacement of metakaoline in the proportion of 5%, 10% while the mass ratio of sand to metakaoline to activator was kept as 2.2:1:1. The samples of $40 \times 40 \times 160$ mm^3 prism specimens were obtained according to EN 1015-11, which was cast and cured at room temperature. And finally the rest amount of data was collected from Pacheco-Torgal et al. [24]. It had tungsten mine waste as base material activated by mixing of two alkali solutions like 24M NaOH and sodium silicate keeping 1:2.5 as mass ratio. 10% of $Ca(OH)_2$ was used as percentage substitution of tungsten mine waste of $50 \times 50 \times 50$ mm^3 cube sample was prepared by mixing the solution with a dry mix of sand, mine waste mud, and $Ca(OH)_2$ with the ratio of mine waste mud to activator as 1:1. Extra water (7, 10%) was added to improve the workability of the mix. In it, water to dry solid binder was 3.6%. Compressive strength was obtained from those three papers which followed the ASTM C109 [1]. Nazari et al. [18] developed ANN models and found to better compare to other prediction models [19]. It may be mentioned here that the developed ANN model had two hidden layers with 12 and 10 number of neurons in hidden layer 1 and 2, respectively, hence number of parameters (weights and biases) were more and the model was not comprehensive. The ANN model also suffers from a lack of a comprehensive procedure for testing the robustness and generalization ability and attains local minima. In the recent decade, AI techniques such as genetic programming (GP), multivariate adaptive regression spline (MARS), functional network (FN) have shown very promising results in over-coming the above-discussed drawbacks.

Usually, in machine learning major portion of the data is used for training and a smaller portion for testing through random sampling of the data to ensure that the testing and training sets are similar for minimising the effects of data discrepancies and to better understand the characteristics of the model and also to limit problems like overfitting and to have an insight on how the model will generalize to an independent dataset [6]. Therefore, the model is trained for a number of times to reduce this effect.

Also in prediction type modelling identification of the controlling parameters is important, as the inclusion of all the features/parameters increases the complexity of the model with a small increase in predictive capability of the developed model. Thus, researchers are constantly looking for reliable predictive models which are not only low in complexity but also high in its predictive capability. One such algorithm, feature selection (FS) algorithm not only minimises the number of fea-tures but also maximises the predictive accuracy (minimisation of error) of the model. The above-described objectives are mutually conflicting in nature, a decrease in one result in an increase in the other. Therefore, multi-objective evo-lutionary algorithms (MOEA) can be implemented, which simultaneously min-imises all the objective functions involved. Feature selection (FS) algorithm is of three types: wrapper, filter and embedded. In wrapper technique, a predictive model is used to evaluate each feature subsets. Each new subset is used to train a model and tested and then ranked based on their accuracy rate or error rate. In filter technique, a proxy measure is used which is fast to compute. Some of the measures used in filter technique are mutual information [12], pointwise mutual information [34], Pearson product-moment correlation coefficient, inter/intra class distance or the scores of significance tests for each class/feature combinations [9, 34]. Filter selects a feature set, which is not tuned to a particular type of model thus resulting into be more general as compared to wrapper technique. Embedded technique uses a catch-all group method performing feature selection as a part of the modelling process. LASSO algorithm [2, 35] is one such technique where during linear modelling the regression coefficients are penalized with an L1 penalty, shrinking many of them to zero. In terms of computational complexity embedded technique is in between filters and wrappers. Implementation of evolutionary algorithms for FS has been made using differential evolution (DE) [13], genetic algorithms (GA) [36], genetic programming (GP) [21], and particle swarm optimisation (PSO) [5, 32, 33].

In this paper prediction models have been developed using multi-gene genetic programming (MGGP), MARS and FN to predict the compressive strength of geopolymers (alkali activated tungsten waste and metakaoline) based on the database available in the literature. A novel type of algorithm known as multi-objective feature selection (MOFS) is also implemented in this paper. In this proposed MOFS (wrapper type approach), artificial neural network (ANN) is combined with non-dominated sorting genetic algorithm (NSAG II) [8], where ANN acts as the learning algorithm and NSGA II performs the feature subset selection and minimises the errors for the developed AI model at the same time. By using three objectives for minimisation (a subset of features, training error, and testing error), a variant of MOEA (modified non-dominated sorting genetic algorithm or NSGA II) is applied to investigate if a subset of features exists with cent percent correct predictions for both training and testing datasets. The features fed to the MOFS are represented in binary form where 1 indicates selection of the feature and 0 indicates its non-selection. The performance of the AI model is evaluated in terms of mean square error; which NSGA II minimises during the multi-objective optimisation process.

2 Methodologies

In the present study artificial intelligence techniques, FN, MARS, and MGGP have been used for development of prediction models. As these techniques are not very common to professional engineering a brief introduction to the above techniques is presented as follows. The feature selection algorithm MOFS is also presented in this section.

2.1 Functional Network (FN)

Functional network (FN) proposed by Castillo et al. [3, 4] is a recent technique, which is being used as an alternate tool to ANN. In FN, network's preliminary topology is derived, centred, around the modelling properties of the real domain, or in other words, it is related to the problems of the domain knowledge, whereas in ANN, by the use of trial and error approach, the required number of hidden layers and neurons are determined, so that a good fitting model to the dataset can be obtained. After the availability of initial topology, functional equations are utilized to reach a much simpler topology. Therefore, functional networks eliminate the problems of artificial neural networks by utilizing together the data knowledge and the domain knowledge from, which the topology of the problem is derived. By the help of domain knowledge FN determines the network structure and from the data, it estimates the unknown neuron function. Initially, arbitrary neural functions are allocated with an assumption that the functions are of multi-argument type and vector valued in nature.

Functional networks can be classified into two types based on their learning methods. They are:

1. Structural learning: In this stage, the preliminary topology of the network is built on the assets obtainable to the designer and further simplifying is done by the help of functional equations.
2. Parametric learning: In this stage estimation of the neuron function is based on the combination of functional families, which is provided initially and then from the available data the associated parameters are estimated. It is similar to the estimation of the weights of the connections in artificial neural networks.

2.1.1 Working with Functional Networks

The main elements around which a functional network is built can be itemized as:

1. Storing Units

   - The inputs—x_1, x_2, and x_3 … require 1 input layer of storing unit.
   - The outputs—f_4, f_5… require 1 output layer of storing unit.
   - Processing units containing 1 or several layers, which evaluates the input from the preceding layers to deliver it to the succeeding layer, f_6.

2. Computing unit's layer, $f1$, $f2$, $f3$: In this computing unit, the neuron evaluates the inputs coming from the preceding layer to deliver the outputs to the succeeding layer.

3. Directed links set: Intermediary functions are not random in nature but it depends on the framework of the networks, such as $x_7 = f_4(x_4, x_5, x_6)$.

All the elements described above together form the functional network architecture. The network architecture defines the topology of the functional network and determines the functional capabilities of the network.

The steps for working with the functional network are as follows:

Step1: Physical relationship of inputs with outputs.
Step2: Preliminary topology of the functional network depends on the dataset of the problem. Artificial neural network selects the topology by trial and error approach, whereas functional networks select the topology on the properties of the data, which ultimately leads to a solo network structure.
Step3: Functional equation simplifies the initial network structure of FN. It is done by constantly searching for a simpler network in comparison to the existing one, which will predict the same output from the same set of inputs. Once a simpler network is found the complicated network is replaced with the simpler one.
Step4: A sole neuron function is selected for the specific topology, which yields a set of outputs.
Step5: In this step data is collected for the training of the network.
Step6: On the basis of the data, which is acquired from Step5, and a blend of the functional families, the neuron function is estimated. Learning stage of the network can be linear or non-linear in nature, which directly depends on the linearity of the neuron function.
Step7: Once a model has been developed it is checked for error rate and also is validated with a different set of data.
Step8: If the model is found to be satisfactory in the cross-validation process, it is prepared to be used.

In FN the learning method is selected on the basis of the neural function, which depends on the type of data $U = \{I_i, O_i\}, \{i = 1, 2, 3, 4, \ldots, n\}$. Learning procedure involves minimisation of the Euclidean norm of the error function and it is represented as:

$$E = \frac{1}{2} \sum_{i=1}^{m} (O_i - F(i))^2 \tag{1}$$

Estimated neural functions fi(x) can be arranged in the following order:

$$f_i(x) = \sum_{j=1}^{m} a_{ij} \phi_{ij}(X) \tag{2}$$

where ϕ is the shape function, having algebraic expressions, exponential functions, and/or trigonometry functions. A set of linear or non-linear algebraic equations is obtained by the help of associative optimisation functions.

Previous information about the functional equation is vital for working with the functional network. The functional equation can be defined as a set of functions, unknown in nature, which excludes the integral and differential equations. Cauchy's functional equation is the most common instance for the functional equations and it is as follows:

$$f(x+y)=f(x)+f(y); \quad x,y \in R \tag{3}$$

For more details, readers can refer Das and Suman [7].

2.2 Multivariate Adaptive Regression Splines (MARS)

MARS correlates between a set of input variables to an output variable through adaptive regression method. In MARS, a non-linear, non-parametric approach is used to develop a prediction model without any prior assumption of any relationship between the input (independent variables) and the output (dependent variable). MARS algorithm creates these relationships by using sets of coefficient and basis functions from the dataset as discussed above. Due to this, MARS is favourable over other learning algorithms where the numbers of inputs (independent variables) are more in number i.e. high.

The backbone of MARS algorithm is founded on divide and conquer strategy in, which the dataset is split into a number of groups of piecewise linear segments known as splines, which varies in gradient. MARS is comprised of knots, which are basically the end points of splines and the functions (piece-wise linear function/piece-wise cubic function) between these knots are called as basis function (BF). In this paper for the case of simplicity of the model, only piece-wise linear basis functions are used.

MARS algorithm proposed by Friedman [10] is a 2-step process to fit data's and is explained below:

i. Forward stepwise algorithm: Basis functions are added in this step. First, the model is developed only by the help of an initial intercept known as β_o. Then in each successive step, a basis function, which shows the greatest decrease in the training error, is annexed. Like this, the whole operation is continued until the number of basis functions reaches its maximum value which has been predetermined beforehand. As a result, an over-fitted model is obtained. Searching of knots among all the variables are done by the adaptive regression algorithm

ii. Backward pruning algorithm: Elimination of the over-fitting of data is done in this phase. The terms in the model are snipped (one by one removal of the terms) in this operation. The best viable sub model is obtained by removing the least effective term. Then the subset of models is equated among themselves by means of the generalized cross-validation (GCV) process.

For a better understanding of MARS algorithm (refer [7]), examine a dataset, which contains an output y for a set of inputs $X = \{X_1, X_2, X_3, \ldots, X_p\}$, which consists of p input variables. MARS generates a model of the form:

$$y = f(X_1, X_2, X_3, \ldots, X_p) + e = f(X) + e \tag{4}$$

where, e = distribution of error; $f(x)$ = a function which is approximated by BFs (piece-wise linear function/piece-wise cubic function).

For the case of simplicity, only piece-wise linear functions have been discussed in this paper for its easy interpretability. The piece-wise linear function is represented as $\max(0, x - t)$ where t is the location of the knot. Its mathematical form is,

$$\max(0, x - t) = \{x - t, \; \text{if } x > t \; or \; 0, \; otherwise\} \tag{5}$$

And finally, $f(x)$ = linear combination of BFs, and interactions between them is defined as,

$$f(X) = \beta_0 + \sum_{i=1}^{M} \beta_m \lambda_m(X) \tag{6}$$

where, λ_m = basis function, which is a single spline or product of 2 or more than 2 splines; β = coefficients of constant values calculated by least square method.

An illustration containing 22 data samples as inputs with an output is taken. Random numbers between one and twelve comprised the input matrix $\{X\}$ with a single output $\{Y\}$, which is obtained as per the equation, given below:

$$Y_i = \frac{1}{\sin(X_i)} - \frac{1}{\cos(X_i)} \tag{7}$$

Also, the data samples are normalized in the range of zero to one and MARS analysis is conducted. The MARS model developed for this dataset is represented as:

$$\hat{Y} = -0.143 + 4.066BF1 - 5.336BF2 + 1.852BF3 \tag{8}$$

where $\hat{Y}$ = predicted values;

$$BF1 = \max(0, X_i - 0.40) \tag{9}$$

$$BF2 = \max(0, X_i - 0.65) \tag{10}$$

$$BF3 = \max(0, 0.65 - X_i) \qquad (11)$$

In this MARS model the knots are situated at, $x = 0.65$ and $x = 0.40$. The R value for this model is 0.805. Proper care should be taken to use normalized values of X_i (Eqs. 9–11) and the denormalized values of the predicted Y_i can be obtained as per Eq. 12.

$$\hat{Y}_{denorm} = \hat{Y}_{norm}\left(X_{i(max)} - X_{i(min)}\right) + X_{i(min)} \qquad (12)$$

Therefore, models developed using MARS algorithm has not only better efficiency but also simplifies the complex equations just like Eq. 7 to a simple linear equation.

2.3 Multi-gene Genetic Programming (MGGP)

Multi-gene genetic programming (MGGP) is a variation of GP where a model is built from the combination of several GP trees. Each tree is composed of genes, which represents a lower non-linear transformation of input variables. The output is created from a weighted linear combination of these genes and is termed as 'multi-gene'. For an MGGP model, the model complexity and accuracy can be controlled by controlling the maximum depth of GP tree (d_{max}) and the maximum allowable number of genes (G_{max}). With the decrease in G_{max} and d_{max} values, the complexity of the MGGP model decreases, whereas its accuracy is hampered. Thus, there exist optimum values of G_{max} and d_{max} which gives fairly accurate results with the relatively compact model [27] for a given problem. The linear coefficients (c_1 and c_2) termed as weights of the gene and bias (c_0) of the model are got by ordinary least square method on the training data.

First, population initialization is done by creating a number of randomly evolved genes with lengths varying from 1 to G_{max}. Then, for each generation, a new population is chosen from the initial population as per their merit and then implementation of reproduction, followed by crossover, followed by mutation operations are performed on the function and terminal sets of the selected GP trees. In subsequent runs, the population is generated by addition and deletion of genes using traditional crossover mechanisms from GP and special MGGP crossover mechanisms. Few distinctive MGGP crossover mechanisms [27] are briefly described below.

2-Point High Level Crossover

The process of mating between two individual parents to swap genes between them is called as a 2-point high level crossover. Suppose there are 2 trees having four genes and three genes respectively marked by G_i to G_n. Assume that the G_{max} value for the model is five. A crossover point, represented by {...} is selected for each individual.

$$[G_1, \{G_2, G_3, G_4\}], [G_5, G_6, \{G_7\}]$$

Genes enclosing the crossover points are interchanged and thus, 2 new offspring are formed as shown below.

$$[G_1, \{G_7\}], [G_5, G_6, \{G_2, G_3, G_4\}]$$

The number of genes in any individual is not allowed to be more than G_{max}. But if it exceeds then, randomly genes are selected and eliminated till each individual has G_{max} genes. This process leads to the creation of fresh genes for both the individuals, as well as the deletion of some genes.

2-Point Low Level Crossover

Standard crossover of GP sub-trees in MGGP algorithm is known as 2-point low level crossover. First, a gene is arbitrarily chosen from each of the individuals and then exchanging of the sub-trees under the selected nodes is done. The newly created trees swap the parent trees in an otherwise unchanged individual in the subsequent generation. There are 6 types of mutations, which can be performed on this stage [11]. For achievement of best MGGP model probability of reproduction, crossover and mutation have to be given, such that the sum of the probability of these operations should not exceed 1.

2.4 Multi-objective Feature Selection (MOFS)

2.4.1 Non-dominated Sorting Genetic Algorithms (NSGA II)

NSGA II [8] is an elitist non-dominated sorting genetic algorithm and is very popular in the application of multi-objective optimisation. Not only does it adopts an elite preservation strategy but also uses the explicit diversity preservation technique. In this first the parent population is initialized, from which the offspring population is created and then both the population are combined and finally classified based on non-dominated sorting. After the completion of non-dominated sorting, filling of the new population starts with the best non-dominated front with the assignment of rank as 1 and this continues for successive fronts and assignment of ranks simultaneously. Along with the non-dominated sorting, another niching strategy adopted is the crowding distance sorting in which the distance reflects the closeness of a solution to its neighbours, greater the distance better is the diversity of the Pareto front. Offspring population is created from parent population by using crowded tournament selection, crossover and mutation operators and this whole operation continue until a termination criterion is met. More details of the algorithm can be found in Deb et al. [8].

2.4.2 NSGA II with ANN for Feature Selection

In this study to solve the feature selection problem wrapper type approach is implemented where binary chromosomes are used to represent the features with a value of 0 and 1, 0 indicating that the required feature is not selected and 1

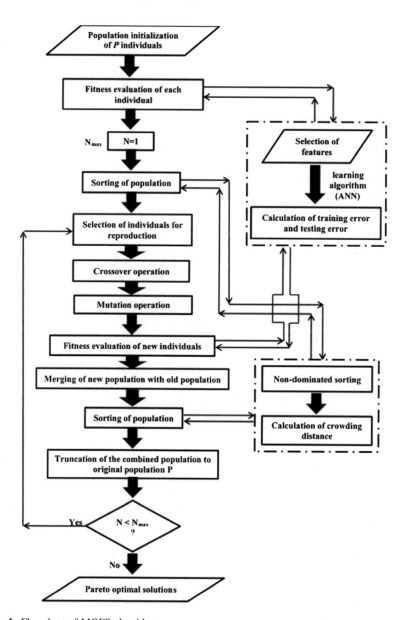

Fig. 1 Flowchart of MOFS algorithm

indicating that the required feature is selected. Three objectives are defined in the NSGA II algorithm, first being the minimisation of the number of selected features, second being the minimisation of training error rate and third being the minimisation of testing error rate in the learning algorithm. The training error and testing error are calculated based on mean square error. Learning algorithm used is feed-forward artificial neural networks (ANNs). Basic flowchart of the MOFS algorithm is presented in Fig. 1.

3   Database and Pre-processing

In this study, 384 data samples were taken with eight parameters from the literature Pacheco-Torgal et al. [23–25]. In those papers tungsten mine waste and metakaoline were used to develop geopolymer activating by alkali solution and extra admixture like superplasticizers as well as calcium hydroxide contents were added with different percentages. Seven variables i.e. curing time in days (T), percentage content of calcium hydroxide by weight (C), superplasticizer percentage by weight (S), NaOH concentration (N), mould type (M), type of geopolymer (G) and H_2O/Na_2O ratio (H) are taken as input parameters and compressive strength (Q_m) is the output. Table 1 presents the statistical values of the dataset used and Fig. 2 shows the variation of input and output parameters of the dataset.

It can be observed (Fig. 2) that when the curing time of the geopolymer is maximum, its compressive strength is minimum and vice versa. Also, with the addition of superplasticizers compressive strength increases but up to a certain extent. Also, when the molar ratio (H) and Ca(OH)$_2$ content is high, compressive strength is less.

The training (288 data samples) and testing (96 data samples) dataset were normalized between 0 and 1 for its implementation in FN and MGGP. For MGGP 500 was taken as the population size and 200 as maximum number of generation keeping 15 as tournament size. Crossover and mutation probability were considered as 0.84 and 0.14 respectively. For MARS modelling 70% data were used for

Table 1 Statistical values of the dataset

Variables	Range	Mean	Std. dev (σ)
Curing time (days) (T)	1.0–90.0	32.67	31.26
Ca(OH)$_2$ content (wt%) (C)	0.0–22.5	12.86	7.61
Superplasticizer (wt%) (S)	0.0–3.0	1.50	1.12
NaOH concentration (N)	6.0–24.0	12.86	5.54
Mold type (M)	1.0–2.0	1.50	0.50
Geopolymer type (G)	1.0–3.0	2.00	0.82
H_2O/Na_2O molar ratio (H)	8.9–19.1	14.63	2.86
Compressive strength (MPa) (Q_m)	1.5–79.0	30.81	16.85

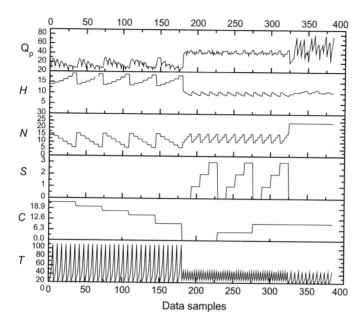

Fig. 2 Variation of input and output parameters of the dataset

training and rest 30% data were used for testing for the normalized values of the dataset in the range of 0–1.

And for the MOFS algorithm ANN training function used was Levenberg-Marquardt type consisting of 3 hidden neurons and performance of the neural network was based on MSE. 70% of the data samples were used for training and the remaining 30% for testing. Data were normalized in the range [0, 1]. In NSGA II uniform crossover technique was applied where replacement of the genetic material of the two selected parents takes place uniformly at several points. Conventional mutation operator was used on each bit separately and changing randomly its value. Parameters used in NSGA II were population size = 50, crossover probability = 0.95, mutation probability = 0.1 and mutation rate = 0.1.

4   Results and Discussion

Statistical comparisons of all the AI models developed in this study was done in terms of average absolute error (AAE), root mean square error (RMSE), correlation coefficient (R) and Nash-Sutcliff coefficient of efficiency (E) and are presented in Table 5. Also, the overfitting ratio, which is the ratio between the RMSE of testing and training was found out and presented in Table 6. Overfitting ratio indicates the generalization of the prediction models. Cumulative probability of the developed models can be expressed as the ratio between the predicted compressive strength

(Q_p) to the measured compressive strength (Q_m) of the geopolymer. The ratio Q_p/Q_m are sorted in ascending order and its respective cumulative probability is found out from Eq. 13.

$$P = \frac{i}{n+1} \tag{13}$$

where; i = order number for the respective Q_p/Q_m and n = total number of data samples. From the cumulative probability distribution (Fig. 8) P_{50}, the ratio of Q_p/Q_m corresponding to 50% probability and P_{90} corresponding to 90% probability are found out. For P_{50} less than one, under prediction is inferred and for greater than one over prediction is implied, with the best model being exactly equal to one. P_{50} and P_{90} values for all the four AI models are given in Table 6. Also, residual plots (residual error between the measured and the predicted values) of all the 4 AI models developed in this research has been presented in Figs. 4, 5, 6 and 7 for the testing dataset (performance on the testing dataset indicates the robustness and generalization capability of the prediction model). If the residuals appear to behave randomly (equally distributed on both sides of the zero line), it suggests that the model fits the data well otherwise it is a poorly fitted model.

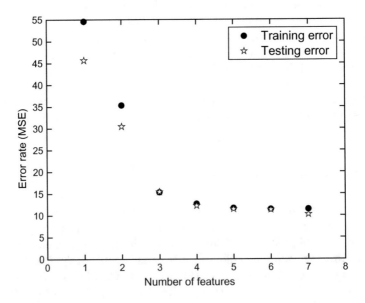

Fig. 3 Pareto optimal solutions

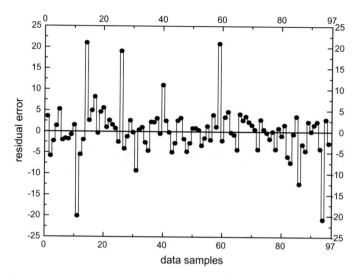

Fig. 4 Residual error of FN model (testing)

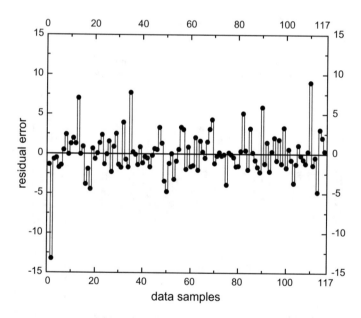

Fig. 5 Residual error of MARS model (testing)

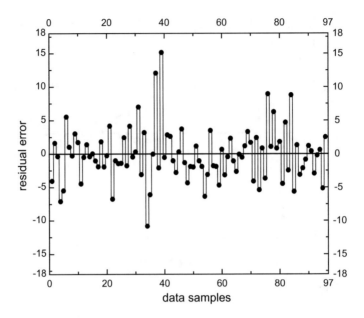

Fig. 6 Residual error of MGGP model (testing)

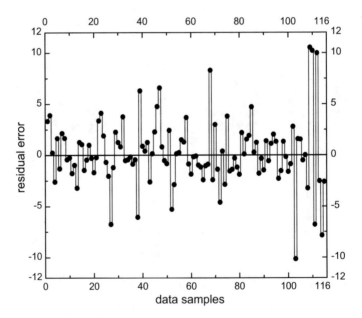

Fig. 7 Residual error of MOFS (ANN) model (testing)

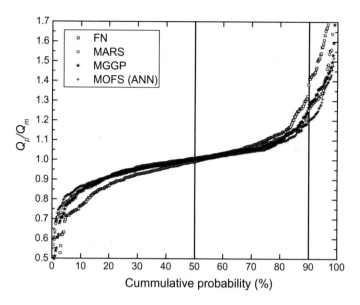

Fig. 8 Cumulative probability distribution of training and testing data

4.1 FN Model

FN models were developed from randomly selected 288 data samples, which were normalized in between 0 and 1. Its prediction value was obtained from the following equation.

$$y = a_0 + \sum_{i=1}^{m} \sum_{j=1}^{m} f_i(x_j) \tag{14}$$

where, n = no. of variables and m = degree of variables. The best model was found to be of associative type with 25 numbers of degree and tanh BF. As the degree of the model was very high, therefore, it was found to be unsuitable for developing a comprehensive model equation. Figure 4 shows the residual error plot between the measured and the predicted compressive strength of geopolymer for the testing data. It can be seen from Fig. 4 that the model fits well along with a maximum deviation of 20 MPa on both sides of the zero line. It can be seen from Table 5 that the values of R in training and testing are same i.e. 0.941, which indicates a strong correlation between predicted and observed values according to Smith [29]. Generally, R is a biased estimate for the prediction models [6]. So another indicator for the goodness of the model can be presented by the help of E. The values of E (Table 5) for training and testing are 0.885 and 0.885 respectively. RMSE and AAE for the FN model as shown in Table 5 are 5.669 MPa, 5.841 MPa and 3.867 MPa and 3.726 MPa for training and testing respectively. The overfitting ratio (Table 6) for the FN model is 1.030, which indicates that the FN model developed is well generalized. Also, P_{50} and P_{90} (Fig. 8) as indicated in Table 6 are 0.992 and 1.411, implying that the model is slightly under predictive.

4.2 MARS Model

In MARS modelling, the best model was obtained corresponding to 11 basis functions and the equivalent model equation is given below.

$$
\begin{aligned}
Q_{p(n)} = 0.49 + 0.5 \times BF1 - 0.61 \times BF2 - 3.09 \times BF3 - 0.33 \times BF4 \\
- 10.1 \times BF5 - 0.05 \times BF6 - 4.06 \times BF7 + 0.36 \times BF8 \\
- 2.45 \times BF9 + 0.74 \times BF10 + 2.22 \times BF11
\end{aligned}
\tag{15}
$$

Details of the respective BF are presented in Table 2. De-normalized value of $Q_{p(n)}$ can be obtained from the following equation:

$$
Q_p = Q_{p(n)}(78.38 - 1.55) + 1.55
\tag{16}
$$

The residual error plot of the MARS model for testing is shown in Fig. 5. It can be seen that the scatter of the error around the zero line is random with a maximum error of approx. −13 MPa from the measured value. The values of R and E in training and testing for the MARS model are 0.963; 0.926 and 0.988 and 0.975 respectively as indicated in Table 5. RMSE and AAE for training and testing (Table 5) are 4.602 MPa; 3.155 MPa and 2.639 MPa; 1.794 MPa respectively. From Table 6 it can be inferred that the MARS model is under-fitted (overfitting ratio = 0.573) and the developed MARS model is good for prediction as its P_{50} value is 1.009 which is nearly same as one.

4.3 MGGP Model

In MGGP model the modelling equation can be developed as follows:

Table 2 Details of the BFs for the MARS model

BF1	max(0, $G - 0.5$)	BF7	max(0, $0.02 - T$)
BF2	max(0, $0.5 - G$)	BF8	$BF1 \times$ max(0, $0.7 - H$)
BF3	$BF1 \times$ max(0.15 $- T$)	BF9	max(0, $N - 0.22$) $\times$ max(0, $C - 0.89$)
BF4	max(0, $0.22 - N$)	BF10	max(0, $N - 0.22$) $\times$ max(0, $H - 0.08$)
BF5	$BF1 \times$ max(0.14 $- H$)	BF11	max(0, $N - 0.22$) $\times$ max(0, $0.08 - H$)
BF6	max(0, $T - 0.02$)		

$$Q_p(n) = 18.8 \times N - 0.142 * T + 0.826 \times H - 19.4 * \exp(N \times H)$$
$$+ 2.16 \times \exp(M) + 30.5 \times \exp(H) - 7.42 \times N$$
$$+ 0.215 \times (T)^{1/4} - 41.7 \times H + 0.826 \times (H)^{1/2} \qquad (17)$$
$$- 0.142 \times (H)^4 + 0.533 \times (T)^{1/2} \times (G)^4 - 25.9$$

where the predicted value of Q_p can be obtained using Eq. 16. Figure 6 shows that the maximum error in prediction for the MGGP model is around 15 MPa on either side of the zero error line for the testing dataset. R between measured and predicted values of compressive strength for the MGGP model as per Table 5 is 0.979 and 0.976 respectively for training and testing. Also from Table 5, E, RMSE and AAE are given as 0.958, 3.405 MPa, 2.449 MPa and 0.950, 3.934 MPa, 2.868 MPa respectively for training and testing. The values of overfitting ratio, P_{50} and P_{90} for the MGGP model as indicated in Table 6 are 1.155, 1.004 and 1.281 respectively. MGGP model is also a good model for prediction as its P_{50} value is close to unity.

4.4 MOFS (ANN) Model

Pareto optimal solutions given by MOFS algorithm are presented below with the number of input parameters used for modelling and the error rate of training and testing in terms of MSE. Results of the multi-objective optimisation are presented in Fig. 3 and the details of the Pareto front are given in Table 3. From Table 3 it can be inferred that the most influential features responsible for the compressive strength of the geopolymer are curing time (T) and molar ratio (H), as these two parameters are selected for a maximum number of times.

Table 3 Details of the Pareto front obtained from the MOFS (ANN) model

Selected features									MSE
							H	Training	54.563
								Testing	45.658
T							H	Training	35.341
								Testing	30.439
T						G	H	Training	15.350
								Testing	15.431
T		S	N				H	Training	12.697
								Testing	12.260
T		S	N			G	H	Training	11.704
								Testing	11.441
T		S	N	M		G	H	Training	11.469
								Testing	11.303
T	C	S	N	M		G	H	Training	11.472
								Testing	10.205

Table 4 Connection weights and biases of the MOFS (ANN) model

Neuron (hidden)	Weights (w_{ik})								Biases	
	Input							Output		
	T	C	S	N	M	G	H	Q_p	b_{hk}	b_0
k_1	−0.318	−0.031	−19.694	−5.261	22.643	1.404	−9.094	736.668	5.164	−169.285
k_2	−0.320	−0.037	0.005	−5.273	1.447	1.410	−9.101	−736.554	5.173	−
k_3	14.642	−0.781	17.010	−0.775	0.966	0.164	0.557	169.499	4.507	−

Figure 3 clearly shows that MSE for training and testing decreases with increase in the number of input variables/features. Also, the difference between the training error and testing error which indicates the generalization of a model (small difference means more generalized is the model) is almost negligible when number of input features is 3 followed closely by when number of features selected is 6; ANN model developed in this paper is for 7 input features (details given in Table 3).

It is evident from Table 5 that model is well generalized as the R values for both training and testing are nearly same (0.981 and 0.979). Thus, the model developed has a good generalized fit between the independent (input parameters) and dependent variables (output). E, RMSE, and AAE of the MOFS (ANN) model are 0.962, 3.387 MPa, 2.395 MPa and 0.958, 3.195 MPa, 2.221 MPa for training and testing dataset respectively (Table 5). From Fig. 7 it can be observed that the MOFS (ANN) model is a good fit model with a maximum residual error of 10 MPa. The input weights, layer weights, and biases of the selected MOFS (ANN) model are given in Table 4. Based on the connection weights and biases (Table 4) of the MOFS (ANN) model, equation is formulated as follows:

$$A_1 = 5.164 - 0.318T - 0.031C - 19.694S \\ - 5.261N + 22.643M + 1.404G - 9.094H \tag{18}$$

$$A_2 = 5.173 - 0.32T - 0.037C + 0.005S \\ - 5.273N + 1.447M + 1.41G - 9.101H \tag{19}$$

Table 5 Statistical values of AI models used in this study

		FN	MARS	MGGP	MOFS (ANN)
R	Training	0.941	0.963	0.979	0.981
	Testing	0.941	0.988	0.976	0.979
RMSE (MPa)	Training	5.669	4.602	3.405	3.387
	Testing	5.841	2.639	3.934	3.195
E	Training	0.885	0.926	0.958	0.962
	Testing	0.885	0.975	0.950	0.958
AAE (MPa)	Training	3.867	3.155	2.449	2.395
	Testing	3.726	1.794	2.868	2.221

Table 6 Overfitting ratio and cumulative probability of the AI models

	Overfitting ratio	P_{50}	P_{90}
FN	1.030	0.992	1.411
MARS	0.573	1.009	1.264
MGGP	1.155	1.004	1.281
MOFS (ANN)	0.943	1.003	1.204

$$A_3 = 4.507 + 14.642T - 0.781C + 17.01S$$
$$- 0.775N + 0.966M + 0.164G + 0.557H \tag{20}$$

$$Q_p = 76.83 \left[\begin{array}{c} -169.285 + 736.668 \text{ tanh } (A_1) - \\ 736.554 \text{ tanh } (A_2) + 169.499 \text{ tanh } (A_3) \end{array} \right] + 1.55 \tag{21}$$

The input values of the variables used in Eqs. 18–20 are normalized values in the range [0, 1]. Table 6 shows that the MOFS (ANN) model is slightly under fitted (overfitting ratio = 0.943) and the model is good in prediction as the P50 value is 1.003 (close to one).

Hence, it can be easily concluded that out of the 4 AI models developed, MOFS (ANN) model is best followed closely by MGGP model as indicated in the statistical comparison (Table 5). However, the model equation developed by MOFS (ANN) model is quite complex (not comprehensive), so for practical on field use MGGP model equation can be utilized, but again in the MGGP model not all the parameters of the geopolymer are used ($Ca(OH)_2$ and superplasticizer content are absent). Thus, it really depends on the user on the choice of AI model to be used. Also from Table 6 all the AI models are good in prediction except the FN model (graphical representation is given in Fig. 8).

5 Conclusion

The present study deals with the compressive strength of geopolymers based on the experimental database available in the literature using different AI methods. Identification of the subset of features responsible for the predictive capacity of the model is addressed here by considering it as a multi-objective optimization problem. Based on different statistical parameters like R, E, RMSE and AAE values, MOFS (ANN) algorithm is found to be more efficient as compared to other AI techniques. The R, E, RMSE and AAE values of the present ANN model, are 0.981, 0.962, 3.387 MPa and 2.395 MPa, respectively, for training and 0.979, 0.958, 3.195 MPa and 2.221 MPa, respectively, for testing data. The model equations are also presented, which can be used by quality control professional engineers to identify the proper proportion of different constituent and the condition of different curing etc. for a desired compressive strength. It was observed that though, the model equation as per the MGGP model is comprehensive, but out of seven parameters of the geopolymer, two important parameters ($Ca(OH)_2$ and superplasticizer content) are not part of the model equation. But, the MOFS (ANN) model is best and though the model equation is not comprehensive, but the model equation is presented in a tabular form. The model equation will help the professional engineers particularly at the initial level to predict the compressive strength of geopolymer, which is a very complex phenomenon.

References

1. ASTM. (2013). International Standard Test Method for Compressive Strength of Hydraulic Cement Mortars (Using 2-in. or [50-mm] Cube Specimens). (ASTM C109/C109M) West Conshohocken, PA 19428-2959. United States.
2. Bach, F. R. (2008). Bolasso: Model consistent Lasso estimation through the bootstrap. In A. McCallum & S. T. Roweis (Eds.), *Proceedings of 25th International Conference on Machine Learning, (ICML2008)*, Helsinki, Finland (pp. 33–400).
3. Castillo, E., Cobo, A., Gutierrez, J. M., & Pruneda, E. (1998). *An introduction to functional networks with applications.* Boston: Kluwer.
4. Castillo, E., Cobo, A., Manuel, J., Gutierrez, J. M., & Pruneda, E. (2000). Functional networks: A new network-based methodology. *Computer-Aided Civil and Infrastructure Engineering, 15,* 90–106.
5. Cervante, L., Xue, B., Zhang, M., & Shang, L. (2012). Binary particle swarm optimisation for feature selection: A filter based approach. In *Proceedings of Evolutionary Computation (CEC), 2012 IEEE Congress*, Brisbane, QLD (art. no. 6256452, pp. 881–888).
6. Das, S. K. (2013). Artificial neural networks in geotechnical engineering: Modeling and application issues, Chapter 10. In X. Yang, A. H. Gandomi, S. Talatahari & A. H. Alavi (Eds.), *Metaheuristics in water, geotechnical and transport engineering* (pp. 231–270). London: Elsevier.
7. Das, S. K., & Suman, S. (2015). Prediction of lateral load capacity of pile in clay using multivariate adaptive regression spline and functional network. *The Arabian Journal for Science and Engineering., 40*(6), 1565–1578.
8. Deb, K., Pratap, A., Agarwal, S., & Meyarivan, T. (2002). A fast and elitist multiobjective genetic algorithm: NSGA-II. *IEEE Transactions on Evolutionary Computation, 6*(2), 182–197.
9. Forman, G. (2003). An extensive empirical study of feature selection metrics for text classification. *Journal of Machine Learning Research, 3,* 1289–1305.
10. Friedman, J. (1991). Multivariate adaptive regression splines. *Annals of Statistics, 19,* 1–141.
11. Gandomi, A. H., & Alavi, A. H. (2012). A new multi-gene genetic programming approach to nonlinear system modeling. Part II: Geotechnical and Earthquake Engineering Problems. *Neural Computing and Applications, 21*(1), 189–201.
12. Guyon, I., & Elisseeff, A. (2003). An introduction to variable and feature selection. *The Journal of Machine Learning Research., 3,* 1157–1182.
13. He, X., Zhang, Q., Sun, N., & Dong, Y. (2009). Feature selection with discrete binary differential evolution. In *Proceedings of International Conference on Artificial Intelligence and Computational Intelligence, AICI 2009*, Shanghai (Vol. 4, art. no. 5376334, pp. 327–330).
14. http://www.geopolymer.org/faq/alkali-activated-materials-geopolymers/.
15. Juenger, M. C. G., Winnefeld, F., Provis, J. L., & Ideker, J. H. (2011). Advances in alternative cementitious binders. *Cement and Concrete Research Cement and Concrete Research, 41,* 1232–1243.

16. Kutyłowska, M. (2016). Comparison of two types of artificial neural networks for predicting failure frequency of water conduits. *Periodica Polytechnica Civil Engineering*. https://doi.org/10.3311/ppci.8737.

17. Mehta, P. K. (2004). High-performance, high-volume fly ash concrete for sustainable development. In *Proceedings of the International Workshop on Sustainable Development and Concrete Technology*, Beijing, China (pp. 3–14).

18. Nazari, A., Hajiallahyari, H., Rahimi, A., Khanmohammadi, H., & Amini, M. (2012). Prediction compressive strength of Portland cement-based geopolymers by artificial neural networks. *Neural Computing and Applications*, 1–9.

19. Nazari, A., & Pacheco-Torgal, F. (2013). Predicting compressive strength of different geopolymers by artificial neural networks. *Ceramics International, 39*, 2247–2257.

20. Nazari, A., & Riahi, S. (2012). Prediction of the effects of nanoparticles on early-age compressive strength of ash-based geopolymers by fuzzy logic. *International Journal of Damage Mechanics, 22*(2), 247–267.

21. Neshatian, K., & Zhang, M. (2009). Pareto front feature selection: Using genetic programming to explore feature space. In *Proceedings of 11th Annual conference on Genetic and Evolutionary Computation, GECCO'09* (pp. 1027–1034). New York, NY, USA: ACM.

22. Pacheco-Torgal, F., Abdollahnejad, Z., Camões, A. F., Jamshidi, M., & Ding, Y. (2012). Durability of alkali-activated binders: A clear advantage over Portland cement or an unproven issue? *Construction and Building Materials, 30*, 400–405.

23. Pacheco-Torgal, F., Castro-Gomes, J., & Jalali, S. (2008). Alkali-activated binders: A review. Part 2. About materials and binders manufacture. *Construction and Building Materials, 22*(7), 1315–1322.

24. Pacheco-Torgal, F., Castro-Gomes, J., & Jalali, S. (2007). Investigations about the effect of aggregates on strength and microstructure of geopoly-meric mine waste mud binders. *Cement and Concrete Research, 37*, 933–941.

25. Pacheco-Torgal, F., Moura, D., Ding, Y., & Jalali, S. (2011). Composition, strength and workability of alkali-activated metakaolin based mortars. *Construction and Building Materials, 25*, 3732–3745.

26. Rashad, A. M. (2014). A comprehensive overview about the influence of different admixtures and additives on the properties of alkali-activated fly ash. *Materials and Design, 53*, 1005–1025.

27. Searson, D. P., Leahy, D. E., & Willis, M. J. (2010). GPTIPS: An open source genetic programming toolbox from multi-gene symbolic regression. In *Proceedings of the International Multi Conference of Engineers and Computer Scientists*, Hong Kong (Vol. 1, no. 3, pp. 77–80).

28. Singh, B., Ishwarya, G., Gupta, M., & Bhattacharyya, S. K. (2015). Geopolymer concrete: A review of some recent developments. *Construction and Building Materials, 85*, 78–90.

29. Smith, G. N. (1986). *Probability and statistics in civil engineering: An introduction*. London: Collins.

30. Tarawneh, B., & Nazzal, M. D. (2014). Optimization of resilient modulus prediction from FWD results using artificial neural network. *Periodica Polytechnica Civil Engineering, 58*(2), 143–154. https://doi.org/10.3311/ppci.2201.

31. Ünes, F., Demirci, M., & Kisi, Ö. (2015). Prediction of Millers Ferry Dam reservoir level in USA using artificial neural network. *Periodica Polytechnica Civil Engineering, 59*(3), 309–318. https://doi.org/10.3311/ppci.7379.

32. Xue, B., Cervante, L., Shang, L., Browne, W. N., & Zhang, M. (2012). A multi-objective particle swarm optimisation for filter based feature selection in classification problems. *Connection Science, 24*(2–3), 91–116.

33. Xue, B., Cervante, L., Shang, L., Browne, W. N., & Zhang, M. (2014). Binary PSO and rough set theory for feature selection: A multi-objective filter based approach. *International Journal of Computational Intelligence and Applications, 13*(2), art. no. 1450009.

34. Yang, Y., & Pedersen, J. O. (1997). A comparative study on feature selection in text categorization. *Proceedings of Fourteenth International Conference on Machine Learning (ICML'97)* (Vol. 97, pp. 412–420), Nashville, Tennessee, USA.
35. Zare, H., Haffari, G., Gupta, A., & Brinkman, R. R. (2013). Scoring relevancy of features based on combinatorial analysis of Lasso with application to lymphoma diagnosis. *BMC Genomics, 14,* art. no. S14.
36. Zhu, Z., Ong, Y. S., & Dash, M. (2007). Wrapper-filter feature selection algorithm using a memetic framework. *IEEE Transactions on Systems, Man, and Cybernetics. Part B, Cybernetics, 37*(1), 70–76.

Application of Big Data Analysis to Operation of Smart Power Systems

Sajad Madadi, Morteza Nazari-Heris, Behnam Mohammadi-Ivatloo
and Sajjad Tohidi

Abstract The volume of data production is increased in smart power system by growing smart meters. Such data is applied for control, operation and protection objectives of power networks. Power companies can attain high indexes of efficiency, reliability and sustainability of the smart grid by appropriate management of such data. Therefore, the smart grids can be assumed as a big data challenge, which needs advanced information techniques to meet massive amounts of data and their analytics. This chapter investigates the utilization of huge data sets in power system operation, control, and protection, which are difficult to process with traditional database tools and often are known as big data. In addition, this paper covers two aspects of applying smart grid data sets, which include feature extraction, and system integration for power system applications. The application of big data methodology, which is analyzed in this study, can be classified to corrective, predictive, distributed, and adaptive approaches.

Keywords Power systems · Big data · C-means algorithm · Operation

1 Introduction

Due to installing smart grid infrastructure, the power companies are confronted with new challenges. Such challenges generally refer to manage smart grid data and use them for improving decision-making. The smart meters generally send information every 15 min, and the processing smart centers face with the massive amount of data. The volume of sent data to data processing centers is approximated about 220 million TB per a day [1]. Big data in the electric power industry can include data with large volume, high velocity, variant, or all three characteristics. The volume of data has been growing significantly according to introduction of new metering devices. Velocity refers to the temporal constraints on collecting, processing and

S. Madadi · M. Nazari-Heris · B. Mohammadi-Ivatloo (✉) · S. Tohidi
Faculty of Electrical and Computer Engineering, University of Tabriz, Tabriz, Iran
e-mail: mohammadi@ieee.org

© Springer Nature Singapore Pte Ltd. 2018
S. S. Roy et al. (eds.), *Big Data in Engineering Applications*,
Studies in Big Data 44, https://doi.org/10.1007/978-981-10-8476-8_17

analyzing data, which is the case with synchrophasor data. Real-time operation is necessary for efficient condition-based asset management and outage prevention, which requires fast processing of large volumes of data. Variety refers to data coming from many different sources that are not necessarily part of the traditional electric utility data.

On the other hand, the power system operators meet with various problems such as challenges in operational efficiency and cost control [2], challenges in stability and reliability [3] and renewable energy management [4]. Big data systems are defined as an effective tool containing a set of means and process for reaching, storing and processing data [5]. Some of the researchers focus on data measurement tools and transfer ways to data centers. In [6], a cloud based power system operational technique is proposed. An approach to keep the security of anonymizing frequent (for example, every few minutes) electrical metering data sent by a smart meter is presented in [7]. A smart meter data security technique for consumer is studied in [8]. However, the researchers are more studying on the application of big data in power system problems. Application of a big data system to specific aims in smart grids requires a significant efforts since the mentioned data are form numerous and independent sources such as phasor measurement units (PMUs) and smart homes controllers [9]. The main application of the decision-making framework can be abbreviated on the exploration of innovative computational concepts to allow novel applications [10]:

(a) Corrective: Such technique is capable to facilitate actions, which aim to remove fault conditions with high rate of expansion in power networks. Risk-based asset management can be defined as an instance of corrective applications [11].

(b) Predictive: Providing exact data estimation of the power network condition such as wind generation and load demand. Short-term renewable energy source forecasting can be introduced as examples of Predictive. Another example of predictive applications are participation rate of load demand response (DR) consumers in DR programs options, and dynamic line rating forecasting, which require a significant amount of data and measurement instruments. In [12], the utilization of load classification, including big data identification and correction, load forecasting and tariff setting, are investigated.

(c) Distributed: the aim of distributed big data approaches is estimation state based on distributed processing. This concept takes advantages of fast control decisions, which can be executed locally. An example for application of this technique is online assessment of voltage stability.

(d) Adaptive: operators monitor occurring faults very closely allowing adjustments in operating strategy. An example of such application is enhanced disturbance detection, and on-line outage management.

The remainder of this chapter is organized as: Sect. 2 discusses asset management. Operations-planning convergence is studied in Sect. 3, and fault detection/protection is analyzed in Sect. 4. In addition, stability margin prediction using PMU measurements is proposed in Sect. 1. Finally, the chapter is concluded in Sect. 5.

2   Asset Management

Smart grids integrate large data sets, which are received from on-line measurement tools such as PMU units. Such data should be processed to obtain suitable response to each request or network condition. Therefore, application of big data needs an asset management. Figure 1 illustrates a framework for handling smart grid big data. The framework includes five stages to model the flow of the smart grid data from data generation to data analytics. The summary of each stage is studied in following sections.

2.1   Data Generation Stage

Smart meters generate data, which are installed in the smart grid. The generated data may be sent from a supplier site such as power plants, wind farms and solar panels or a demand site such as residential homes and factories. As well as, some data can be sent from operation center to consumers such as electrical price, which is generally used to optimize electrical cost by big consumers and demand response

Fig. 1 Framework for handling smart grid big data

Data Analytics

Data Querying

Data Storing/ Processing

Data Acquisition

Data generation

signals. Weather condition is other data which is revived at weather station units. Such data is applied to predict electrical demand and output power of renewable energy sources (wind farms units and solar panels).

2.2 Data Acquisition

In the smart grid application, data acquisition stage can be classified into three steps. This stage includes data collection, data transmission, and data pre-processing. At first, centralized/distributed agents collect the data generated from different data supplies. After data collection, the collected data should be sent to a master node. Once the primary data are integrated, they are transferred to a data storage structure for subsequent processing. The last step in data acquisition stage is pre-processing. The collected data generally have different types and formats since such data are generated by various sources. In the pre-processing step, combining data from diverse sources provides a unified view of the data. In addition, the characteristic of the data such as the timestamp, smart meter's ID, generated/consumed power and location may be inaccurate or incomplete. Therefore, in the preprocessing of data, inaccurate and incomplete data are modified to improve the quality of data.

Flume is known as a suitable function of data acquisition. The collection, integration and transmission steps are covered by utilization of flume. Whereas, a Flume unit stores received data into one or more channels. Each channel memorizes the received data at the first time. Then it removes the channel data and sends it into the external storages. In the external storage units, the format of Flume data is changed into an appropriate format. Consequently, the data preprocessing step is completed and a unified view of the data is obtained.

2.3 Data Storing and Processing

After the acquisition stage, the data is classified and stored for future processing. Such stage includes a single name node for managing file system big data, and sets of data nodes, which is used to save the actual data. On the other hand, transmitted data is decomposed into one or more sections, and these sections are stored in collections of data nodes.

2.4 Data Querying

In this stage, smart grid data is reed from a smart grid data repository. Then especial data is selected. Such data can belong to load demand of an area or the power generated by renewable supplies. In the data querying stage, cluster sets are generally used to obtain prompt results.

2.5 Data Analytics

The smart grid data must be used for optimal operation of smart grids. For instance, this data can be applied to schedule maintenance plans, or exactly future demand/renewable generation prediction. In addition, such data can be used to online protection of smart grids such as detection of high impedance faults. Moreover, utilization of such data can reduce controlling tools and improved traditional response approaches such as wide area small signal stability.

3 Operations-Planning Convergence

In the operation and planning convergence application, the certain models are applied to estimate actual physical state of power systems. Operators generally use the results of big data processing models for getting long-term and short-term decisions. Such decisions include expansion of transmission line, which is classified to long-term process, and balance clearing market, which is short-term decision making process. On the other hand, the operators are capable to find out the future power system conditions with high accuracy according to processing big data received by smart meters.

In the power system network, the stochastic variables such as wind speed, air radiation and demand of load are forecasted. After prediction stochastic variables, the power generated by thermal units is scheduled in day-ahead market. Renewable power generation forecasting and load demand prediction are used in power market operator before big data processing. However, such forecasting based on big data processing has high accuracy because methods account more data and feature to reach future conditions of power system.

The major role of big data in power system is better operation of power system components. An instance for this role is dynamic line rating. The capacity of power line transmission is determined by worth scenario of weather condition. Therefore, such values are generally less than real line capacity. Data measured by smart meters can be used to estimate of real time capacity of lines. The framework of operation is illustrated in Fig. 2.

3.1 Big Data and Dynamic Line Rating (DLR)

In traditional concepts, the capacity of overhead line is calculated by worst scenario of weather conditions since such scenario is not capable to measure online. This method for determining the capacity of line is known as static line rating (SLR). Big data process takes advantages of providing required data to accomplish overhead line projects. Dynamic line rating is introduced in the following.

Fig. 2 The framework of operation

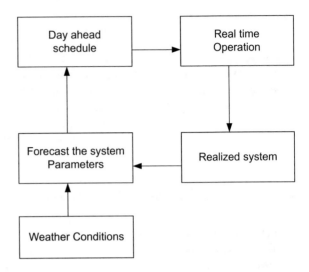

A model for calculating dynamic capacity of overhead lines is presented in this chapter. This model is based on the IEEE standard 738-bus system [13]. IEEE model is used steady-state heat balance equation for calculating the current of overhead conductors.

$$
\begin{cases}
R(T_C)I^2 + Q_s = Q_r + Q_c \\
Q_s = \gamma D S_i \\
Q_r = S_B \pi D K_r (T_c^4 - T_1^4) \\
Q_c = \lambda N_u (T_2 - T_1) \pi \\
N_u = 0.65 \mathrm{Re}^{0.2} + 0.23 \mathrm{Re}^{0.61} \\
\mathrm{Re} = 1.644 \times 10^9 V D [T_1 + 0.5(T_c - T_1)]^{-1.78}
\end{cases}
\tag{1}
$$

where, Q_s, Q_r, Q_c are the solar heating, radiative cooling, and convective cooling, respectively. In addition, N_u is the Nusselt number, and Re is the Reynolds number that shows the impact of wind speed on the capacity of overhead lines. The wind speed is demonstrated by indicator V. Considering the above equations, the maximum current can be obtained using the following equation.

$$
I_{\max} = \max \left(\begin{cases}
\sqrt{\dfrac{\left[1.01 + 0.0371 \times \left(\dfrac{D\rho_f V}{\mu_f}\right)^{0.52}\right] \times \left[K_f \times K_{angle} \times \Delta T\right]}{R(T_c)}} \\[4ex]
\sqrt{\dfrac{\left[0.0119 \times \left(\dfrac{D\rho_f V}{\mu_f}\right)^{0.52}\right] \times \left[K_f \times K_{angle} \times \Delta T\right]}{R(T_c)}}
\end{cases} \right.
\tag{2}
$$

where, ΔT is $T_c - T_2$, in which T_c and T_2 are conductor temperature and ambient air temperature, respectively. Moreover, D is the conductor diameter, and ρ_f is the density of air at temperature of $\frac{T_c+T_2}{2}$. In addition, v is the speed of air stream at conductor, and μ_f is the dynamic viscosity of air at temperature of $\frac{T_c+T_2}{2}$. Beside, K_f is the thermal conductivity of air at temperature of $\frac{T_c+T_2}{2}$, and K_{angle} is a parameter representing the angle between wind speed and the conductor axis. Capacity of each sag points of overhead line is calculated and minimum capacity of these points is selected for capacity of overhead line. This model for calculating dynamically capacity of overhead lines is difficult, and it often is used for determining static capacity of overhead lines. In determining static capacity of overhead lines, the worst case scenario such as minimum wind speed and highest ambient temperature are used. The value obtained using the worst case scenario is called static line rating. SLR can be computed by using the following equation.

$$
I_{max}^{SLR} \approx \frac{\left[1.01+0.0371 \times \left(\frac{D\rho_f V^{SLR}}{\mu_f}\right)^{0.52}\right] \times \left[K_f \times K_{angle} \times \Delta T^{SLR}\right]}{R(T_c)} \tag{3}
$$

where, V^{SLR} is wind speed at static line rating condition, and ΔT^{SLR} is temperature difference. High ambient temperature and high conductor temperature is selected for calculating temperature difference. These values may change at seasonal condition; however, it is assumed constant for all seasons in this work. Other parameters can be found at design standards of overhead lines. In this chapter, wind speed is set to 0.5 m/s and ambient temperature is set to 50 for SLR calculation.

DLR can be estimated with real time weather condition. This method is based on impact of weather condition on capacity of overhead lines. Real time weather conditions consist of wind speed and air temperature at each sag points of overhead line. In [14], a simplified method for DLR calculation using SLR is presented. This method neglected the correlation between air temperature and wind speed and investigated their effect on capacity of overhead lines. The ratio of η is defined to model the impact of wind speed and air temperature on capacity of overhead lines.

$$
I_{max}^{DLR} = \max \left(\begin{cases} I_{max}^{SLR} \\ \left(\frac{v}{v_{SLR}}\right)^{0.26} \sqrt{\frac{T_c-T}{T_c-T_{SLR}}} I_{max}^{SLR} \\ \frac{0.566}{v_{SLR}^{0.26}} \left(\frac{\rho_f}{\mu_f}\right)^{0.04} D^{0.04} v^{0.3} \sqrt{\frac{T_c-T}{T_c-T_{SLR}}} I_{max}^{SLR} \end{cases} \right) \tag{4}
$$

Capacity of each sag point of overhead lines is estimated by (4) and minimum capacity is considered as dynamic line rating for studies period. This method used separate equation for calculating DLR at low wind speed scenario and high wind speed scenario. DLR should not be lower than SLR. The measured data is applied to compute line capacity by (4). However, power system is generally scheduled in day ahead. Therefore, the system operators need an exact forecasted DLR to use this option. Data Analytics step in big data process can obtain the value of DLR. Different techniques are presented in recent studies for forecasting data required for power system planning. In this chapter, the authors utilized radial basis function

Fig. 3 A graphical
representation of RBF

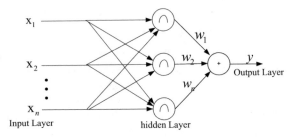

(RBF) neural networks in order to predict DLR. Figure 3 illustrates a RBF neural network that inputs of RBF network is designated by $\{x_1, x_2, \ldots, x_n\}$ and y is the output [15, 16]. This RBF network has a hidden layer of basis function or neurons. The output of each neurons calculation is based on distance between the neuron center and the input vector. The RBF network is formed by a weighted sum of the neurons outputs that are shown by Eqs. 7 and 8. ϕ_{ij} is radial basis function. Gaussian function, multi-quadric function and linear function can be used for radial basis function but the most commonly used basis function is Gaussian function that is shown by (7). Where, x_i is training data, σ is width of Gaussian functions, x is the central point of function, and $\|x - x_i\|$ is distance of sample from the central function. In this chapter, Gaussian function is used as basic function and gradient descent algorithm is used for training of the RBF network. In this algorithm, the weights are updated in a direction of the negative of the gradient, which is shown by (8).

$$y = \sum_{j=1}^{N} w_j \phi_j (\|x_i - x_j\|) \tag{5}$$

$$y = \sum_{j=1}^{N} w_j \phi_{ij} \tag{6}$$

$$\phi_i(\|x - x_i\|) = exp(- \frac{\|x - x_i\|^2}{2\sigma^2}) \tag{7}$$

$$w_j(t+1) = w_j(t) - \eta \nabla E(w_j(t)) \tag{8}$$

3.2 The Performance of Big Data Processing (DLR Calculation) in System Operation

In this subsection, a simple instance is investigated to show the impact of big data processing (DLR calculation) in system operation. The system includes three thermal generation units, two wind farms, a hydro unit, and a pumped storage plant. The schematic of the studied test system is demonstrated in Fig. 4.

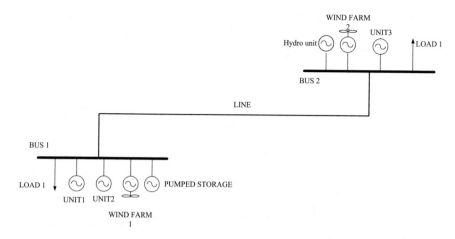

Fig. 4 The schematic of the studied test system

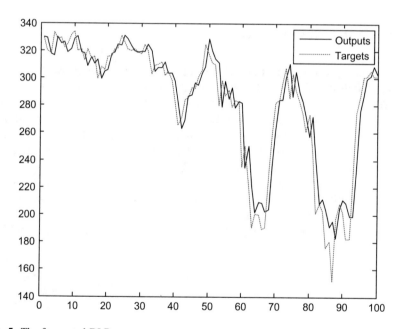

Fig. 5 The forecasted DLR

Table 1 Data for thermal generation units

Generation unit/characteristics	Unit 1	Unit 2	Unit 3
$P_{\max}(MW)$	50	100	110
$P_{\min}(MW)$	5	10	10
$C(\$/MWh)$	10	30	35
$C^{RU}(\$/MWh)$	16	13	10
$C^{RD}(\$/MWh)$	15	12	9

Table 2 The operation cost

	System with SLR	System with DLR
Schedule	18235	16850
Reserve	4750	3833
Total	13485	13017

The data for thermal generation units is provided in Table 1. The lower and upper bounds of generation units, and cost functions and reserve cost of the units are reported in this table.

The mathematical models for this optimization problem is presented in [17]. Here, we study the impact of DLR calculated by big data processing in the operation cost. The forecasted value of DLR is shown in Fig. 5. The RBF neural network is employed to predict the parameters of DLR. Then, the value of DLR is calculated by (4). The operation cost for a system with static line capacity, which is 180 MW and a system with DLR is reported in Table 2.

4 Fault Detection/Protection

The big data process approach can be used for fault detection by the system operator. High impedance fault generally happens when the power network components such as distribution line or high impedance surface touches a high impedance object for instance trees. Detection of this type of fault is very hard because such types of fault has high impedance in the fault point [18]. Hence, the common protection devices can not properly operate. The electric arcing (harmonic and non-harmonic components) is generally deployed to detect such fault types. However, some of the power system operation conditions such as air switching operation, nonlinear load, capacitor bank operation, power factor correction, voltage profile management and losses minimization, have a same behavior to high impedance faults. Therefore, high impedance fault detection approach should be

Fig. 6 Framework of high impedance fault detection

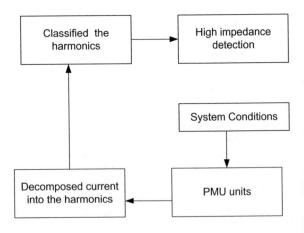

able to classify such operation and fault conditions [19, 20]. In this section, the impact of big data is studied, which receives from the smart meter tools of fault detection. The framework of this problem is shown by Fig. 6. The current harmonics are used to detect high impedance fault and distinguish fault with other similar operations. In addition, the researchers propose different machine learning approaches to solve such problem. In this study, the Group Method of Data handing (GMDH) is deployed to recognize high impedance faults occurring in power systems. The following subsection introduces the GMDH approach.

4.1 GMDH

GMDH can be classified into the learning machine based on the rule of heuristic self-organized map. A schematic of this network is shown in Fig. 7. Such type of machine learning is presented by Ivankenko [21]. Such method includes several steps similar to gardening steps [22]. The abbreviated method contains seeding, rearing, crossbreeding and the sifting of seed. In GMDH networks, a model is built by a set of neurons which are connected together in each layer and producing a neuron in the next layer. A mathematical model of GMDH is illustrated by (9). In such formulation, parameter $\hat{f}$ is an approximation function to determine input class and output of the model for input vector of $X = \{x_1, x_2, \ldots, x_n\}$ is represented by $\hat{y}$.

$$\hat{y} = \hat{f}(x_1, x_2, \ldots, x_m) \tag{9}$$

In the GMDH, general correlation between the class number and input variables is found. Such correlation is based on mathematical equations. The parameters of such equations are determined for minimizing objective function, which is illustrated as:

$$OBJ_{\min} = \sum_{i=1}^{M} \left[\hat{f}(x_{i1}, x_{i2}, \ldots, x_{im}) - y_i \right]^2 \tag{10}$$

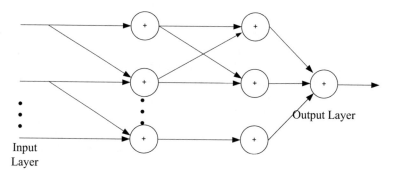

Fig. 7 Schematic of GMDH network

where objective function is shown by *OBJ*. The number of training data is determined by M. $\hat{f}(x_{i1}, x_{i2}, \ldots, x_{im})$ shows estimation class which is obtained by
GMDH. Actual class of such input variables is represented by y_i.

A complicated discrete form of the Volterra function is generally used to represent impact of input and output variables. Such function is shown in (11) ω is the
GMDH parameter and x, y are the input and output variables respectively.

$$y = \omega_0 + \sum_{i=1}^{n} \omega_i x_i + \sum_{i=1}^{n} \sum_{j=1}^{n} \omega_{ij} x_i x_j + \sum_{i=1}^{n} \sum_{j=1}^{n} \sum_{k=1}^{n} \omega_{ijk} x_i x_j x_k + \ldots \tag{11}$$

This chapter uses Levenberg Marquardt (LM) algorithm to train GMDH. Such
algorithm is similar to iterative and the goal of method is minimizing a function
illustrated in (12). The mathematical model for this learning method is:

$$E = \frac{(y - \hat{y})^2}{2} \tag{12}$$

$$z_1 = W_{11}^T . X_{11} \tag{13}$$

$$z_2 = W_{12}^T . X_{12} \tag{14}$$

$$W_{21} = \left\{\omega_{21}^0, \omega_{21}^1, \omega_{21}^2, \omega_{21}^3, \omega_{21}^4, \omega_{21}^5\right\}^T \tag{15}$$

$$W_{11} = \left\{\omega_{11}^0, \omega_{11}^1, \omega_{11}^2, \omega_{11}^3, \omega_{11}^4, \omega_{11}^5\right\}^T \tag{16}$$

$$W_{12} = \left\{\omega_{12}^0, \omega_{12}^1, \omega_{12}^2, \omega_{12}^3, \omega_{12}^4, \omega_{12}^5\right\}^T \tag{17}$$

$$X_{21} = \left\{1, z_1, z_2, z_1 z_2, z_1^2, z_2^2\right\}^T \tag{18}$$

$$X_{11} = \left\{1, x_1, x_2, x_1 x_2, x_1^2, x_2^2\right\}^T \tag{19}$$

$$X_{12} = \left\{1, x_1, x_3, x_1 x_3, x_1^2, x_3^2\right\}^T \tag{20}$$

where input variable is determined by x_k and z_k is intermediate variables. Vector of
neurons is denoted by X_{ts}. In such symbol, the number of layers and number of
neurons in each layer are specified by t and s, respectively, and W_{ts} is weight vector.

The Jacobian matrix as the partial differentiation taken for the error function,
based on the chain rule is defined as:

$$[J_{2s}] = \left\{\frac{\partial E}{\partial W_{2s}}\right\}^T = \frac{\partial E}{\partial \hat{y}} \left\{\frac{\partial z_s}{\partial \omega_{2s}^0}, \frac{\partial z_s}{\partial \omega_{2s}^1}, \frac{\partial z_s}{\partial \omega_{2s}^2}, \frac{\partial z_s}{\partial \omega_{2s}^3}, \frac{\partial z_s}{\partial \omega_{2s}^4}, \frac{\partial z_s}{\partial \omega_{2s}^5}\right\} \tag{21}$$

$$[J_{1s}] = \left\{\frac{\partial E}{\partial W_{1s}}\right\}^T = \frac{\partial E}{\partial \hat{y}} \left\{\begin{array}{cc} \frac{\partial \hat{y}}{\partial X_{2s}} \cdot \frac{\partial X_{2s}}{\partial z_s} \cdot \frac{\partial z_s}{\partial \omega_{1s}^0} & \frac{\partial \hat{y}}{\partial X_{2s}} \cdot \frac{\partial X_{2s}}{\partial z_s} \cdot \frac{\partial z_s}{\partial \omega_{1s}^1} \\ \frac{\partial \hat{y}}{\partial X_{2s}} \cdot \frac{\partial X_{2s}}{\partial z_s} \cdot \frac{\partial z_s}{\partial \omega_{1s}^2} & \frac{\partial \hat{y}}{\partial X_{2s}} \cdot \frac{\partial X_{2s}}{\partial z_s} \cdot \frac{\partial z_s}{\partial \omega_{1s}^3} \\ \frac{\partial \hat{y}}{\partial X_{2s}} \cdot \frac{\partial X_{2s}}{\partial z_s} \cdot \frac{\partial z_s}{\partial \omega_{1s}^4} & \frac{\partial \hat{y}}{\partial X_{2s}} \cdot \frac{\partial X_{2s}}{\partial z_s} \cdot \frac{\partial z_s}{\partial \omega_{1s}^5} \end{array}\right\}^T \tag{22}$$

According to such equations updated weight matrixes are obtained as:

$$W_{2s}^{new} = W_{2s}^{old} + [J_{2s}^T J_{2s} + \mu.I] J_{2s}^T E \tag{23}$$

$$W_{1s}^{new} = W_{1s}^{old} + [J_{1s}^T J_{1s} + \mu.I] J_{1s}^T E \tag{24}$$

where μ is the learning rate between 0 and 1. Furthermore, by adjusting the weighting coefficients by Levenberg–Marquardt, the corresponding quadratic polynomial neurons are presented as by:

$$\begin{aligned} (S/D)_2^1 = & -0.1734 - 0.0266N_s + 0.0488\theta \\ & + 0.0129(N_s)(\theta) + 0.0182(N_s)^2 + 0.0488(\theta)^2 \end{aligned} \tag{25}$$

$$\begin{aligned} (S/D)_8^1 = & -0.0785 + 0.0213KC \\ & + 1.1734 \times 10^{-5}\text{Re} + 3.3563 \times 10^{-7}(KC)(Re) \\ & - 1.1385 \times 10^{-4}(KC)^2 \\ & + 4.0551 \times 10^{-7}(Re)^2 \end{aligned} \tag{26}$$

$$\begin{aligned} (S/D)_9^1 = & -0.0306 - 0.0419KC + 6.3156\text{Re}_d \\ & + 1.8427(KC)(Re) + 7.7577 \times 10^{-4}(KC)^2 \\ & - 226.2307(Re_d)^2 \end{aligned} \tag{27}$$

$$\begin{aligned} (S/D)_1^2 = & -0.012 + 0.1195(S/D)_2^1 + 0.09(S/D)_5^1 \\ & + 0.0142(S/D)_2^1(S/D)_5^1 + 0.0074\left((S/D)_2^1\right)^2 \\ & + 0.0067\left((S/D)_5^1\right)^2 \end{aligned} \tag{28}$$

The number of layers and number of neurons in the layer is shown by (S/D).

4.2 Simulation Results

4 sets of prescribed events are deployed for constructing the big data based on GMDH model with the target of high impedance faults detection. They are defined as follows:

- Set 1: tripping of the nonlinear load
- Set 2: operating an air switching
- Set 3: operating a capacitor bank
- Set 4: occurring a high impedance fault

Each set of such events is simulated under different smart grid operating states. Such operating states are normal system loading, minimum system loading and maximum system loading. A part of training data is illustrated in Table 3. In such table class 1 demonstrates the fault occurring condition and other condition is shown by class 0. Figure 8 shows the result of accuracy of the proposed technique.

Table 3 A part of training data

The first harmonic	3th harmonic	5th harmonic	7th harmonic	9th harmonic	Class
55.875	0.06	19.7521	6.7813	0.0519	0
62.897	0.0575	56.8906	12.7226	0.0715	0
81.2775	0.2668	88.2769	16.236	0.0837	0
56.3028	0.0693	24.7787	7.9388	0.0928	0
66.5219	0.2616	73.3183	17.1846	0.0393	0
100.4955	0.1788	90.987	17.2103	0.3523	0
62.8311	0.0574	56.6858	12.6978	0.0702	0
58.0326	0.0781	37.6233	10.2805	0.1079	0
376.6222	0.3133	74.5632	16.4449	0.468	0
160.1868	0.6673	62.1958	13.185	0.4715	0
434.8189	0.7889	70.6437	16.0099	1.0728	0
308.784	1.4628	0.7841	0.4299	308.784	1
327.2618	1.4614	0.7938	0.44	327.2618	1
438.2918	1.4036	0.8013	0.4645	438.2918	1
455.3458	1.3903	0.7973	0.4645	455.3458	1
168.8984	1.2423	0.5377	0.248	168.8984	1
85.295	0.6179	0.1831	0.0556	85.295	1
95.9708	0.728	0.2344	0.0798	95.9708	1
164.6647	2.8955	10.2766	7.8294	164.6647	1
192.1416	2.6278	7.5285	6.927	192.1416	1
123.5071	2.35E-04	2.37E-04	2.40E-04	2.42E-04	0
152.6619	2.85E-04	2.85E-04	2.85E-04	2.84E-04	0
194.9956	3.04E-04	3.09E-04	3.17E-04	3.30E-04	0
248.9097	1.33E-04	1.46E-04	1.64E-04	1.90E-04	0
266.7952	2.24E-04	2.33E-04	2.47E-04	2.68E-04	0
375.0032	1.28E-04	1.46E-04	1.73E-04	2.09E-04	0

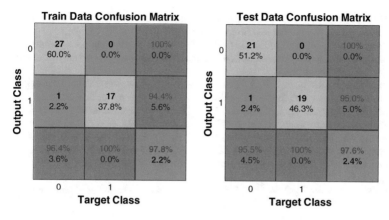

Fig. 8 The accuracy of high impedance detection method

5 Conclusion

In this chapter, the application of big data analysis to smart grids is studied. The first step in big data processing is the asset management, which provides data for the certain analysis related to applications of power systems. Two various utilization cases of big data are investigated in the chapter. The first case is operation of a test system considering dynamic line rating (DLR), which is forecasted by implementation of the big data received from the smart meters. The obtained results proves that the big data can be effective in reducing total cost for a small case study. Accordingly, it can be proved that a significant reduction of total cost for big cases can be obtained. The second case study is a decision making based on big data for detecting high impedance fault occurrence in the power system, which is defined as worthy and difficult subjects in the power systems protection.

References

1. Zhou, K., Chao, F., & Yang, S. (2016). Big data driven smart energy management: From big data to big insights. *Renewable and Sustainable Energy Reviews, 56,* 215–225.
2. Momoh, J. A. (2009). Smart grid design for efficient and flexible power networks operation and control. In *2009 Power Systems Conference and Exposition,. PSCE'09. IEEE/PES.* IEEE.
3. Amin, M. (2008). Challenges in reliability, security, efficiency, and resilience of energy infrastructure: Toward smart self-healing electric power grid. In *2008 IEEE Power and Energy Society General Meeting-Conversion and Delivery of Electrical Energy in the 21st Century.* IEEE.
4. Hossain, M. S., et al. (2016). Role of smart grid in renewable energy: An overview. *Renewable and Sustainable Energy Reviews, 60,* 1168–1184.
5. Wu, X., Zhu, X., Wu, G.-Q., & Ding, W. (2014). Data mining with big data," *IEEE Transactions on Knowledge and Data Engineering, 26,* 97–107.

6. Zhou, K., & Yang, S. (2015). A framework of service-oriented operation model of China' s power system. *Renewable and Sustainable Energy Reviews, 50,* 719–725.
7. Efthymiou, C., & Kalogridis, G. (2010). Smart grid privacy via anonymization of smart metering data. In *2010 First IEEE International Conference on Smart Grid Communications (SmartGridComm).* IEEE.
8. McKenna, E., Richardson, I., & Thomson, M. (2012). Smart meter data: Balancing consumer privacy concerns with legitimate applications. *Energy Policy, 41,* 807–814.
9. Hu, H., Wen, Y., Chua, T.-S., & Li, X. (2014). Toward scalable systems for big data analytics: A technology tutorial. *IEEE Access, 2,* 652–687.
10. Kezunovic, M., Xie, L., & Grijalva, S. (2013). The role of big data in improving power system operation and protection. In *2013 IREP Symposium on Bulk Power System Dynamics and Control-IX Optimization, Security and Control of the Emerging Power Grid (IREP)* (pp. 1–9).
11. Dalal, G., Gilboa, E., & Mannor, S. (2016). Distributed scenario-based optimization for asset management in a hierarchical decision making environment. *Power Systems Computation Conference (PSCC), 2016,* 1–9.
12. Yang, S.-l., & Shen, C. (2013). A review of electric load classification in smart grid environment. *Renewable and Sustainable Energy Reviews, 24,* 103–110.
13. IEEE. (2006). IEEE standard for calculating the current temperature of bare overhead conductors.
14. Wallnerstrom, C. J., Huang, Y., & Soder, L. (2015). Impact from dynamic line rating on winpower integration. *IEEE Transactions on Smart Grid, 6*(1), 343–350.
15. Chen, D. (2017). Research on traffic flow prediction in the big data environment based on the improved RBF neural network. *IEEE Transactions on Industrial Informatics.*
16. Shetty, R.P., Sathyabhama, A., & Adarsh Rai, A. (2016). Optimized radial basis function neural network model for wind power prediction. In *2016 Second International Conference on Cognitive Computing and Information Processing (CCIP).* IEEE.
17. Morales, J.M., et al. (2013). *Integrating renewables in electricity markets: Operational problems* (Vol. 205). Springer Science & Business Media.
18. Sulaiman, M., Adnan, T., & Ibrahim, Z. (2013). Using probabilistic neural network for classification high impedance faults on power distribution feeders. *World Applied Sciences Journal, 23*(10), 1274–1283.
19. Mor, V., & Vaghamshi, A. (2016). Review on fault detection, identification and localization in electrical networks using fuzzy-logic.
20. Kagan, N., et al. (2016). Computerized system for detection of high impedance faults in MV overhead distribution lines. In *2016 17th International Conference on. Harmonics and Quality of Power (ICHQP).* IEEE.
21. Dag, O., & Yozgatligil, C. (2016). GMDH: An R package for short term forecasting via GMDH-type neural network algorithms. *The R Journal, 8*(1), 379–386.
22. Xiao, J., et al. (2016). Churn prediction in customer relationship management via GMDH-based multiple classifiers ensemble. IEEE Intelligent Systems, *31*(2), 37–44.

A Structural Graph-Coupled Advanced Machine Learning Ensemble Model for Disease Risk Prediction in a Telehealthcare Environment

Raid Lafta, Ji Zhang, Xiaohui Tao, Yan Li, Mohammed Diykh and Jerry Chun-Wei Lin

Abstract The use of intelligent and sophistic technologies in evidence-based clinical decision making support have been playing an important role in improving the quality of patients' life and helping to reduce cost and workload involved in their daily healthcare. In this paper, an effective medical recommendation system that uses a structural graph approach with advanced machine learning ensemble model is proposed for short-term disease risk prediction to provide chronic heart disease patients with appropriate recommendations about the need to take a medical test or not on the coming day based on analysing their medical data. A time series telehealth data recorded from patients is used for experimentations, evaluation and validation. The Tunstall dataset were collected from May to October 2012, from industry collaborator Tunstall. A time series data is segmented into slide windows and then mapped into undirect graph. The size of slide window was empirically determined. The structural properties of graph enter as the features set to the machine learning ensemble classifier to predict the patient's condition one day in advance. A combination of three classifiers—Least Squares-Support Vector Machine, Artificial Neural Network, and Naive Bayes—are used to construct an ensemble framework to classify the graph features. To investigate the predictive ability of the graph with the ensemble

R. Lafta · J. Zhang (✉) · X. Tao · Y. Li · M. Diykh
Faculty of Health, Engineering and Sciences, University of Southern
Queensland, Toowoomba, Australia
e-mail: ji.zhang@usq.edu.au

R. Lafta
e-mail: raidluaibi.lafta@usq.edu.au

X. Tao
e-mail: xtao@usq.edu.au

Y. Li
e-mail: yan.li@usq.edu.au

M. Diykh
e-mail: mohammed.diykh@usq.edu.au

J. C.-W. Lin
School of Computer Science and Technology, Harbin Institute of Technology
Shenzhen Graduate School China, Shenzhen, China
e-mail: jerrylin@ieee.org

© Springer Nature Singapore Pte Ltd. 2018
S. S. Roy et al. (eds.), *Big Data in Engineering Applications*,
Studies in Big Data 44, https://doi.org/10.1007/978-981-10-8476-8_18

classifier, the extracted statistical features were also forwarded to the individual classifiers for comparison. The findings of this study shows that the recommendation system yields a satisfactory recommendation accuracy, offers a effective way for reducing the risk of incorrect recommendations as well as reducing the workload for heart disease patients in conducting body tests every day. A 94% average prediction accuracy is achieved by using the proposed recommendation system. The results conclusively ascertain that the proposed system is a promising tool for analyzing time series medical data and providing accurate and reliable recommendations to patients suffering from chronic heart diseases.

1 Introduction

The chronic diseases such as heart disease have developed and become one of the major public health problems which accounting for 50% of disease burden worldwide [22]. According to World Health Organization (WHO), these diseases were caused more than 60% of all death in 2005 [1]. Nowadays, many of people around the world are suffering from different chronicle diseases because the lack of used diseases prediction tools. Therefore, the survival rates have been noticeably increased due to using sophisticated techniques to predict diseases in a right time.

Recommendation systems are computer-based information systems designed to support and assist medical practitioners in implementation evidence-based practices and improved decision-making [10, 31]. The recommendation systems can help in minimizing medical errors and providing more detailed data analysis in shorter time [38].

Telehealth systems offer a real time and quick way that is enable healthcare practitioners and chronic diseases patients to exchange information easily [11, 45], and subsequently have enjoined fast developments in many countries due to fast service delivery and its low-cost. Most telehealth services are conveyed through Web-based applications which utilize Internet and Web browsers, together with sensors, wearable devices and mobile. Given the significance of disease risk prediction in the medical field [48] as well as the urgency of acquiring more effective analytic techniques for disease risk prediction, great endeavors are expected to enhance the quality of evidence-based decisions and recommendations in the telehealth environment. In telehealth system, patients with chronic heart disease require taking daily medical tests to monitor their heart health conditions. Yet, carrying out various necessary medical tests every day for chronical disease patients in the current practice brings lots of inconvenience and even burden to the patients and adversely affect their life quality. Generating accurate intelligent recommendations to guide their daily medical tests can significantly decrease their workload in taking those tests while keeping the associated health risk in a worthy low level.

In many cases, an accurate medical recommendation is based upon the prediction of patients' short-term disease risk, which is one of the most important functions in telehealth systems. A set of disease risk prediction models have become available in the medical literature using statistical analysis tools and approaches based on data mining tools. These models have been utilized for different healthcare and medical issues [7, 9, 17, 21, 30, 33, 36, 39, 46, 47]. However, most of the existing work only focus on the long-term medical prediction. Nevertheless, the short-term prediction, which is studied in our work, has turns to be more challenging than the long-term prediction as patients' conditions may experience more dramatic and abrupt changes during the short-term timeframe.

In this work, we utilize a structural graph to process the time series medical data of heart disease patients to facilitate the subsequent data analytics to produce the accurate prediction and recommendations. Graphs can be mathematically defined as abstract representations of networks that consist from set of nodes linked by edges [28, 35]. In last years, graph theory has been widely increasingly used in analysing and classification of the complex networks relationships such as, social networks, biological and brain networks, signal and image processing. It is used in neuroscience research to analyse and study the brain diseases [32, 43, 44]. Some studies [16, 26, 37] showed that graph theory can be considered as a one of robust tools to characterize the functional topological properties of brain networks for both normal and abnormal brain functioning [25, 41]. It is also used in image processing as a powerful tool to analyse and classify digital images [34]. The time series of EEG signals are converted into graphs by [12, 13] for EEG sleep stages classification.

The intelligent, accurate medical recommendations in our work rely on the use of classification approaches to produce reliable prediction of the short-term medical risks of the patients. By nature, this is a classification problem which involves using classification methods (called classifiers) to predict the necessity of taking body test of a given medical measurement.

There are several reasons that pushed us to construct the ensemble classifier. First, it provides an efficient solution for building a single model for applications of which the amount of data may be very large [40]. Second, it has also been proven to be an effective tool thanks to its ability to improve the overall accuracy of the prediction model. Empirical results showed that machine learning ensembles are often more accurate than the individual classifiers that make them up [3, 42]. Bagging aggregation is a machine learning ensemble algorithm designed to enhance the accuracy and stability of machine learning algorithms [8], which was proposed by Breiman in the mid-1990s [40]. It has been proven to be a very popular, efficient and effective method for building an ensemble model.

Due to the ensemble outperforms individual classifiers, a combination of three classifiers—Least Squares-Support Vector Machine, Artificial Neural Network, and Naive Bayes—are used to construct an ensemble framework in this work.

The contributions of this work can be summarized as follows:

- First, the time series medical data of a given patient will be segmented into smaller overlapped sliding windows based on the size of the sliding window used in the data analysis;
- Then, each sliding window is mapped into undirect graph in order to extract the structural properties of each graph;
- Finally, the extracted structural properties of each graph are then input into our ensemble learning model to produce a binary recommendation concerning whether that patient needs to take a medical test on the coming day for a certain medical measurement such as the heart rate or blood pressure.

In this paper, a novel short-term recommendation system for chronic heart disease patients is proposed. This system is developed using a structural graph with a machine learning ensemble model to provide patients in a telehealth environment with appropriate recommendations for the necessity of taking a medical body test on the coming day. Such recommendations are established based on the prediction of their heart conditions using their time series medical data from the past few days.

To verify the performance of the proposed model, the metrics of accuracy, workload saving and risk are used and experimental evaluations are conducted on a real-life time series data collected from a pilot study on a group of heart failure patients. The experimental results demonstrate that the proposed model yields a reasonably good recommendation accuracy and can effectively reduce the workload required in medical tests for the patients. It also can effectively reduce the risk of incorrect recommendations. We believe that this analytic model is promising in risk assessment and management associated with heart failure and other similar diseases.

The remainder of this paper is organised as follows. Section 2 explains the details of the proposed methodology including describes the machine learning classifiers that constructing the proposed ensemble model. Section 3 discussed in details the experimental evaluation results and the used dataset and also compared the results of the proposed method with other results of common methods. Finally we conclude the paper and highlight the future work in Sect. 4.

2   Methodology

Figure 1 illustrates the overall architecture of our recommendation system used for chronic heart disease patients in the telehealth environment. In this section, we present in details the architecture of the recommender system.

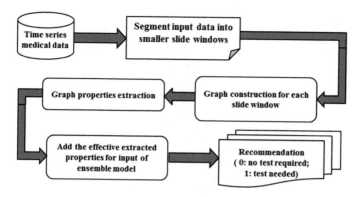

Fig. 1 An overview of the proposed methodology

2.1 Time Series Segmentation

In our system, the input time series data, represented as $X = \{y_1, y_2, y_3, \ldots, y_n\}$ which contains n data, is segmented into a set of overlapped sub-segments based on a predefined value of parameter k that specifies the size of the sliding window, corresponding to each sub-segment. In this work, many experiments are conducted with different numbers of slide window sizes (k). It is important to divide the time series data into several windows because each slide window will be mapped into a separated graph and then extract the effective features from graphs to represent the slide windows.

2.2 Graph Construction and Structural Graph Similarity

Each slide window was mapped as an undirect graph. A graph is a pair of sets $G = (V, E)$, where V is a set of nodes (vertices, or points) so that each node represents the value of a test measurement for that day and E is a set of connections among the nodes of graphs. Therefore, each pair of nodes in a graph are connected by a link if there is a relationship between them [4, 5, 29]. The Euclidean distance has been widely used as a similarity measuring method [6, 18, 19]. Let $D_{ij} = 1, 2, 3, \ldots, M$ be the set of time series of M test measurements in each slide window. Each test measurement in a slide window is assigned to be a node in an undirected graph. Lets n_1 and n_2 be nodes in an undirected graph. They are connected if the distance (d) between them are less or equal to a determined threshold [13]:

$$(n_1, n_2) \in E, if \quad d(n_1, n_2) \leq \theta \qquad (1)$$

where θ is a determined threshold. An example of an undirect graph is shown in Fig. 2. A graph G can be described by giving a square matrix $N \times N$ called adjacency or connectivity matrix A to describe the connections among the nodes of graph.

Fig. 2 An example of an undirect graph

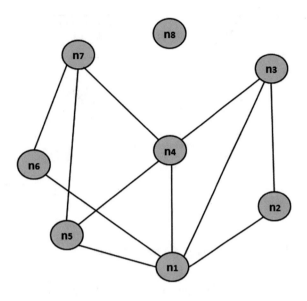

Table 1 The adjacency matrix of a graph G

	n_1	n_2	n_3	n_4	n_5	n_6	n_7
n_1	0	1	1	1	1	1	0
n_2	1	0	1	0	0	0	0
n_3	1	1	0	1	0	0	0
n_4	1	0	1	0	1	0	1
n_5	1	0	0	1	0	0	1
n_6	1	0	0	0	0	0	1
n_7	0	0	0	1	1	1	0

The adjacency matrix contains zeros in it's diagonal and thus it is a symmetric matrix. The adjacency matrix is qual to one if there is a connect between two nodes, and zero otherwise [6].

$$A(n_i, n_j) = \begin{cases} 1 & \text{if } (n_i, n_j) \in E, \\ 0 & \text{otherwise.} \end{cases} \quad (2)$$

For example, Table 1 shows the adjacency matrix of a graph G that consists from 7 nodes. We can notice that each element a_{ij} in an adjacency matrix A is equal to 1 when the connection exists, and zero otherwise. The diagonal of matrix A is still zero for all it's elements.

2.3 Graph Features

The adjacency matrix of a graph G can be used to extract the statistical features of a graph G [13, 14, 26]. The statistical features of a graph can be used for prediction in this study. The following sections illustrate the most common extracted features from a graph G.

2.3.1 Degree Distributions of the Graph

The degree distribution, that denoted by $P(k)$, refers to the proportion of nodes with degree k divided by the total number of nodes in the graph [13]. It can be mathematically defined as follow:

$$P(k) = \frac{|\{n \mid d(n) = k\}|}{N} \tag{3}$$

where $d(n)$ refers to the degree of node n, N is the total number of nodes in the graph.

2.3.2 The Clustering Coefficient of the Graph

Clustering coefficient (CC) is one of the most important measures used to characterize the local and global structures of a graph [13, 26, 28]. Let n_i be a node in a graph G. Thus the local clustering coefficient of a given node n_i is computed as the proportion of links among n_i's neighbours which are actually realised compared with the total number of possible connections. For example, the clustering coefficient of a node n_3 in Fig. 2 is 1 because the node n_3 has three neighbours, which can have a maximum of 3 connections among them and all of them are realised. The overall level of clustering in a graph is measured as the average of the local clustering coefficients of all the nodes:

$$C' = \frac{1}{N} \sum_{i=1}^{N} C_{ni} \tag{4}$$

where, N is the number of nodes in a graph G and C_{ni} is the local clustering coefficient of the node n_i.

2.3.3 Jaccard Coefficient of the Graph

Jaccard Coefficient (it also called Jaccard Index) is a statistical tool that used to measure the similarity and diversity between two nodes of a graph [20]. Let n_i and n_j are two nodes in a graph G. Thus the Jaccard coefficient $\Gamma(n_i, n_j)$ is defined as the ratio of the set of the neighboring intersections between those two nodes to the set of the neighboring unions for the two nodes. It can be mathematically defined as follows:

$$\Gamma(n_i, n_j) = \frac{|N(n_i) \cap N(n_j)|}{|N(n_i) \cup N(n_j)|} \tag{5}$$

where $N(n_i)$ is the set of neighbors of the node n_i that have an edge from n_i to them, and $N(n_j)$ is the set of neighbors of the node n_j that have an edge from n_j to them.

2.3.4 Average Degree

The average degree (AD) points out to the average number of links connecting in a node n_i to the other nodes in the graph [2]. The average degree of a graph can be defined as the total number of links for each node divided by the number of nodes in a graph [12]:

$$AD = \frac{1}{N} \sum_{i=1}^{m} K_i \tag{6}$$

where K_i is the degree of node n_i and N is the total number of nodes in a graph.

For example, we can easily calculate the degree of each node for a graph G shown in Fig. 2 and then calculate the average degree (AD) as follows:

$K(n_1) = 5, K(n_2) = 2, K(n_3) = 3, K(n_4) = 4, K(n_5) = 3, K(n_6) = 2, K(n_7) = 3, K(n_8) = 0$, and $AD = 2$.

2.4 Bootstrap Aggregation (Bagging)

An ensemble approach is a very effective method that combines the decisions of multiple base classifiers in order to overcome the limited generalization performance of each base classifier and generate more accurate predictions than individual base classifier. Bootstrap aggregation, a.k.a bagging, is a machine learning ensemble algorithm designed to enhance the accuracy and stability of machine learning algorithms [15, 27]. In the bootstrap method, the classifiers are trained independently and then aggregated by an appropriate combination strategy. Specifically, our ensemble model can be divided into two phases. In the first phase, the model uses bootstrap sampling to generate a number of training sets. In the second phase, the training of the three base classifiers, i.e., Least Square-Support Vector Machine, Neural Network and Naive Bayes, is performed using the bootstrap training sets generated during the first phase. Figure 3 shows an example of the bagging algorithm which involves the three classifiers to build our ensemble model. In this study, the training set was divided into multiple datasets using the bootstrap aggregation approach, and then the classifiers were individually applied to these datasets to generate the final prediction.

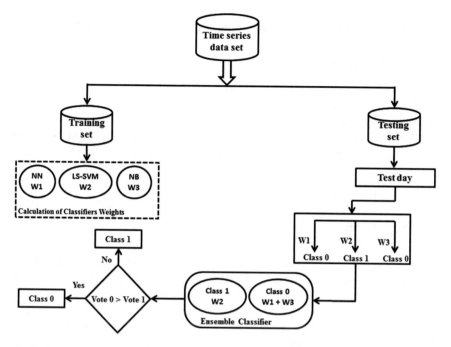

Fig. 3 An example of a bagging algorithm

It is noted that different individual classifier in the bagging approach may perform differently. Therefore, we assign a weight to each classifier's vote, based on how well the classifier performs. The classifier's weight is calculated based on its error rate. The classifier that has a lower error rate is considered more accurate and is therefore assigned a higher weight. The weight of classifier C_i's vote is calculated as follows:

$$w(C_i) = \log \frac{1 - error(C_i)}{error(C_i)}, 1 \le C_i \le 3 \tag{7}$$

The following example is presented to facilitate the understanding of our weighted bagging ensemble model:

1. Least Square-Support Vector Machine, Neural Network, and Naive Bayes are used as individual base classifier in the ensemble model. Suppose that the classifier training is performed on the training data and the error rate is calculated for each base classifier as 0.14 for LS-SVM, 0.25 for NN, and 0.30 for NB;
2. As per Eq. (7), the weight 0.78 is assigned to LS-SVM, 0.47 to NN, and 0.36 to NB;
3. Suppose that the three base classifiers generate the following predictions for a coming testing day: LS-SVM predicts 0, NN predicts 1, and NB predicts 1 (Here,

0 means no test is required on the testing day for a medical measurement; 1 means a test is required otherwise);

4. The ensemble classifier will use the weighted vote to generate the following prediction results:

   Class 1: NN + NB $\longrightarrow$ 0.47 + 0.36 $\longrightarrow$ 0.83,

   Class 0: LS-SVM $\longrightarrow$ 0.78.

5. Finally, according to the weighted vote, the class 1 has a higher value than class 0. Therefore, the ensemble classifier will classify this testing day as being in Class 1, suggesting that the patient in question need to take the test on that day for a medical measurement.

3 Experiment Result

This study aims at short-term risk assessment in chronic heart diseases patients based on analytic of a patient's historical medical data using structural graph similarity and machine learning-based ensemble classifier. As mentioned above, the time series slide windows were converted into undirect graphs. Then, the suitable features from graphs were extracted and entered as input features set for the ensemble classifier. The detailed experimental results are discussed in the following sub sections.

3.1 Performance Assessment

In this section, we present the details concerning the design of our experimental evaluation including datasets and performance metrics.

As the predictive performance of the proposed model is quite important, assessment of potential predictions is critically dependent on the quality of the used dataset. For this reason, telehealth data from Tunstall dataset will be conducted in this work. We use a real-life dataset obtained from our industry collaborator Tunstall to test the practical applicability of the proposed model. A Tunstall dataset obtained from a pilot study has been conducted on a group of heart failure patients and the resulting data were collected for their day-to-day medical readings of different measurements in a tele-health care environment. The Tunstall database employed in the development of the algorithm consists of data from six patients with a total of 7,147 different time series records. Data were acquired between May and January 2012, using a remote telehealth collaborator. The dataset is by nature in a time series and contains a set of measurements taken from the patients on different days. Each record in the dataset consists of a few different meta-data attributes about the patients such as patient-id, visit-id, measurement type, measurement unit, measurement value, measurement question, date and date-received. The characteristics of the features of the dataset are shown in Table 2.

Table 2 Characteristic features of the dataset

Feature name	Feature type
Id	Numeric
Id-patient	Numeric
HCN	Numeric
Visit-id	Numeric
Measurement type	Nominal
Measurement unit	Nominal
Measurement value	Numeric
Measurement question	Nominal
Date	Numeric
Date-received	Numeric

In addition, each record contains a few medical attributes including Ankles, Chest Pain, and Heart Rate, Diastolic Blood Pressure (DBP), Mean Arterial Pressure (MAP), Systolic Blood Pressure (SBP), Oxygen Saturation (SO_2), Blood Glucose, and Weight. Ethical clearance was obtained from the University of Southern Queensland (USQ) Human Research Ethics Committee (HREC) prior to the onset of the study. This dataset is used as the ground truth result to test the performance of our proposed model. The recommendations produced by our proposed model will be compared with the actual readings of the measurement in question recorded in the dataset to see how accurate our recommendations are.

Since a patient's historical medical data often have the class-imbalance problem (i.e., the number of normal data is much larger than that of the abnormal data), we carefully dealt with the class-imbalance problem when training the classifiers. The over-sampling and under-sampling methods have been used as good means to address this problem.

The selected input data were divided into two groups as the training and the testing sets. The slide windows time series data have been randomly divided into about 75% for the training of ensemble's classifier and 25% for the testing purpose. Several of experiments were designed and conducted to evaluate the proposed model using a real-life Tunstall database. Different sizes of slide windows were used to determine the best selected features set and the best size for each slide window as well. All the experimental results were conducted using MATLAB (R2015) on a desktop computer with the configurations of a 3.40 GHz Intel core i7 CPU processor with 8.00 GB RAM.

The performance of proposed method was evaluated by calculating the *accuracy*, *workload saving*, and *risk*. Accuracy refers to the percentage of correctly recommended days against the total number of days that recommendations are provided; workload saving refers to the percentage of the total number of days when recommendations are provided against the total number of days in the dataset, while risk refers to the percentage of incorrectly recommended days that recommendations are no test needed. Mathematically, Accuracy, workload saving and risk are defined as follows [24]:

$$Accuracy = \frac{NN}{NN + NA} \times 100\% \tag{8}$$

$$Saving = \frac{NN + NA}{|\mathcal{D}|} \times 100\% \tag{9}$$

$$Risk = \frac{NR}{|\mathcal{D}|} \times 100\% \tag{10}$$

where NN denotes the number of days with correct recommendations, NA denotes the number of days with incorrect recommendations, NR denotes the number of days with incorrect days that recommendations are no test needed, and $|\mathcal{D}|$ refers to the total number of days in the dataset. Here, a correct recommendation means that the model produces the recommendation of "no test required" for the following day and the actual reading for that day in the dataset is normal. If this is a case, the recommendation is considered accurate.

3.2 Prediction Accuracy with Different Number of Features

We first carried out experiments to evaluate the recommendation performance of our system under different sets of statistical features extracted from the siding windows of the dataset. Several experiments are carried out to determine the best set of the graph features by which the original time series can be represented with the best form. The four graph features were tested separately to evaluate the prediction accuracy of the proposed system. Figure 4 shows the ranking of the statistical features based on their performance where the features were sorted in a descending order based on their effectiveness in predicting patient's condition.

3.2.1 Two-Features Set

To determine the best combination of the two graph features, a set of experiments was designed. In this experiment, at each time, a two features set of graphs was picked up from the ordered list in Fig. 4 and sent to ensemble classifier. The number of permutation of two graph features that was tested in this paper was six cases. Figure 5 shows the performance of the proposed method based on the graphs features. Based on the obtained results, it was observed that the combination of Jaccard coefficient and degree distribution recorded the highest accuracy of 81% compared to other combinations. We found that those two features were able to give the promising prediction. However, the lowest accuracy of 56% rate was recorded by the pair of

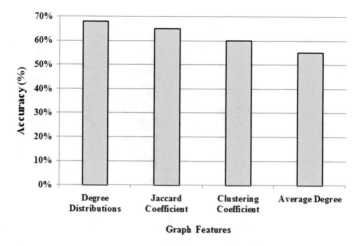

Fig. 4 Ranking of the graph features based on their accuracy performance

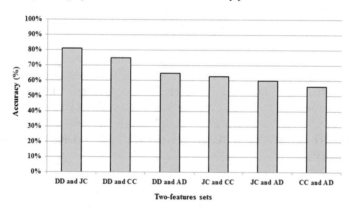

Fig. 5 Accuracy based on two-features sets (*Note DD* Degree Distribution; *JC* Jaccard Coefficient; *CC* Clustering Coefficient; *AD* Average Degree)

clustering coefficients and average degree. For further investigation, three features set was tested in the next experiment.

3.2.2 Three-Features Set

To assess the method ability to predict the status of patient with a high accuracy, the proposed method was tested using three features set. The first three graph features in Fig. 4 were selected. The three features were degree distribution, Jaccard coefficient and clustering coefficient. Figure 6 shows the performance of the proposed method using three and four features sets. The most noticeable results from this experiment were that the prediction accuracies were exceeded 94% compared with the sets of

Fig. 6 Accuracy based on three and four features sets

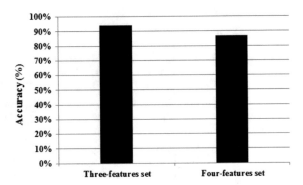

two features. For more accurate results, different experiments were designed with different data size. The results showed that there is a stability in the performance of the proposed method. Another three features set was also tested in this paper, however, the results confirmed that the three features set of degree distribution, Jaccard coefficient and clustering coefficient was the best combination of the graphs features to provide the recommendation accurately.

Four features was also tested and investigated in this paper. Based on the results in Fig. 6, the prediction accuracy of the proposed method was achieved a low rate compared with three features set. It was archived 87% using all the graph features including degree distribution, Jaccard coefficient, clustering coefficient and average degree. In this paper, the combination of the first three features for degree distribution, Jaccard coefficient and clustering coefficient was considered as they achieved the best accuracy.

3.3   Prediction Accuracy with Different Size of Slide Windows

The second influence in this work is associated to the size of window. In this experiment, the best window size is investigated to obtain the desired prediction accuracy. From the obtained results, it is clear that there is a positive relationship between the selected size of slide window and the predictive performance of the proposed system. It was found that when the number of nodes in a graph is increased due to the increasing the size of a slide window, the proposed method generates more accurate recommendations. To determine the optimum size of a slide, a set of experiments were conducted with different sizes of windows. It found that the model performance is improved by increasing the size of slide window (the number of nodes). This is because the characterises of time series data are clearly presented when the number of graph nodes is increased. Therefore, we tested our proposed model with different sizes of window and started with 7, 10, 15 and 20 days. In these experiments, the three features set of degree distribution, Jaccard coefficient and clustering coefficient were considered. The four Medical attributes including Heart Rate, Diastolic Blood

Table 3 Performance evaluation based on slide windows of 7 days

Measurement	Accuracy (%)	Saving (%)	Risk (%)
Heart rate	86.37	60.34	05.21
DBP	85.30	57.18	05.40
MAP	87.70	61.33	05.00
SO$_2$	84.30	55.44	05.90

Pressure (DBP), Mean Arterial Pressure (MAP), and Oxygen Saturation (SO$_2$) were used in the following experiments.

3.3.1 Slide Window of 7 Days

A slide windows of 7 days were used to test the predictive performance of the proposed method. Each day in the slide window was represented by a node in a graph. The three structural properties of the graphs were extracted and considered the key features to represent each window. The metrics of accuracy, workload saving and risk for all the graphs were calculated to verify the performance of proposed method. Table 3 presents the metrics of accuracy, workload saving and risk for each measurement in the Tunstall dataset.

Based on the obtained results, it was noticed that the performance of the proposed method was not good enough to predict the patient's condition due to the number of the graph nodes was not enough to reflect the behaviours of the time series data. To tackle this issue, the number of nodes in each graph was increased by considering a new size window. In the next experiment, the influence of using a window size of 10 days was discussed.

3.3.2 Slide Window of 10 Days

The time series data were segmented into windows by using a slide window of 10 days and then each window was transferred into a graph. As mentioned before, 10 days slide windows were considered to improve the accuracy of the proposed method and to make more accurate recommendations. One of the interesting findings in this paper, the proposed method yielded a high performance using 10 days slide window compared with the window size of 7 days. It can be noticed that the performance of proposed method significantly improved due to the number of graph nodes increased. It was found that the graphs nodes reflect big differences between the patient states which include whether he/she requires medical test or not. Table 4 shows the obtained results by the proposed method after considering the window size of 10 days.

Table 4 Performance evaluation based on slide windows of 10 days

Measurement	Accuracy (%)	Saving (%)	Risk (%)
Heart rate	92.75	59.80	03.95
DBP	91.40	58.77	04.50
MAP	90.55	60.65	04.80
SO$_2$	91.60	55.50	04.20

Table 5 Performance evaluation based on slide windows of 15 days

Measurement	Accuracy (%)	Saving (%)	Risk (%)
Heart rate	94.80	62.30	02.60
DBP	93.80	59.50	03.60
MAP	93.90	61.40	03.40
SO$_2$	94.60	61.80	02.90

Based on the obtained results in Table 4, It is interesting to note that the accuracies, for all the measurements, improved by more than 5% compared to the results in Table 3. In addition, using window size of 10 days did not considerably affect the performance of workload saving although the accuracy and risk are increased.

3.3.3   Slide Window of 15 Days

For further investigation, a window size of 15 days were adopted to test the performance of the proposed method. In this experiment, the size of window was increased into 15 day. Table 5 represents the metrics of accuracy, workload saving and risk for all the measurements using slide window size of 15 days. Based on results in Table 5, the average of accuracy of the proposed method were exceeded 94% across different measurements. The obtained results proved that the size of the window has a significantly potential on the accuracy of the prediction for all the measurements. One of the most important observations, the graphs characteristics were became significant to exhibit different behaviours when the patient state change from required test to not required test. We found that the connectivity among the graph nodes (clustering coefficients) are strong enough to reveal the difference between time series data.

Different sizes of window including 20, 25 and 30 days were also tested and evaluated in this study. It was noticed that there are no significant differences compared with the obtained results using the 15 days slide windows. Thus, the optimal window size was 15 days because it reflects the actual behavior of the time series data, on the basis of observation on the obtained results.

3.4 Comparative Study

To investigate the performance of our recommendation system, two performance comparisons were conducted in this section. In the first experiment, the performance of the recommendation system was evaluated based on individual classifier as well as machine learning-based ensemble classifier. In the second experiment, the proposed system was compared with some of our previous approaches. All the obtained results were recorded and evaluated.

3.4.1 Performance Evaluation Based on a Single Classifier as Well as Ensemble Model

In this experiment, we evaluate the performance of our system under 15 day slide windows and the three graph features set based on the previous information. Table 6 shows the results of comparison among the ensemble classifier and the individual classifiers. Based on the results, the system performance using individual classifier was between 80 and 85% across different measurements. The maximum accuracy of 85% was obtained by LS-SVM, while the minimum accuracy of 80% was gained by Nave Byes. We can notice that although the proposed system was conducted with different classifiers, there is no a big fluctuation in its performance and the accuracies of those classifiers are quite closer. One of the gold solution to improve the performance of the proposed method and to decrease the error rate is to combine multi-classifiers to classify the extracted features.

However, in this paper, an ensemble machine learning was used to classify the graphs features. Our recommendation system achieved a better prediction accuracy compared with the individual classifiers with an increase of 12%. As mentioned above, each classifier is trained and conducted with the dataset separately and then they combined according to an appropriate criteria. By comparing the results in Table 6, we can observe that the performance of the proposed sytem was escalated when the ensemble machine learning was adopted.

For more investigation, the execution time of the proposed model was calculated based on the ensemble classifier as well as individual classifiers. Figures 7 and 8 show the complexity time for each individual classifier and the ensemble model. we observed that the ensemble model takes more time to complete the training and

Table 6 Performance evaluation based on a single classifier and an ensemble model

Classifier	Accuracy (%)	Saving (%)	Risk (%)
LS-SVM	85.30	61.10	04.30
Neural network	83.50	60.80	05.10
Naive bayes	80.40	60.95	04.90
Ensemble model	94.27	61.25	03.01

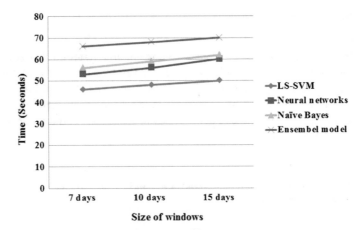

Fig. 7 Comparison of the execution time between the classifiers and the ensemble model under different slide windows

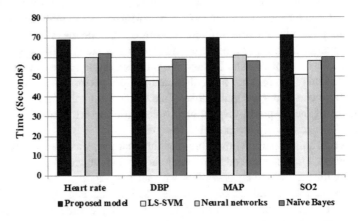

Fig. 8 Comparison of the execution time between the classifiers and the ensemble model under different measurements

prediction than the individual base classifier. This is reasonable as the ensemble model needs to aggregate the results from the base classifiers to generate the weights for them and produce the final recommendation. The ensemble model sacrifices a little on the execution time for achieving better recommendation effectiveness for patients. Additionally, the training stage can be performed off-line so that it will not adversely affect the efficiency in generating recommendations for patients during the prediction stage.

Table 7 Prediction accuracy comparison with other methods

Tunstall dataset

Method	Size of window	Techniques used	Accuracy (%)
Raid et al. [23]	5 days	Basic heuristic algorithm	86
Raid et al. [24]	5 days	Basic heuristic algorithm, Regression-based algorithm and Hybrid algorithm	91
Proposed method	15 day	Structural graph similarity and ensemble model	**94**

3.4.2 Effectiveness Comparison with Previous Approaches

To evaluate the performance of the proposed method, the prediction results were compared with some of our previously proposed methods that tackle the exactly same problem as we do in this paper using the same Tunstall dataset for a fair comparison. Table 7 represents the performances comparison among the two other reported methods and our proposed method. Based on results, the proposed model is the best among the three methods. Raid et al. [23] used a innovative time series prediction algorithm to provide recommendations to heart disease patients in the tele-health environment. The best accuracy was achieved using slide windows of 5 days. The average of the accuracy for all patients they achieved was 86% across all measurements. An intelligent recommender system, supported by three innovative predictive algorithms, was proposed by Raid et al. [24] for short-term risk assessment on patients in telehealth environment. The size of slide window was empirically detected by 5 days as the best accuracy in this study. The average of accuracy results obtained was 91% for all measurements. It clearly seems from the above results in Table 7 that the proposed model yielded the highest accuracy compared with the two others methods using the same dataset.

4 Conclusions and Future Research Directions

In this work, we propose a recommendation system supported by the structural graph properties and advanced machine learning ensemble for short-term disease risk prediction and medical test recommendation in the telehealth environment for patients suffering from chronic heart disease. This study applies the structural graph, which effectively represents the medical time series data and input the extracted statistical features to the ensemble model to generate the accurate, reliable recommendations for chronic heart disease patients. Three popular and capable classifiers, i.e., Least Square-Support Vector Machine, Neural Network, and Naive Bayes are used to construct the ensemble framework.

The experimental results show that the proposed system using slide windows of 15 day with the optimal statistical features set produced by the structural graph properties yields a better predictive performance for all measurements. The results also show that our system using the ensemble classifier with optimal features set can correctly predict up to 94% of the subjects across all measurements. It is also observed that our system is more effective than the individual base classifiers used in the ensemble model and outperforms the previously proposed approaches to solve the same problem. Our evaluation establishes that our recommendation system is effective in improving the quality of clinical evidence-based decisions and help reduce the time costs incurred by the chronic heart disease patients in taking their daily medical test, whereby improving their overall life quality.

There are several directions for our future research work in this study. First, we want to evaluate our proposed system using additional appropriate datasets which preferably have a large number of data records. We are also interested in applying other ensemble techniques, such as boosting and Adaboost, to produce recommendations and conducting a comparative study on those different ensemble models. Finally, given the generality of our proposed model in dealing with medical time series data, we will explore the possibility to apply our system to support telehealth care for patients suffering from other type of diseases.

References

1. Abegunde, D. O., Mathers, C. D., Adam, T., Ortegon, M., & Strong, K. (2007). The burden and costs of chronic diseases in low-income and middle-income countries. *The Lancet, 370,* 1929–1938.
2. Artameeyanant, P., Sultornsanee, S., & Chamnongthai, K. (2015). Classification of electromyogram using weight visibility algorithm with multilayer perceptron neural network. In: *7th International Conference on Knowledge and Smart Technology (KST)* (pp. 190–194). Chonburi, Thailand: IEEE.
3. Bashir, S., Qamar, U., & Khan, F. H. (2015). BagMOOV: A novel ensemble for heart disease prediction bootstrap aggregation with multi-objective optimized voting. *Australasian Physical and Engineering Sciences in Medicine, 38,* 305–323.
4. Bernhardt, B. C., Bonilha, L., & Gross, D. W. (2015). Network analysis for a network disorder: The emerging role of graph theory in the study of epilepsy. *Epilepsy and Behavior, 50,* 162–170.
5. Blondel, V. D., Gajardo, A., Heymans, M., Senellart, P., & Van Dooren, P. (2004). A measure of similarity between graph vertices: Applications to synonym extraction and web searching. *SIAM Review, 46,* 647–666.
6. Boccaletti, S., Latora, V., Moreno, Y., Chavez, M., & Hwang, D.-U. (2006). Complex networks: Structure and dynamics. *Physics Reports, 424,* 175–308.
7. Braamse, A. M., Jean, C. Y., Visser, O. J., Heymans, M. W., van Meijel, B., Dekker, J., et al. (2016). Developing a risk prediction model for long-term physical and psychological functioning after hematopoietic cell transplantation. *Biology of Blood and Marrow Transplantation, 22,* 549–556.

8. Breiman, L. (1996). Bagging predictors. *Machine learning, 24*, 123–140.
9. Chang, C. D., Wang, C. C., & Jiang, B. C. (2011). Using data mining techniques for multi-diseases prediction modeling of hypertension and hyperlipidemia by common risk factors. *Expert Systems with Applications, 38*, 5507–5513.
10. Chen, D., Jin, D., Goh, T. T., Li, N., & Wei, L. (2016). Context-awareness based personalized recommendation of anti-hypertension drugs. *Journal of Medical Systems, 40*, 202.
11. Dewar, A. R., Bull, T. P., Malvey, D. M., & Szalma, J. L. (2017). Developing a measure of engagement with telehealth systems: The mHealth technology engagement index. *Journal of Telemedicine and Telecare, 23*, 248–255.
12. Diykh, M., & Li, Y. (2016). Complex networks approach for EEG signal sleep stages classification. *Expert Systems with Applications, 63*, 241–248.
13. Diykh, M., Li, Y., & Wen, P. (2016). EEG sleep stages classification based on time domain features and structural graph similarity. *IEEE Transactions on Neural Systems and Rehabilitation Engineering, 24*, 1159–1168.
14. Fang, Z., & Wang, J. (2014). Efficient identifications of structural similarities for graphs. *Journal of Combinatorial Optimization, 27*, 209–220.
15. Gao, H., Jian, S., Peng, Y., & Liu, X. (2016). A subspace ensemble framework for classification with high dimensional missing data. *Multidimensional Systems and Signal Processing*, 1–16.
16. He, Y., & Evans, A. (2010). Graph theoretical modeling of brain connectivity. *Current Opinion in Neurology, 23*, 341–350.
17. Huang, F., Wang, S., & Chan, C.C. (2012). Predicting disease by using data mining based on healthcare information system. In: *Granular Computing (GrC)*, (pp. 191–194). Hangzhou, China: IEEE.
18. Huang, X., & Lai, W. (2006). Clustering graphs for visualization via node similarities. *Journal of Visual Languages and Computing, 17*, 225–253.
19. Jain, A. K., Murty, M. N., & Flynn, P. J. (1999). Data clustering: A review. *ACM Computing Surveys (CSUR), 31*, 264–323.
20. Kogge, P. M. (2016). Jaccard Coefficients as a Potential Graph Benchmark. In *Parallel and Distributed Processing Symposium Workshops* (pp. 921–928) Chicago: IEEE.
21. Krishnaiah, V., Narsimha, D. G., & Chandra, D. N. S. (2013). Diagnosis of lung cancer prediction system using data mining classification techniques. *International Journal of Computer Science and Information Technologies, 4*, 39–45.
22. Kuh, D., & Shlomo, Y. B. (2004). *A Life Course Approach to Chronic Disease Epidemiology.* Oxford University Press.
23. Lafta, R., Zhang, J., Tao, X., Li, Y., & Tseng, V. S. (2015). An intelligent recommender system based on short-term risk prediction for heart disease patients. In *Web Intelligence and Intelligent Agent Technology (WI-IAT)* (pp. 102–105). Singapore: IEEE.
24. Lafta, R., Zhang, J., Tao, X., Li, Y., Tseng, V. S., Luo, Y., et al. (2016). An intelligent recommender system based on predictive analysis in telehealthcare environment. *Web Intelligence, 4*, 325–336.
25. Lang, S. (2017). Cognitive eloquence in neurosurgery: Insight from graph theoretical analysis of complex brain networks. *Medical Hypotheses, 98*, 49–56.
26. Li, X., Hu, X., Jin, C., Han, J., Liu, T., Guo, L., et al. (2013). A comparative study of theoretical graph models for characterizing structural networks of human brain. *International journal of biomedical imaging*, 2013.
27. Li, S., Tang, B., & He, H. (2016). An imbalanced learning based MDR-TB early warning system. *Journal of Medical Systems, 40*, 164.
28. Micheloyannis, S., Pachou, E., Stam, C. J., Vourkas, M., Erimaki, S., & Tsirka, V. (2006). Using graph theoretical analysis of multi channel EEG to evaluate the neural efficiency hypothesis. *Neuroscience Letters, 402*, 273–277.
29. Miraglia, F., Vecchio, F., & Rossini, P. M. (2017). Searching for signs of aging and dementia in EEG through network analysis. *Behavioural Brain Research, 317*, 292–300.
30. Mohktar, M. S., Redmond, S. J., Antoniades, N. C., Rochford, P. D., Pretto, J. J., Basilakis, J., et al. (2015). Predicting the risk of exacerbation in patients with chronic obstructive pulmonary

disease using home telehealth measurement data. *Artificial Intelligence in Medicine, 63,* 51–59.

31. Njie, G. J., Proia, K. K., Thota, A. B., Finnie, R. K., Hopkins, D. P., Banks, S. M., et al. (2015). Clinical decision support systems and prevention: A community guide cardiovascular disease systematic review. *American Journal of Preventive Medicine, 49,* 784–795.

32. Panzica, F., Varotto, G., Rotondi, F., Spreafico, R., & Franceschetti, S. (2013). Identification of the epileptogenic zone from stereo-EEG signals: A connectivity-graph theory approach. *Frontiers in Neurology, 4.*

33. Polat, K., & Gne, S. (2007). Breast cancer diagnosis using least square support vector machine. *Digital Signal Processing, 17,* 694–701.

34. Sarsoh, J. T., Hashem, & K. M. (2012). Classifying of human face images based on the graph theory concepts. *Global Journal of Computer Science and Technology, 12.*

35. Schaeffer, S. E. (2007). Graph clustering. *Computer Science Review, 1,* 27–64.

36. Snchez, A. S., Iglesias-Rodrguez, F. J., Fernndez, P. R., & de Cos, Juez F. (2016). Applying the K-nearest neighbor technique to the classification of workers according to their risk of suffering musculoskeletal disorders. *International Journal of Industrial Ergonomics, 52,* 92–99.

37. Stam, C. J., & Reijneveld, J. C. (2007). Graph theoretical analysis of complex networks in the brain. *Nonlinear Biomedical Physics, 1,* 3.

38. Thong, N. T. (2015). HIFCF: An effective hybrid model between picture fuzzy clustering and intuitionistic fuzzy recommender systems for medical diagnosis. *Expert Systems with Applications, 42,* 3682–3701.

39. Tuffry, S. (2011). *Data Mining and Statistics For Decision Making,* Wiley Chichester.

40. Valentini, G., Masulli, F. (2002). Ensembles of learning machines. In *Italian Workshop on Neural Nets* (pp. 3–20). Heidelberg: Springer.

41. Vecchio, F., Miraglia, F., Piludu, F., Granata, G., Romanello, R., Caulo, M., et al. (2017). Small World architecture in brain connectivity and hippocampal volume in Alzheimers disease: A study via graph theory from EEG data. *Brain Imaging and Behavior, 11,* 473–485.

42. Verma, L., Srivastava, S., & Negi, P. (2016). A hybrid data mining model to predict coronary artery disease cases using non-invasive clinical data. *Journal of Medical Systems, 40,* 1–7.

43. Vural, C., & Yildiz, M. (2010). Determination of sleep stage separation ability of features extracted from EEG signals using principle component analysis. *Journal of Medical Systems, 34,* 83–89.

44. Wang, J., Qiu, S., Xu, Y., Liu, Z., Wen, X., Hu, X., et al. (2014). Graph theoretical analysis reveals disrupted topological properties of whole brain functional networks in temporal lobe epilepsy. *Clinical Neurophysiology, 125,* 1744–1756.

45. Wang, J., Qiu, M., & Guo, B. (2017). Enabling real-time information service on telehealth system over cloud-based big data platform. *Journal of Systems Architecture, 72,* 69–79.

46. Yang, J. G., Kim, J. K., Kang, U. G., & Lee, Y. H. (2014). Coronary heart disease optimization system on adaptive-network-based fuzzy inference system and linear discriminant analysis (ANFISLDA). *Personal and Ubiquitous Computing, 18,* 1351–1362.

47. Yeh, D. Y., Cheng, C. H., & Chen, Y. W. (2011). A predictive model for cerebrovascular disease using data mining. *Expert Systems with Applications, 38,* 8970–8977.

48. Zhang, J., Lafta, R., Tao, X., Li, Y., Zhu, X., Luo, Y., et al. (2017). Coupling a fast fourier transformation with a machine learning ensemble model to support recommendations for heart disease patients in a telehealth environment. *5,* 10674–10685.

Printed in the United States
By Bookmasters